PROCEEDINGS OF THE 30TH INTERNATIONAL GEOLOGICAL CONGRESS
VOLUME 14

STRUCTURAL GEOLOGY AND GEOMECHANICS

Proceedings of the 30th International Geological Congress

Proceedings of the
30th International Geological Congress

Beijing, China, 4 - 14 August 1996

VOLUME 14

Structural Geology and Geomechanics

EDITORS:
Yadong Zheng
Department of Geology, Peking University, Beijing, China
G.A. Davis
University of Southern California, Los Angeles, CA, USA
An Yin
University of Southern California, Los Angeles, CA, USA

CRC Press
Taylor & Francis Group
Boca Raton London New York

CRC Press is an imprint of the
Taylor & Francis Group, an **informa** business

First published 1997 by VSP BV Publishing

Published 2019 by CRC Press
Taylor & Francis Group
6000 Broken Sound Parkway NW, Suite 300
Boca Raton, FL 33487-2742

© 1997 by Taylor & Francis Group, LLC
CRC Press is an imprint of Taylor & Francis Group, an Informa business

First issued in paperback 2019

No claim to original U.S. Government works

ISBN 13: 978-0-367-44822-6 (pbk)
ISBN 13: 978-90-6764-249-1 (hbk)

**Visit the Taylor & Francis Web site at
http://www.taylorandfrancis.com**

**and the CRC Press Web site at
http://www.crcpress.com**

CONTENTS

Proc. 30th Int'l. Geol. Congr., Vol. 14
Zheng *et al.* (Eds)
© VSP 1997

Preface

The 30th IGC was held at the China World Trade Centre of Beijing, from 4th to 14th August 1996. 366 abstracts were submitted for Symposia 5, Structural geology and Geomechanics, and 118 oral presentations were given at its eight sessions on the following subjects:

5-1. Deformation and rheology of continental lithosphere: theory and experiment (convened by M.S. Paterson, S. Suo and Z. Wang);

5-2. Mechanism of rock deformation (convened by R. Liu, Y. Sun, T. Shimamoto and C.J.L. Wilson);

5-3. Tectonophysical simulation and digital modeling (convened by M. Dragoni and H. Hong);

5-4. Extensional structures and metamorphic core complexes (convened by G.A. Davis and H. Song);

5-5. Nappes, decollements, and thrust structures (convened by A. Yin and Y. Zheng);

5-6. Strike-slip structures and inversion tectonics (convened by M. Faure and J. Xu);

5-7. Application of geomechanics to the evaluation of the stability of major engineering works; structural control of ore minerals and prediction of occurrence of minerals resources (convened by Q. Chen, S. Cui and J. Kutina);

5-8. Tectonic concepts other than plate tectonics (convened by J.M. Dickins, S. Maruyama and Q. Xiao).

This volume contains 23 papers recommended by the conveners and is divided into six parts based on the subjects. We thank the contributors for their cooperation and the reviewers for their quick and careful work. They helped us considerably to meet a desired mid-1997 publication, less than year after the conference.

Beijing
March 1997

Yadong Zheng
Peking University
Gregory A. Davis
University of Southern California
An Yin
University of California, Los Angeles

PART 1

MECHANISM OF DEFORMATION

Proc. 30th Int'l. Geol. Congr., Vol.14, pp. 3 -15
Zheng *et al.* (Eds)
© VSP 1997

Netlike and Homogeneous Plastic Flows in the Crust and Mantle and the Flow Law

SHENG-ZU WANG

Institute of Geology, State Seismological Bureau, Beijing 100029, China.

Abstract

Netlike flow (*NF*) and homogeneous flow (*HF*) are two basic types of ductile solid-state flow. *NF* is a viscoplastic flow, i.e. a viscous flow accompanied by localized ductile shear deformation, forming plastic-flow belts and their conjugate network, i.e. plastic-flow network (*PFN*), which is dominant in the lower crust and mantle lid, playing an important role for controlling intracontinental tectonic deformation, stress field, seismicity, and other geological-geophysical processes. *HF*, or called convective flow (*CF*), is a viscous flow without solid deformation localization but similar to liquid-state flow with heterogeneity in thermal, density and velocity structures, which exists in the asthenosphere and deeper mantle. Based on a viscoplastic model in which *NF* and *HF* are taken into account, a "power / linear-binomial" combined flow law is proposed for describing the rheological behavior of steady-state creep. The netlike-flow coefficient, β, is used in it for measuring the development level of *PFN* in specimen and the conventional power flow law is only a special state of the combined flow law with β equal to zero. The combined flow law provides a theoretical basis for understanding the rheological behavior of the Earth's interior and confirms further the inevitability of *PFN* in some depth ranges of the crust and mantle.

Keywords: Crust, Mantle, Netlike flow, Homogeneous flow, Combined flow law

INTRODUCTION

The author's previous studies [21-24] indicate that the mechanical behavior of rocks depends not only on microscopic mechanisms, but is also associated with macroscopic structures of deformation involving the development of localized "weakness planes", such as brittle fractures or ductile shear belts and their conjugate networks. In the light of plastic flow with or without shear network, i.e. plastic-flow network (*PFN*), it is divided into two types, netlike flow (*NF*) and homogeneous flow (*HF*).

Conjugate shear belts or slip lines as a localized deformation phenomenon in the ductile field have long been recognized since slip bands or the so-called Lüders' bands were observed on the surfaces of forged pieces [11]. It was later mentioned by a number of researchers that localized deformation traces similar to the Lüders' bands are also seen on deformed specimens of single crystals, such as NaCl and KCl [15,16,18], and rocks, such as marble, limestone and sandstone [4,5,13]. All these phenomena were, however, regarded only as a consequence of deformation rather than considering the effects of them on the mechanical behaviors of rocks.

As a matter of fact. *PFN* is not only a consequence of ductile shear deformation, but plays an important role in controlling rheological behavior of rocks. It was inferred that *NF* may exist in the lower lithosphere, including the lower crust and lithospheric mantle, where temperature and confining pressure conditions are suitable for it [23,25,26]. In this paper the types of plastic flow and the evidences of *NF* are discussed and. in particular, a new combined flow law is suggested instead of the conventional power flow law for describing the rheological behavior of *NF* and *HF* comprehensively.

TWO TYPES OF PLASTIC FLOW

NF and *HF* are two basic types of ductile solid-state flow. *NF* is a viscoplastic flow. i.e. a viscous flow accompanied by localized ductile shear deformation, forming plastic-flow belts and their conjugate network, *PFN*. *HF* is characterized by relatively homogeneous deformation without *PFN* structure. *HF* can also be called convective flow (*CF*) because it exists in the mantle where convection takes place. Note that *CF*, as large-scale convection in the mantle, is similar to liquid-state flow which is heterogeneous in thermal, density and velocity structures.

It is inferred in consideration of the transition of deformation from brittle to ductile in the crust and mantle that *NF* is dominant in the lower lithosphere and *HF* in the asthenosphere and deeper mantle [23,24]. For instance, a *PFN* system manifested by the netlike distribution of earthquakes exists in the central/eastern Asian continent (Fig.1) [23]. As is well known, intracontinental earthquakes occur dominantly in the upper crust, forming the so-called seismogenic layer. However, the distribution of earthquakes is controlled by *PFN* in the lower lithosphere, because the energy necessary for earthquakes comes mainly from *NF* in the underlying layer, as a result, forming a "plastic-flow / seismic" two-layer network system. This network system is driven by the compression of the Indian plate at the Himalayan arc and is distributed over a large area of the central-eastern Asian continent. Furthermore, it is easy to estimate the directions of the maximum compressive stresses in the lower crust, as shown by the arrows in Fig.1, using the method of bisecting conjugate angles of *PFN*.

NF in the lower lithosphere is also responsible for the netlike distributions of magmatic and volcanic activities. velocity structures. etc., because the shear displacement along plastic-flow belts and its frictional-thermal effect result in the decrease of density and strength of intrabelt media, promoting the upwelling of magma and the variations of velocity in the lithosphere. Moreover, the effects of *PFN* on the temporal-spatial distribution of seismicity opens possibility to use the hypothesis of *PFN* for earthquake prediction. In summary, the existence of *PFN*s in the lithosphere and the effects of them on continental dynamic processes and tectonic deformation have so far been confirmed by the following evidences:

1. Geological and geophysical evidences
(1) Netlike distribution of earthquakes defining the "plastic-flow / seismic" networks such as those in central/eastern Asia, western Asia and southern Europe [23,25,26].

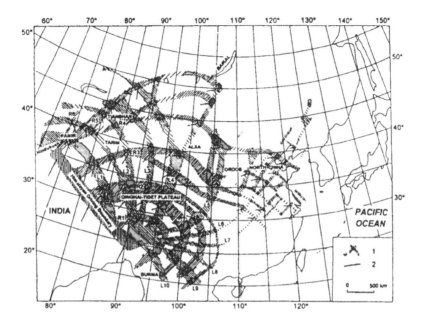

Figure 1. The "plastic-flow / seismic" network system in central-eastern Asia [23]. Symbols: 1, "plastic-flow/seismic" belts; 2, direction of maximum compressive stress; L1, L2, ..., numbers of the left-handed belts; R1, R2, ..., numbers of the right-handed belts.

(2) Netlike distribution of magmatic rocks, i.e. *PFN*-controlled magmatic networks, such as the late-Paleozoic, Mesozoic and Cenozoic magmatic networks in central/eastern Asia [29] and the Precambrian-Paleozoic magmatic network in the southern Appalachian area of North America [27].

(3) Netlike velocity structure, such as that in the Xinjiang and adjacent regions revealed by the seismic tomography, including the high-velocity network at 22km depth and the low-velocity networks at the different depths in the range 50 to 220km, which are identical in general trends with the "plastic-flow / seismic" network in that area [29].

2. Evidences from laboratory experiments

(1) Experimental deformation of minerals, rocks and other solid materials indicates that *PFN*s exist extensively in their deformed specimens under suitable conditions of temperature and confining pressure. The plastic-flow belts, i.e. ductile shear zones, in rocks are transcrystalline and often shown as relatively-dark bands (melanocratic bands) in thin sections under ordinary light [21,22,24].

(2) Physical simulations, using plasticized rosin and dried talcum powder slurry as the analog materials to model the ductile lower layer and brittle upper layer in the lithosphere, respectively, demonstrate *PFN* in the lower layer and the *PFN*-controlled fracture network in the overlying brittle layer corresponding to the upper crust [29].

3. Application to earthquake prediction

A tectonophysical method for intermediate- and long-term earthquake prediction has been developed using the *PFN* hypothesis as one of its theoretical bases [28]. The results of the annual predictions for 1994, 1995 and 1996 indicate that about 80% of earthquakes $M_S \geq 6.0$ occurred in the predicted seismic energy background zones.

In addition to the above-mentioned evidences, the establishment and verification of the combined flow law stated later in this paper will further theoretically confirm the hypothesis.

A COMBINED FLOW LAW FOR STEADY STATE CREEP

As is well known, the power flow law has been used for describing the rheological behavior of steady state creep, showing a straight line with a slope expressed by the stress exponent n in a logarithmic chart of strain rate versus differential stress. It is also indicated by researchers [2,12,17,19,24] that $log\,\dot{\varepsilon}$ versus $log\,\sigma$ tends to include a few segments with different values of n, which increases step by step with strain rate or stress owing to the changes of rheological mechanism. Furthermore, it is noteworthy that some experimental results show that the distribution of data points within a segment of $log\,\dot{\varepsilon}$ versus $log\,\sigma$ is convex instead of linear [19]. Conventionally, all the deviations of data points, including the convex distribution and other kinds of deviation, are attributed to experimental errors. It will be proved, however, that the convexity of the segments is not an accidental phenomenon but an inevitable consequence resulting from the development of *PFN*.

In consideration of viscous flow accompanied by localized ductile shear deformation, a viscoplastic model is suggested as shown in Fig.2. It includes two components, viscous flow and viscoplastic flow, with a parameter β called netlike-flow coefficient ranging from 0 to 1 to indicate the development level of *PFN*. Based on this model, a "power / linear-binomial" combined flow law has been derived for describing the rheological behavior as follows with its schematic chart of $log\,\dot{\varepsilon}$ versus $log\,\sigma$ shown in Fig.2:

$$\dot{\varepsilon} = (1-\beta)\,\dot{\varepsilon}_v + \beta\,\dot{\varepsilon}_{vp} \tag{1}$$

where $\dot{\varepsilon}_v$ and $\dot{\varepsilon}_{vp}$ for different segments are expressed as follows:

Segment " — A " ($\sigma \leq \sigma_A$, $n = n_1 = 1$, $\beta = 0$):

$$\dot{\varepsilon} = \dot{\varepsilon}_v = C_{v1}\,\sigma \tag{2}$$

$$C_{v1} = 1/(2\,\eta_1) = \dot{\varepsilon}_A / \sigma_A \tag{3}$$

Segment "$A — B$ " ($\sigma_A \leq \sigma \leq \sigma_B$, $n = n_2 > 1$, $\beta = \beta_2$):

$$\dot{\varepsilon}_v = \dot{\varepsilon}_{v2} = C_{v2}\,\sigma^{n_2} \tag{4}$$

$$C_{v2} = \dot{\varepsilon}_A / \sigma_A{}^{n_2} \tag{5}$$

$$\dot{\varepsilon}_{vp} = \dot{\varepsilon}_{vp2} = C_{vp2}\,(\sigma - \sigma_{y2}) \tag{6}$$

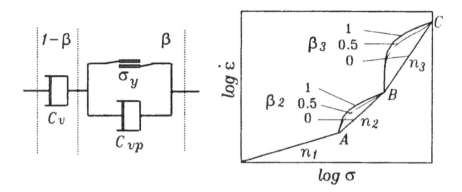

Figure 2. A rheological model for plastic flow and its *log* $\dot{\varepsilon}$ versus *log* σ. Symbols: $\dot{\varepsilon}$, strain rate; σ, differential stress; β, β_2 and β_3, netlike-flow coefficients; C_v and C_{vp}, fluidities of viscous and viscoplastic flows, respectively; σ_y, yield limit; n_1, n_2 and n_3, stress exponents.

$$C_{vp2} = (\dot{\varepsilon}_B - \dot{\varepsilon}_A) / (\sigma_B - \sigma_A) \tag{7}$$

$$\sigma_{y2} = \sigma_A - \dot{\varepsilon}_A / C_{vp2} \tag{8}$$

Segment "$B - C$" ($\sigma_B \leq \sigma \leq \sigma_C$, $n = n_3 > n_2$, $\beta = \beta_3$):

$$\dot{\varepsilon}_v = \dot{\varepsilon}_{v3} = C_{v3}\,\sigma^{n_3} \tag{9}$$

$$C_{v3} = \dot{\varepsilon}_B / \sigma_B^{n_3} \tag{10}$$

$$\dot{\varepsilon}_{vp} = \dot{\varepsilon}_{vp3} = C_{vp3}(\sigma - \sigma_{y3}) \tag{11}$$

$$C_{vp3} = (\dot{\varepsilon}_C - \dot{\varepsilon}_B) / (\sigma_C - \sigma_B) \tag{12}$$

$$\sigma_{y3} = \sigma_B - \dot{\varepsilon}_B / C_{vp3} \tag{13}$$

where $\dot{\varepsilon}$, $\dot{\varepsilon}_v$ and $\dot{\varepsilon}_{vp}$: total, viscous and viscoplastic strain rates, respectively; n : stress exponent; β : netlike flow coefficient; $\dot{\varepsilon}_A$ and $\dot{\varepsilon}_B$: strain rates at inflection points A and B, respectively; σ_A and σ_B : inflection-point stresses, or called apparent yield limits; σ_{y2} and σ_{y3}: yield limits; C_{v1} and η_1 : fluidity and viscosity of viscous flow in the segment "$-A$"; C_{v2} and C_{v3}: fluidities of viscous flow in the segments "$A - B$" and "$B - C$"; C_{vp2} and C_{vp3}: fluidities of plastic-flow belts in segments "$A - B$" and "$B - C$".

EXPERIMENTAL VERIFICATION OF THE COMBINED FLOW LAW

Plastic Flow of WRC-Mixture

WRC is a mixture of wax (W), rosin (R) and cement powder (C) in the ratio W : R : C = 1 : (R/W) : 2.4, in which wax and rosin (colophony) are mixed in their molten state at elevated temperature with cement powder as filling material. The fluidity of the mixture increases with decreasing the content of rosin which is brittle at room temperature. The uniaxial compression creep tests of the specimens with different ratios of R/W were

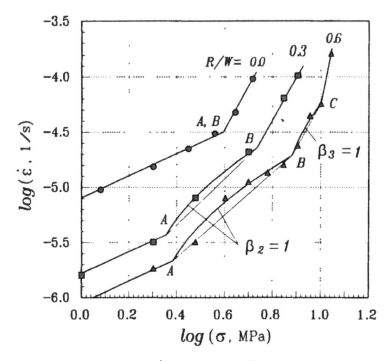

Figure 3. The $log\ \dot{\varepsilon}$ versus $log\ \sigma$ of WRC at room temperature.

Table 1.   The rheological parameters of WRC

R/W		0.0	0.3	0.6
	n_1	1	1	1
A	η_1 (Pa.s)	6.20×10^{10}	2.99×10^{11}	5.56×10^{11}
	n_2		1.98	1.91
	$\dot{\varepsilon}_A$ (s^{-1})	3.18×10^{-5}	3.55×10^{-6}	2.19×10^{-6}
A	σ_A (MPa)	3.95	2.13	2.43
|	σ_{y2} (MPa)		1.51	1.77
B	C_{v2} (MPa$^{-n_2}$s^{-1})		8.01×10^{-7}	4.00×10^{-7}
	C_{vp2} (MPa^{-1}s^{-1})		5.81×10^{-6}	3.33×10^{-6}
	β_2		1	1
	n_3	3.95	3.90	3.96
	$\dot{\varepsilon}_B$ (s^{-1})	3.18×10^{-5}	2.29×10^{-5}	1.95×10^{-5}
B	σ_B (MPa)	3.95	5.46	7.64
|	σ_{y3} (MPa)			6.40
C	C_{v3} (MPa$^{-n_3}$s^{-1})			6.24×10^{-9}
	C_{vp3} (MPa^{-1}s^{-1})			1.58×10^{-5}
	β_3	0	0	1

carried out at room temperature. The size of cylindrical specimens is 38.5mm in diameter and 61~79mm in height.   It has been determined   in terms of the development level of

PFN in the deformed specimens that *NF* takes place at R/W=0.2~0.6 and *HF* at R/W=0; *NF* is partially developed in specimen at R/W=0.1. The experimental data [10] of *log ε̇* versus *log σ* and the curves fitted for them using the combined flow law are shown in Fig.3 and the rheological parameters are given in Table 1.

The value of n_1, n_2 and n_3 are equal to 1, 1.98~1.91 and 3.90~3.96, respectively. The values of β_2 at R/W=0.6 and 0.3 as well as β_3 at R/W=0.3 are equal to 1 and, in general, the development level of *PFN* in specimens tends to diminish with the decrease of ratio R/W and to vanish at R/W=0.

Plastic Flow of Limestone

The stress relaxation data of Solenhofen limestone [20] are used for analyzing the feature of *log ε̇* versus *log σ*, including two groups of triaxial compression tests conducted at temperature of 600°C and confining pressure of 200MPa, namely, groups SH-2/4 and SH-3 with the axial pressure applying normal to and parallel to the bedding, respectively. As shown in Fig.4, the experimental results are fitted quite well using the combined flow law with the rheological parameters given in Table 2.

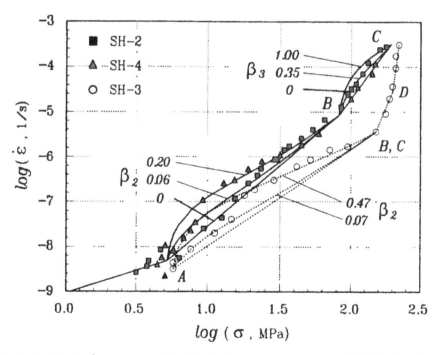

Figure 4. The *log ε̇* versus *log σ* of Solenhofen limestone at temperature of 600°C and confining pressure of 200MPa. (The experimental data are cited from [20])

The values of n_2 and n_3 are equal to 2.15~2.71 and 4.21, respectively. Comparing them with those of Solenhofen limestone at the similar temperature and confining pressure, $n_2=1.66$ and $n_3=4.70$, estimated previously using the power flow law [17], it is indicated that the underestimation of n_2 and overestimation of n_3 in the previous study may result from neglecting the convexity of the segments "$A — B$" and "$B — C$" and that the slopes of the upper and lower portions of an ascending convex curve must be slower and steeper than its slope on the average, respectively.

The various values of β_2 and β_3 indicate the convexity and deviation of experimental data point distribution to be associated with the different development levels of NF in specimens. It is, therefore, therefore inferred that the plastic-flow belts and networks would be observed as melanocratic traces under the ordinary light from the thin sections which cross the conjugate slip planes. Note that a few of data points within the segments $A — B$ and $B — C$ of the group SH-2,4 are distributed along the lines with β_2 and β_3 equal to zero, implying no PFN to exist in those specimens, that is to say, homogeneous flow may be non-Newtonian.

Table 2.   The Rheological Parameters of Solenhofen Limestone

	Group of Tests		SH-2/4	SH-3
A	n_1		1	
	η_1	(Pa.s)	5.36×10^{14}	
A | B	n_2		2.71	2.15
	$\dot{\varepsilon}_A$	(s^{-1})	4.90×10^{-9}	3.16×10^{-9}
	σ_A	(MPa)	5.25	5.78
	σ_{y2}	(MPa)	5.20	5.65
	C_{v2}	(MPa$^{-n_2}$ s^{-1})	5.47×10^{-11}	7.25×10^{-11}
	C_{vp2}	(MPa^{-1} s^{-1})	1.11×10^{-7}	2.46×10^{-8}
	β_2		0.00; 0.06; 0.20	0.07; 0.47
B | C	n_3		4.21	
	$\dot{\varepsilon}_B$	(s^{-1})	8.51×10^{-6}	2.57×10^{-6}
	σ_B	(MPa)	82.22	146.2
	σ_{y3}	(MPa)	79.18	
	C_{v3}	(MPa$^{-n_3}$ s^{-1})	7.28×10^{-14}	
	C_{vp3}	(MPa^{-1} s^{-1})	2.79×10^{-6}	
	β_3		0.00; 0.35; 1.00	

The apparent yield limit, σ_A, is equal to 5.25~5.78 MPa, corresponding to the critical shear stress $\tau_{y2}= \sigma_A / 2 =2.63~2.89$ MPa. Comparing τ_A with the critical resolved shear stress, τ_c, of calcite single crystal under uniaxial compression at temperature of 600°C, namely, τ_c = 7.5 [12] or 16.0 MPa [30] for r-slip and τ_c = 18.0 MPa [30] for f-slip, it is indicated that the values of τ_A estimated using the combined flow law are considerably less than those of τ_c of single crystals. The strain rate as low as less than 10^{-8}/s may be responsible for the low value of σ_A.

Plastic Flow of Olivine

The experimental data cited from paper [8] include three parts: olivine single crystals tested under uniaxial compression at temperatures of 1428~1650°C [8]: Mt.Burnette dunite (dry) under triaxial compression at 1100~1350°C and confining pressure of 1.5GPa [1]: Mt.Burnette dunite (dry) at 1100°C and 1.5GPa [6]. These data were adjusted to 1400°C by multiplying the experimental strain rate by $\exp(Q/RT-Q/RT_o)$, where $Q=125$ kcal/mol and $T_o=1400$°C [8]. Some researchers used these adjusted data for studies of the rheological behaviors of olivine on the basis of the power flow law and referred to $\log \dot{\varepsilon}$ versus $\log \sigma$ as a continuous curve [8], a straight line [19]. or a straight-curved line [12]. It should be mentioned that the deviations between the data from the different researchers. for example. between those from Carter et al.[1.8] and Kirby et al.[6,8]. are so considerable that they cannot be regarded only as accidental errors.

Based on the combined flow law. the data points are fitted quite well as shown in Fig.5 and the rheological parameters are given in Table 3. The previous studies indicated that the n values range over 3.0~3.7 for olivine single crystals [3,7,8,9] and over 3.0~4.8 for dry dunite [1,6,7,14]. Both are greater than n_2 (=2.27~2.39) and less than n_3 (=5.60~5.24), because the appropriate segmentation of $\log \dot{\varepsilon}$ versus $\log \sigma$ and the effect of the ductile deformation localization were neglected. The values of β_2 (=0, 0.20 and 1.0) and β_3 (=0 and 0.39~1.0) indicate the different development levels of NF and it is inferred that the deviations between the data from Carter et al.[1,8] and Kirby et al.[6,8] may be caused by the differences in β value.

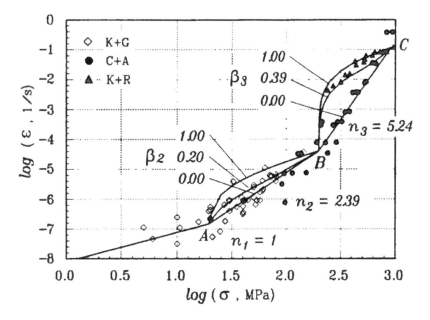

Figure 5. The $\log \dot{\varepsilon}$ versus $\log \sigma$ of olivine at temperature of 1400°C. Symbols: K+G, Kohlstedt et al.,1974 [8]: C+A, Carter et al.,1970 [1,8]; K+R, Kirby et al.,1973 [6,8].

Table 3.   The Rheological Parameters of Olivine

A	n_1		1
A	η_1	(Pa.s)	6.41×10^{13}
A | B	n_2		2.39
	$\dot{\varepsilon}_A$	(s^{-1})	1.51×10^{-7}
	σ_A	(MPa)	19.41
	σ_{y2}	(MPa)	18.72
	C_{v2}	$(MPa^{-n_2} s^{-1})$	1.26×10^{-10}
	C_{vp2}	$(MPa^{-1} s^{-1})$	2.20×10^{-7}
	β_2		0.00; 0.20; 1.00
B | C	n_3		5.24
	$\dot{\varepsilon}_B$	(s^{-1})	3.98×10^{-5}
	σ_B	(MPa)	199.53
	σ_{y3}	(MPa)	199.29
	C_{v3}	$(MPa^{-n_3} s^{-1})$	3.46×10^{-17}
	C_{vp3}	$(MPa^{-1} s^{-1})$	1.66×10^{-4}
	β_3		0.00; 0.39; 1.00

DISCUSSIONS: IMPLICATIONS FOR RHEOLOGICAL PARAMETERS

According to the "power / linear-binomial" combined flow law, rheological behavior is characterized by the following basic parameters: (a) stress exponent, n, inflection- point stress, σ_A, σ_B or σ_C, and fluidity of matrix, C_v, for viscous flow component; (b) yield limit, σ_y, and fluidity of intrabelt media, C_{vp}, for viscoplastic flow component; (c) netlike-flow coefficient, β, for measuring the fraction of viscoplastic flow component.

The stress exponent, n, as it is known, depends mainly on microscopic mechanism of plastic flow. It is thought for limestone and olivine that $n=n_1$ for segment " — A " corresponds to the diffusion-controlled mechanisms; $n=n_2$ for segment "$A — B$ " depends dominantly on the dislocation-glide mechanism; and $n=n_3$ for segment "$B — C$" may reflect some changes in mechanisms, for instance, the occurrence of second slip bands or kink bands perpenndicular to the primary slips, the formation of wavy slip bands resulting from cross-slips, or, even, the plastic flow accompanied partially with fracturing [12,21].

The yield limit, σ_{y2}, or, approximately, the inflection-point stress, σ_A, characterizes the starting point of ductile shear deformation localization, beyond which the deformation transits from homogeneously-distributed to localized. Note that the variations of n value mentioned above are a manifestation of intracrystalline changes in mechanisms, while the yield limit indicates a threshold above which transcrystalline, localized deformation is possible to occur. Also, σ_{y3} and σ_{y4}, or, σ_B and σ_C, characterize some changes in deformation localization.

The experimental results of WRC indicate that the strain rate tends to decrease with decreasing the content of brittle material (rosin) and the segment "$A — B$" tends to vanish as the ratio R/W approximate to zero. Similarly, such kinds of changes can also resulting from the elevation of temperature. The fluidity C_v or C_{vp} is proportional to exp(-Q/RT) [7.19], where Q is the activation energy, R is the gas constant and T is the absolute temperature, and the strain rate is therefore changed as a function of temperature. In addition, the locations of segments "$A — B$" and "$B — C$" would be changed due to the increase of inflection-point stresses, σ_A, σ_B and σ_C, as elevating temperature and the segments "$A — B$" or "$B — C$" would vanish as $\sigma_A = \sigma_B$ or $\sigma_B = \sigma_C$.

The netlike-flow coefficient, β_2 or β_3, measuring the development level of *PFN* or shear deformation localization, can be considered to depend on the fraction of the volume distributed with localized deformation in specimen. It is indicated by the experimental and fitting results of limestone and olivine that $\beta_2 = 0$ or $\beta_3 = 0$, as an extreme state of ductile deformation localization, can exist in rocks, where no transcrystalline, localized shear deformation takes place, or in single crystals, where the scattered dislocation glides or small-scale slips are not connected to be continuous slip planes.

CONCLUSIONS

1. The rheological behavior of rocks and other materials depends not only on microscopic mechanisms, but also the macroscopic structure, e.g. belt-like, conjugately netlike, or other kinds of structure. In the light of plastic flow with or without the netlike structure, it is divided into two basic types, netlike flow (*NF*) and homogeneous flow (*HF*). *NF* is a viscoplastic flow, i.e. a viscous flow accompanied by the localization of ductile shear deformation or the development of plastic-flow network (*PFN*). *NF* or *PFN*s exist extensively in the lower crust and mantle lid, playing an important role for controlling intracontinental tectonic deformation, stress field, seismicity, and other geological-geophysical processes. *HF*, i.e. convective flow, is a viscous flow without solid deformation localization but with heterogeneity in thermal, density and velocity structure for the large-scale convection in the mantle.

2. Based on a viscoplastic model in which the two basic types of plastic flow are taken into account, a "power / linear-binomial" combined flow law is proposed for describing the rheological behavior of steady-state creep. The netlike-flow coefficient, β, is used in the combined flow law for measuring the development level of *PFN* in specimen and the conventional power flow law is only a special state of the combined flow law, in which the value of β is equal to zero. The combined flow law provides a new approach to rheological data processing so that the errors resulting from the curve fitting of $log\,\dot{\varepsilon}$ versus $log\sigma$ using the conventional power flow law can be avoided. Moreover, in addition to the parameters for the power flow law, more information about rheological behavior can be obtained, including the netlike-flow coefficients, β_2 and β_3, the yield limits, σ_{y2} and σ_{y3}, and the fluidities of intrabelt media, C_{vp2} and C_{vp3}, etc.

3. The establishment and verification of the combined flow law from laboratory experiments and theoretical analysis confirm further the inevitability of *PFN* in some depth ranges of the crust and mantle. The combined flow law provides a theoretical basis for understanding the rheological behavior of the Earth's interior. For instance. it is possible. through further improvements in the *PFN* hypothesis and the combined flow law. to estimate the *n* and *β* values of different layers in terms of their development level of *PFN*. or. vice versa. to infer the structure types of different layers in terms of the results of rheological experiments of rocks.

Acknowledgments

I sincerely thank Jian-guo Li, Zong-chun Zhang and Rui-qing Song for helps in the laboratory experiments and data processing and Gregory A.Davis and Yadong Zheng for helpful reviews and suggestions. This study is supported by the Joint Earthquake Science Foundation of China, grant 85080103.

REFERENCES

1. N.L.Carter and H.G.Ave'Lallemant. High temperature flow of dunite and peridotite, *Geol.Soc. Amer.Bull.*, **81**, 2181-2202 (1970).
2. P.N.Chopra. The plasticity of some fine-grained aggregates of olivine at high pressure and temperature. In: *Mineral and Rock deformation: Laboratory Studies, The Paterson Volume*, B.E.Hobbs and H.C.Heard (eds.), Geophys.Monogr. 36, Amer.Geophys.Union, Washington, D.C., pp.25-33 (1986).
3. W.B.Durham and C.Goetze. Plastic flow of oriented single crystals of olivine, I. Mechanical data, *J.Geophys.Res.*, **82**, 5737-5753 (1977).
4. M.Friedman and J.M.Logan. Lüders' bands in experimentally deformed sandstone and limestone, *Bull.Geol.Soc.Am.*, **84**, 1465-1476 (1973).
5. H.C.Heard. Transition from brittle fracture to ductile flow in Solenhofen limestone as a function of temperature, confining pressure, and interstitial fluid pressure. In: *Rock Deformation*, D.Griggs and J.Handin (eds.), Geol.Soc.Am., Memoir 79, pp.193-226 (1960).
6. S.H.Kirby and C.B.Raleigh. Mechanisms of high-temperature, solid state flow in minerals and ceramics and their bearing on creep behavior of the mantle, *Tectonophysics*, **19**, 165-194 (1973).
7. S.H.Kirby and J.W.McCormick. Inelastic properties of rocks and minerals: Strength and rheology. In: *Handbook of Physical Properties of Rocks*, Volume III, R.S.Carmichael (ed.), CRC Press, Boca Raton, Florida, pp.139-280 (1984).
8. D.L.Kohlstedt and C.Goetze. Low-stress hing-temperature creep in olivine single crystals, *J.Geophys.Res.*, **79**(14), 2045-2051 (1974).
9. D.L.Kohlstedt, C.Goetze and W.B.Durham. Experimental deformation of single crystal olivine with application to flow in the mantle. In: *The Physics and Chemistry of Minerals and Rocks*, R.G.J.Strens (ed,), John Wiley & Sons, London, pp.35-49 (1976).
10. J.-g.Li, R.-q.Song and S.-z.Wang. Experimental study of brittle-ductile transition and macro scopic structures of some solid materials, *Progress in Geophysics*, **8**(4), 70-80 (1993). (in Chinese).
11. W.Lüders. Über die äusserung der elasticität an stahlartigen eisenstäben und stahlst-äben über eine beim biegen solcher stäbe beobachtete molekularbewegung, *Dingler's Polytechische Journal*, **155**, 18-22 (1860).

12. A.Nicolas and J.P.Poirier. *Crystalline Plasticity and Solid State Flow in Metamophic Rocks.* John Wiley & Sons, London (1976).

13. M.S.Paterson. *Experimental Rock Deformation - The Brittle Field.* Springer-Verlag, Berlin (1978).

14. C.B.Raleigh and S.H.Kirby. Creep in the upper mantle, *Miner.Soc.Am.Spec.Pap.*, **3**, 113-121 (1970).

15. F.Rinne. Double refraction due to stress, NaCl, *Z.Krist.*, **61**, 389 (1925).

16. E.Schmid and W.Boas. *Plasticity of Crystals.* F.A.Hughes, London (1950).

17. S.M.Schmid, J.N.Boland and M.S.Paterson. Superplastic flow in fine-grained limestone, *Tectonophysics*, **43**, 257-291 (1977).

18. W.Schütze. Stage of deformation, yield points of potassium halides, normal stress law, progress of fracture, *Z.Physik*, **76**, 135 (1932).

19. D.L.Turcotte and G.Schubert. *Geodynamics: Applications of Continuum Physics to Geological Problems.* John Wiley & Sons, New York (1982).

20. A.N.Walker, E.H.Rutter and K.H.Brodie. Experimental study of grain-size sensitive flow of synthetic, hot-pressed calcite rocks. In: *Deformation Mechanisms, Rheology and Tectonics*, Knipe,R.J.(ed.), Geological Society Special Publications, **54**, pp.259-284 (1990).

21. S.-z.Wang. Experimental investigation and mathematical simulation of deformation and failure of salt-lake rocksalt, *Chinese J.Geotech.Engin.*, 4(4), 17-28 (1982). (in Chinese)

22. S.-z.Wang. Experimental investigation on conjugate shear networks in rocks. In: *Research on Recent Crustal Movement* (1). Seismological Press, Beijing, pp.171-178, 190-192 (1985). (in Chinese)

23. S.-z.Wang. Net-like earthquake distribution and plastic-flow network in central and eastern Asia, *Phys.Earth Planet.Inter.*, **77**, 177-188 (1993).

24. S.-z.Wang. Brittle-ductile transition and plastic-flow networks in rocks, *Progress in Geophysics*, 8(4), 25-37 (1993). (in Chinese)

25. S.-z.Wang. Earthquake distribution and plastic-flow networks in the southern Eurasian continent. In: *Continental Earthquakes, IASPEI Publication Series for the IDNDR*, Volume No.3, pp.142-147 (1993).

26. S.-z.Wang. "plastic-flow / Seismic" network systems and tectonic units in central-eastern Asia, *Earthq.Res.China*, 9(3), 321-333 (1995).

27. S.-z.Wang. Netlike magmatite distribution controlled by plastic-flow network tectonics in Asia and North America. In: *IUGG 21th General Assembly, Boulder, Abstacts*, p.B368 (1995).

28. S.-z.Wang. A tectonophysical method for earthquake prediction: Applications of the multi-criterion principle and the "Vor/Net" continental dynamics model, *Seismol.& Geol.*, 18(suppl.), 71-85 (1996).

29. S.-z.Wang. *A study on the relationship between the intraplate lithospheric structure and tectonics and the preparation of strong earthquakes.* The General Report of Project No.85080103 supported by the Joint Earthquake Science Foundation of China 3 (1996).

30. H.-R.Wenk. Carbonete minerals. In: *Preferred Orientation in Deformed Metals and Rocks: An Introduction to Modern Texture Analysis*, H.-R.Wenk (ed.), Academic Press, New York, pp.241-258 (1985).

Proc. 30th Int'l. Geol. Congr., Vol. 14, PP. 16-26
Zheng et al. (Eds)
© VSP 1997

Rheology of Ancient Middle-Lower Continental Crust in the Qinling Orogenic Belt, Central China

SHUTIAN SUO AND ZENGQIU ZHONG
Faculty of Earth Sciences, China University of Geosciences, Wuhan 430074

Abstract

The Qinling Orogenic Belt is a polyphase collisional belt between the North China and Yangtze blocks in central China, in which the Tongbai-Dabie segment is considered to be the eastern partion or root belt of the orogenic belt. The crystalline axial belt in the Tongbai-Dabie segment represents Precambrian middle-lower continental crust exhumed at the surface through syn-orogenic tectonic processes and late-or post-orogenic extensional processes. The rheological features of the ancient middle-lower crust are assessed on the basis of the frictional failure in the brittle regime, and power-law creep in the ductile regime. The present-day and Mid-Late Proterozoic rheological profiles showing variation of rock strength with depth in the crust of the Tongbai-Dabie segment are qualitatively constructed and compared, respectively. The profiles suggest that the crustal texture, composition and deformation have been strongly heterogeneous and partitioned with the rheological stratification since Mid-Late Proterozoic times at least. The formation of the Luotian old metamorphic core complex in the Dabie area is interpreted in terms of mechanical instabilities arising from sialic partial melting in the middle crust in the Mid-Late Proterozoicx.

Keywords: Qinling Orogenic Belt, Rheological stratification, Mechanical instability, Middle-Lower crust, Old metamorphic core complex

INTRODUCTION

The Qinling Orogenic Belt(QOB) is a polyphase collisional belt between the North China and Yangtze blocks in central China, in which the Tongbai-Dabie segment (TDS) is considered as the eastern partion or root belt of the orogenic belt. The crystalline axial belt in the TDS represents Precambrian middle-lower continental crust exhumed at the surface through polyphase syn-orogenic tectonic processes and late- or post-orogenic extensional processes [21, 22, 23]. It has been the focus of intense geological studies for decades because of the discovery of coesite, coesite pseudomorphs and microdiamonds in eclogite blocks within the Precambrian crystalline axial belt [23:P. 59]. The crystalline axial belt comprises the Dabie Complex of Late Archean age (ca. 2650-2500 Ma) and the Early Proterozoic Hong'an Group (ca. 2000-1850 Ma), which are overlain by Mid-Late Proterozoic low-grade metamorphic rocks known as the Suixian Group in the southern margin of the TDS. The Dabie Complex experienced upper-amphibolite to granulite facies metamorphism and the Hong'an Group rocks were metamorphosed to epidote-amphibolite facies. Detailed accounts of the general geology and high- and ultrahigh-pressure metamorphism of the ancient middle-lower continental crust in Early Mesozoic time have been given by others [5, 6, 10, 14, 21, 22, 24, 26]. The purpose of this paper is to examine the rheological features of the exposed ancient middle -lower crust and to provide a geological background of high pressure-ultrahigh pressure rocks on the basis of geological, chronological and geophysical data available in the TDS from a physical standpoint. It is argued that deformation partitioning and partial melting in the continental crust are responsible for the continental heterogeneity and rheological stratification at different crustal levels. The crustal deformation behaviour on a large scale and the transition from compressional to extensional regimes are accommodated by mechanical instabilities and thermal states in the orogenic belt. Consequently, the TDS provides a window into the roots of the QOB that allows direct investigation of structural and lithological relationships at various scales.

STRUCTURE OF THE ANCIENT MIDDLE-LOWER CRUST

The ancient crustal segment has an extremely complicated structure [21, 22]. The most prominent features are the anastomosing tectonic patterns with lenticular or rhomboid domains of low strain separated by high-strain zones or shear zones of different scales. These anastomosing shear zones are composed of eclogite-facies rocks (Fig. 1), intermediate pressure amphibolite facies rocks and lower grade greenschist facies rocks. Commonly, the large-scale shear zones, for example, the Das-

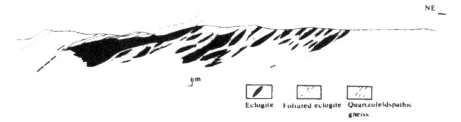

NE

Eclogite   Foliated eclogite   Quartzofeldspathic gneiss

Figure 1. Cross-section across a high-pressure shear zone near Shime, Anhui.

hankou shear zone [21] which is 250 Km long and 5 Km wide (Fig.2), are located along preexisting crustal heterogeneities of different types and have experienced a long rheological evolution. The rocks within the shear zones exhibit well-developed pervasive mylonitic fabrics and reflect plastic flow deformation and dynamic crystallization under different physical conditions (Fig. 3) [15, 21, 22].

Microstructural observations indicate that the intense flow deformation in these shear zones was mainly controlled, respectively, by the behaviour of omphacite, feldspar and quartz [6, 7, 17, 22]. The eclogites in the HP-UHP shear zones are of various types. Their P-T conditions of formation are also different; the P-T parameters for the diamond-bearing eclogite are 4000 Mpa, and 900°C [26]. Physical parameters were estimated through the application of geological thermobarometers and microstructures of mylonites within a number of shear zones in the quartzofeldspathic crust. These parameters indicats that the shear zones formed under lower-amphibolite and greenschist facies conditions in the TDS; they are summarized in Table 1. In the low-strain domains between the shear zones, Late Archean and

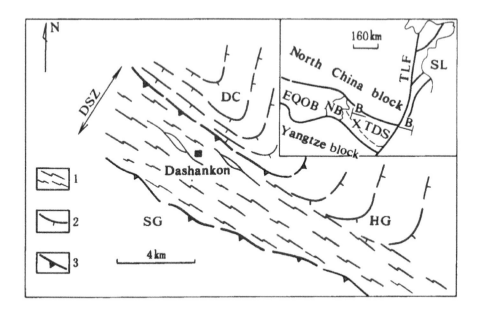

Figure 2. Simplified geological map of the Dashankou area. DSZ shows the approximate extent of the Dashankou shear zone in this area. 1. mylonitic foliation; 2. gneissic foliation; 3. detachment belt; DC, Dabie Complex; HG, Hong'an Group; SG, Suixian Group. Inset shows location of the Tongbai-Dabie segment in central China; EQOB, eastern Qinling orogenic belt; NB, Nanxiang basin; TDS, Tongbai-Dabie segment; SL, Sulu area; TLF, Tan-Lu fault. X marks the Dashankou shear zone in the TDS. B-B indicates direction and location of crustal section shown in Fig. 6.

Figure 3. Photomicrograph of quartzo-feldspathic mylonite from the Haohanpo shear zone, showing a well-developed planar fabric and brittle deformation of plagioclase in a fine-grained dynamically recrystallized quartz matrix. XZ-thin section, crossed nicols, width of the photograph 3 mm.

Early Proterozoic metamorphic fabrics formed under conditions of granulite or amphibolite facies with low to zero deviatoric stress [17] are clearly preserved and recognized (Fig. 4). At least four episodes of previous fold deformation can be distinguished in these domains at the outcrop scale [22]. Under the microscope, however, no intragranular strain features can be recognized. The data imply that the deformation and metamorphic processes in the ancient middle-lower crust have been clearly partitioned at various scales [1,7]. Much of the late strain was localized in ductile deformation zones or shear zones, and the materials outside these zones remained relatively undeformed on the grain scale. The deformation, thus, may have been largely accommodated by the rheology of the rocks in the shear zones rather than the bulk rheology of rocks within the middle-lower crust as a whole. Microscopic evidence and field observations also suggest that the strain localization in the crust was mainly controlled by mechanical and rheological instabilities at different levels and scales [13]. As a consequence, the structure of the ancient middle-lower crust is heterogeneous.

Table 1. Physical parameters for shear zones formed under lower-amphibolite and greenschist facies conditions in the TDS.

Name	Width km	Shear Strain	Temperature ℃	Confining Pressure MPa	Differential Stress Mpa	Strain Rate * S^{-1}
Dashankou	1.0 – 5.0	2.9 – 4.5	350 – 400	350 – 400	30 – 100	8.6×10^{-11} 1.7×10^{-12}
Taibaiding	–	–	350	–	119.7	4.5×10^{-10}
Shilipu	–	–	350	–	151.0	8.62×10^{-13}
Songpa	1.2	–	500	200 – 500	102.7	5.87×10^{-11}
Haohanpo	0.5 – 1.0	–	410	400	101.9	4.3×10^{-14}
Gumial	1.5 – 3.0	1.0 – 6.2	350 – 500	200 – 600	110.3	2.8×10^{-12} 2.4×10^{13}
Wangjialao	0.2 – 0.8	0.3 – 3.8	500	300	103.4	2.5×10^{-13}
Pingtianfan	–	–	–	–	103.8	2.5×10^{-13}
Youfangzhuang	0.1 – 0.2	9.41	600	–	70 – 112	1.2×10^{-11} 7.8×10^{-11}
Miaoxizhuang	–	8.02	450	–	103 – 146	2.5×10^{-13} 10.0×10^{-13}
Daoshiwan	–	0.4 – 3.0	350 – 450	–	103.0	2.5×10^{10}
Huaihedian	–	5.5 – 3.5	350 – 450	–	140.0	2.5×10^{-10} 2.3×10^{-12}
Mozitan	–	–	350 – 450	–	21.7 – 99.6 59.9 – 85.6	0.74×10^{-13} 3.34×10^{-10}

* Reference: Twiss (1977, 1980), Mercier et al (1977), Koch (1983)

The QOB is a polyphase collisional belt between the North China and Yangtze

blocks. A series of tectonic stages elucidates the broad features of the tectonic evolution of the TDS [23]. The sialic crust was eventually formed at about 2650 Ma
to 2500 Ma of Late Archean. During the Mid-Late Proterozoic the continental collisional event occurred and thickened the crust, leading to partial melting and
migmatization. The main collisional orogeny and HP and UHP eclogite facies metamorphism occurred in Early Mesozoic time. The formation and exhumation process
of the HP and UHP rocks in the collisional belt will not be discussed further here.
The present three dimensional texture pattern of the crust represents the result of
preferred Indosinian-Yanshanian tectonic events and late- or post-orogenic extensional collapse [12, 22].

Figure 4. Photomicrograph taken in
cross-polarized light of a relatively undeformed upper amphibolite facies garnet-
bearing amphibolite gneiss with coarse
and isotropic equilibrium texture, field
of view 3 mm.

RHEOLOGICAL PROFILE OF THE PRESENT-DAY CRUST

A present-day profile (Fig. 5) of the ancient middle-lower crustal rheology has been
constructed based on reported geophysical data, heat-flow measurements (Table
2), regional geological observations and estimated physical parameters for deformation [21, 22, 23]. The crustal rheology was calculated on the basis of the frictional failure in the brittle field [9, 19], and the power-law steady-state creep in the
plastic field [8] using an extensional regime and a representative strain rate of $1 \times 10^{-14} S^{-1}$. The profile indicates that the present-day crust contains different
strength layers of predominantly brittle and plastic deformation. The crust, rang-
ing from 32 to 35.5 Km in thickness, can be roughly divided into three layers of
different bulk rheology and integrated yield strength: 1) an upper crust of a relatively high-strength with a thickness of about 14 Km; 2) a middle crust of a lower
strength of fluid crustal layer [25], ranging from 14 to 21 Km in depth, with an
average density of the rocks of about 2.66×10^3 kgm^{-3}, a P-wave velocity of 6.02
km S^{-1} and 10-30 Ωm for the resistivity; and 3) a lower crust that may be substantially stronger than the overlying fluid crustal layer; its physical parameters of
the rocks are inferred: 6.48 to 6.79 Km S^{-1} for the P-wave velocity and 2.84×10^3 kgm^{-3} for the density. The Moho is an important mechanical boundary with a
cuspate-lobate fold or flexure style on the basis of explosion seismology data [22].

Table 2. Surface heat flow data for the Tongbai-Dabie segment and adjacent areas

Location	Geothermal gradient ℃Km⁻¹	Thermal conductivity W.m⁻¹K⁻¹	Surface heat flow mW.m⁻²
Tanghe	26.70	2.08	55.67
Zaoyang	23.33	1.86	43.41
Zaoyang	30.22	1.49	44.90
Tongbai	28.00	4.46	52.33
Tongbai	27.80	4.28	49.81
Tongbai	31.30	3.74	48.98
Queshan	18.17	2.69	49.00
Pingdingshan	27.20	6.35	72.30
Wuchang	17.60	2.64	46.54
Wuhan	27.51	1.66	45.69
Xinzhou	25.60	1.85	47.29
Qinzhou	20.08	3.38	67.84
Yinan	–	–	61.90
Lujiang	–	–	76.90
Huoshan	22.20	3.91	86.70
Huangshi	18.71	3.53	65.97

Sources: Duan Runmu et al. (1990), Wu Qianfan et al. (1990), Liang Ruxin et al. (1992)

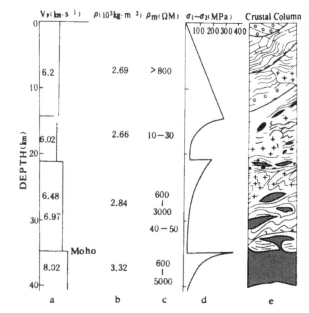

Figure 5. One-dimensional rheological profile for the present-day crust of the Tongbai-Dabie segment: a, velocity profile; b, density; c, resistivity; d, strength profile; e, crustal column. See text for further explanation.

It implies a high strength contrast between the lithospheric mantle and the overlying lower crust.

Geological and geophysical studies show the upper crust profile perpendicular to the TDS is characterized by a positive flower structure [23; Fig. 2]. It has been inferred that the entire brittle upper crust probably floats on the underlying fluid crustal layer in the middle-crust. If so, the fluid crustal layer can be regarded as a large based detachment belt for the upper crust. Thus the compressional and extensional deformation in the upper crust may be strongly influenced or accommodated by the fluid crustal layer in the TDS. The rheological stratification of the present-day crust shown in Figure 5 is also demonstrated by the distribution of a large number of relatively shallow earthquake foci depth in the study area (Fig. 6). The depth distribution of crustal earthquakes delimits two seismogenetic zones which coincide well with the base and top surfaces of the middle crust, respectively. They, in fact, correspond to the brittle-plastic transition zones.

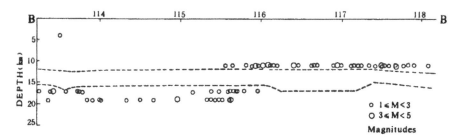

Figure 6. Earthquake foci depth distribution along the TDS, central China occurred in 1970-1979. The seismic activity was largely restricted to the top and base of the middle crust (after Zheng Ye et al. 1989, 1994). See Fig. 2 for direction and location of section (B-B). Dashed lines represent the top and base of a lower velocity layer in the middle crust and are from explosion seismology data (Zheng Ye et al. 1989).

ROLE OF PARTIAL MELTING IN THE SIALIC CRUST IN MID-LATE PROTEROZOIC

Extensive granitic gneisses and migmatites of Mid-Late Proterozoic age have been recognized in the TDS from detailed structural studies and regional mapping at a scale of 1:50000. You et al. [27, 28] have given a detailed account of the petrographic, geochemical and spatial distribution features of these granitic and migmatite bodies. They pointed out that the granitic gneisses comprise a series of metamorphic granites of tonalitic-granodioritic-adamellitic composition intruded into the Late Archean metamorphic supracrustal rocks; the migmatites were mainly formed by anatexis or partial melting of ancient sialic crustal rocks. Following the migmatite nomenclature of Sawyer et al. [20] and Burg et al. [2], the migmatites in the ancient middle-lower crust can be divided into metatexites and diatexites using the extent to which palaeosome fabric is preserved. The metatexites have a dominantly country rock composition and fabric, with a low percentage of leucosome that represents the melt fraction segregated during partial melting. The diatexites consist of foliated xenoliths of the original source-rock dispersed in leucosome material and igneous melts of successive generations [2, 22]. Partial melting under favourable hydration conditions and elevated temperature resulted in a great decrease in the integrated yield strength and density, leading to mechanical instability and the enhancement of the deformability of the rocks similar to the fluid crustal layer in the present-day crust, as mentioned above. Field criteria and microstructural analyses reveal that the metatexites had a dominantly elastic-plastic behaviour. In contrast, the diatexites had an overall viscous behaviour and the ability to develop disharmonic and non-cylindrical folds with variable wavelengths, amplitudes, and non-uniform orientation of fold axes and axial planes [11].

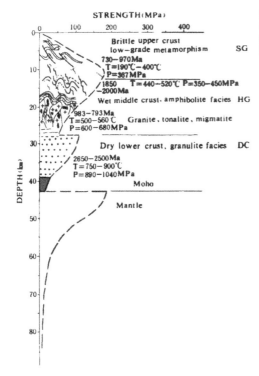

Figure 7. Paleorheological profile dealing with the thickened crust formed under a compressional regime during the Mid-Late Proterozoic. SG, HG, DC as in Fig. 2.

Extracted from melanocratic biotite gneiss inclusions in the tonalitic gneiss and migmatite at Fenghuangguan, Luotian, the colorless idiomorphic zircon single crystal yields the U-Pb dating of 796 Ma, while from the adamellite gneiss at Chinjiahe, Dawu, the zircon U-Pb dating yields an age of 832 Ma. The Guolingzhai fine-grained gneissic granite collected from the Tongbai area yields the Rb-Sr isochron dating of 983 ± 39 Ma. All of these, combined with isotopic ages reported for collision granites, for instance, the Dehe porphyritic biotite two feldspar granite(ca . 889 ± 22 Ma) in the eastern Qinling area by a number of authors [22, 27] indicate the intensively tectono-thermal and crustal partial melting events of Mid-Late Proterozoic age on the QOB. Thermobarometric estimations based on the amphibole-plagiolclase geobarothermometry and mineral assemblage show that the extensive crustal partial melting and migmatization occurred at 600-680 MPa, 500-560℃ and about 22-25 km in depth, indicating a middle crustal environment

[22, 23, 28].

The regional geological study indicates that the migmatites and old granitic gneisses are mainly localized within the area of exposed granulite and upper-amphibolite facies rocks of Late Archean age or the Dabie Complex, and are essentially absent from the overlying Early Proterozoic Hong' an Group rocks of epidote-amphibolite facies. It implies that the basal detachment zone for the Early Proterozoic sequence [22] may be an important mechanical barrier [2]. A paleorheological profile (Fig. 7) of the crust during the Mid-Late Proterozoic has been constructed assuming a compressional regime and using the data described above, together with the low-grade metamorphic and mechanical behavious of the Mid-Late Proterozoic Suixian Group rocks [22, 23]. From the paleorheological profile it is clear that the ancient crustal structure in Mid-Late Proterozoic time was also heterogeneous and rheologically stratified. This profile can be used to explain the tectonic features and origin of the Luotian old metamorphic core complex similar to those described in the North American Cordillera [3, 4, 12, 20]. In the Luotian metamorphic core complex the regional foliation defines a dome shape and contains an outward plunging mineral lineation (Fig. 8). The migmatites and granitic gneisses are characteristic of the thermo-center in their spatial distribution, decreasing in migmatization intensity from the centre of the complex to its margin. Tonalites and granitoid bodies with a flat-syntectonic layering occupy much of the core sector. The contact zone between the migmatitic supracrustal rocks and granitic gneisses belonging to the Dabie Complex and the epidote-amphibolite facies rocks of Early Proterozoic age is a regional low-angle shear or detachmemt zone. The lower formation of the upper plate, for example, the phosphatic and conglomeratic unit of the Early Proterozoic

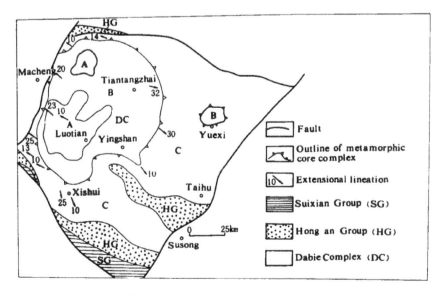

Figure 8. Schematic geological map of the Luotian metamorphic core complex illustrating the main tectonic elements. A, tonalite and granitoid body; B, diatexite; C, metatexite.

sequence is tectonically thinned or is absent altogether.

Partial melting of continental crust is a complex geodynamic and physicochemical process, and can occur in different tectonic settings and ways [2, 17]. Here, we relate the partial melting marked by the migmatization and anatexis of the sialic crust to crustal thickening and injection of mantle-derived hotter basic magma or to lower crustal magmatic underplating [18] demonstrated by the development of basic dyke and sill swarms during the Mid-Late Proterozoic continental collision [21, 23].

Mechanical and thermal instabilities induced by crustal stacking and the accumulation of partial melt of the crust have led to the transition from compressional to extensional regimes and the change from horizontal compression to vertical pure shear extension, which resulted in the formation of the metamorphic core complexes and crustal thinning concomitant with uplift and doming of the middle-lower crust. Of course, such a paleorheological profile of the crust is too simple in view of its great structural and compositional complexity [16]. Some main tectonic features, however, can be explained quite well.

SUMMARY

1. The texture of the ancient middle-lower continental crust in the Qinling Orogenic Belt is heterogeneous. Deformation partitioning is an important characteristic of continental deformation at different scales.

2. The tectonic evolution and behaviour of rocks in the ancient crust have been controlled by rheological stratification since the Mid-Late Proterozoic at least.

3. The crustal thickening by continental collision and the injection of mantle-derived basic magma have led to crustal melting and mechanical instability, and promoted the transition from the compressional regime to the extensional regime that resulted in the development of the old metamorphic core complexes of Mid-Late Proterozoic age.

Acknowledgements

Funding for this research was provided by the National Science Foundation (Nos. 49290100 and 49472147). Collaborations with Professor You Zhendong and Dr. Wang Jianhai over many years have contributed to the development of these ideas. Professors G. A. Davis and Zheng Yadong are thanked for helpful review comments.

REFERENCES

1. T. H. Bell. Deformation partitioning and porphyroblast rotation in metamorphic rocks: a radical reinterpretation. *Jour. metamorphic Geol.* 3, 109-118 (1985).
2. J. P. Burg and O. Vanderhaeghe. Structures and way-up criteria in migmatites, with application to the Velay

dome (French Massif Central), *Jour. Struct. Geol.* 15, 1239-1301 (1993).

3. G. H. Davis. Shear-zone model for the origin of metamorphic core complexes, *Geology.* 11, 342-347 (1983).

4. G. A. Davis, G. S. Lister and S. T. Reynolds. Structural evolution of the Whipple and South mountains shear zones, south-western United States, *Geology.* 14, 7-10 (1986).

5. W. G. Ernst, J. G. Liou and R. G. Coleman. Comparative petrotectonic study of five Eurasian ultrahigh-pressure metamorphic complexes, *International Geology Riview.* 37, 191-211 (1995).

6. K. J. Hsu. Exhumation of high-pressure metamorphic rocks, *Geology.* 19, 107-110 (1991).

7. M. R. Handy. Flow laws for rocks containing two non-linear viscous phases: a phenomenological approach. *Jour. Struct. Geol.* 16, 287-301 (1994).

8. S. H. Kirby. Rheology of the lithosphere , *Reviews of Geophysics and Space Physics.* 21, 1458-1487 (1983).

9. C. Lowe and G. Ranalli. Density, temperature and rheological models for the southeastern Canadian Cordillera: implications for its geodynamic evolution, *Can. Jour. Earth Sci.* 30, 77-93 (1993).

10. J. G. Liou, R. Y. Zhang and W. G. Ernst. An introduction to ultrahigh-pressure metamorphism, *The Island Arc.* 3, 1-24 (1994).

11. E. L. McLellan. Deformational behaviour of migmatites and problems of structural analysis in migmatite terrains, *Geol. Mag.* 121, 339-345 (1984).

12. J. Malavieille. Late orogenic extension in mountain belts: insights from the Basin and Range and the Late Paleozoic Variscan belt, *Tectonics.* 12, 1115-1130 (1993).

13. A. Ord and B. E. Hobbs. The strength of the continental crust, detachment zones and the development of plastic instabilities. *Tectonophysics.* 158, 269-289 (1989).

14. A. I. Okay and A. M. C. Sengor. Tectonics of an ultrahigh-pressure metamoprhic terrane: the Dabie Shan/Tongbai Shan orogen, China, *Tectonics,* 12, 1320-1334 (1993).

15. P. Philippot and H. L. M. van Roermund. Deformation processes in eclogitic rocks: evidence for the rheological delamination of the oceanic crust in deeper levels of subduction zones, *Jour. Struct. Geol.* 14, 1059-1077 (1992).

16. E. H. Rutter and K. H. Brodie. Lithosphere rheology – a note of caution. *Jour. Struct. Geol.* 13, 363-367 (1991).

17. E. H. Rutter and K. H. Brodie. The rheology of the crust. In: *The Geololgy of the Lower Continental Crust.* D. M. Fountain, R. Arculus and R. Kay(Eds.). pp. 201-267. Elsevier, Amsterdam (1992).

18. E. H. Rutter, K. H. Brodie and P. J. Evans. Structural geometry, lower crustal magmatic underplating and lithospheric stretching in the Ivrea-Verbano zone, northern Italy, *Jour. Struct. Geol.* 15, 647-662 (1993).

19. R. H. Sibson. Frictional constraints on thrust, wrench and normal faults, *Nature* (London), 249, 542-544 (1974).

20. E. W. Sawyer and S. K. Barnes. Temporal and compositional differences between subsolidus and anatectic migmatite leucosomes from the Quetico metasedimentary belt, Canada, *Jour. metamorphic Geol.* 6, 437-450 (1988).

21. S. T. Suo and H. W. Zhou. Transpressive deformation across Tongbai-Dabie orogenic belt, *Jour. China Univ. Geosci.* 3, 1-8 (1992).

22. S. T. Suo, L. K. Sang, Y. J. Han, Z. D. You, Z. Q. Zhong, J. H. Wang, H. W. Zhou and Z. M. Zhang. *The Petrology and Tectonics in Dabie Precambrian Metamorphic Terrunes, central China,* Press of China University of Geosciences, Wuhan (1993) (in Chinese with English Abstract).

23. S. T. Suo. Crustal texture and rheological evolution of Tongbai-Dabie orogenic belt, China, *Jour China Univ. Geosci.* 6, 59-63 (1995).

24. X. Wang, J. G. Liou and H. K. Mao. Coesite-bearing eclogites from the Dabie Mountains in central China, *Geology,* 17, 1085-1088 (1989).

25. B. Wernicke. The fluid crustal layer and its implications for continental dynamics. In: *Exposed Cross-Section of the Continental Crust.* M. H. Salibury and D. M. Fountain (Eds.). pp. 509-544. Kluwer Academic Publishers (1990).

26. S. T. Xu, A. I. kay and S. Ji. Diamond from the Dabie Shan metamorphic rocks and its implication for tectonic setting, *Science.* 256, 80-82 (1992).

27. Z. D. You, S. T. Suo, Y. J. Han, Z. Q. Zhong, L. K. Sang and N. S. Cheng. Metamorphic evolution of the east Qinling and Dabieshan tectonic belt, central China, *Jour. South east Asian Earth Sciences.* 9, 397-430 (1994).

28. Z. D. You, Y. J. Han, Z. Q. Zhong, L. K. Sang, N. S. Chen and Z. M. Zhang. Petrogenesis of eclogites in the light of punctuated metamorphic evolution in Dabie terrane, China, *Jour. China Univ. Geosci.* 6, 79-84 (1995).

Proc. 30th Int'l. Congr., Vol. 14, pp. 27-40
Zheng *et al.*. (Eds)
© VSP 1997

Fluidization and Rapid Injection of Crushed Fine-Grained Materials in Fault Zones during Episodes of Seismic Faulting

AIMING LIN

Faculty of Science, Kobe University, Nada-ku, Kobe 657, Japan

Abstract

Injection veins of pseudotachylyte and fault gouge are mainly composed of fine-grained crushed materials, can be linked to seismic faulting; good examples of this linkage occur in the Iida-Matsukawa fault, Nagano Prefecture, central Japan. The pseudotachylytes, which show dense and aphanitic appearances, and fault gouge occur as simple veins (fault vein) along the main fault plane and as complex network veins (injection vein) in the neighboring cataclasite. Powder X-ray diffraction patterns and petrological analysis indicate that both the pseudotachylyte and fault gouge consist entirely of fine-grained angular clasts and that those materials have similar X-ray diffraction patterns with those of the host granite. The similarity of chemical compositions and distribution patterns of grain size also show that the injection veins of pseudotachylyte and fault gouge have the same source material as that of fault veins. Such injection veins, which are mainly composed of fine-grained crushed materials, are also found in the melting-originated and experimentally-generated pseudotachylyte veins. Field occurrences and petrological characteristics, therefore, strongly suggest that the injection veins of pseudotachylyte and fault gouge formed during seismic faulting by fluidization and rapid injection of fine-grained crushed materials generated in the shear zone in a gas-solid-liquid system.

Keywords: rapid injection, fluidization, fault gouge, pseudotachylyte, injection vein

INTRODUCTION

Pseudotachylytes found as simple veins and injected networks in fault zones are widely considered to record "fossil" earthquake, i.e. events of seismic slip along faults. Those rocks generally have some aspects of an igneous rock intruded into fractures, although others resemble a sedimentary breccia cemented by fine-grained matrix. The melting origin of pseudotachylyte has been demonstrated by field and petrologic studies [15,10,4-6]. Experimental results also show that melt-originated pseudotachylyte can be generated by frictional heating at depths as shallow as several tens of meters [4, 7]. However, it has also been reported recently that there are some crushing-originated pseudotachylytes which are dark, aphanitic in appearance, and show the occurrences of simple veins and networks injected in the wallrocks have an origin by crushing, not melting [8,9]. Furthermore, veins of fault gouge injected along fractures in the fault wallrocks have also been found [8,9], as have injection veins composed of angular wallrock fragments [2,17,18].

These crushing-originated pseudotachylyte and fault gouge veins have the general

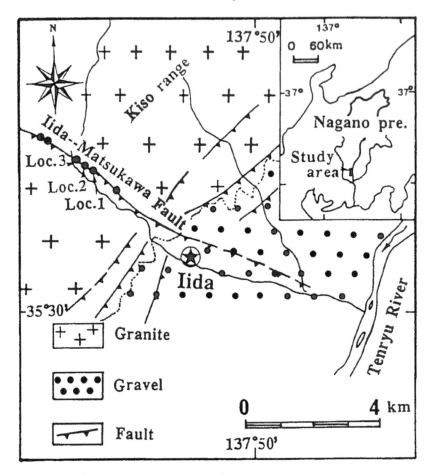

Figure 1. Simplified geological map of the Iida region, southern Nagano Prefecture, central Japan.   Solid circles show the locations of main outcrops   along the Iida-Matsukaga fault.

character of injection into host rocks. Occurrences of injected pseudotachylyte and fault gouge veins along particular faults suggest that the associated fractures   were   related to dynamic rupture propagation and slip during incremental coseismic displacement. The process of earthquake rupture propagation within the seismogenic zone is inherently complex at all scales, but a lack of adequate geologic data inhibits any detailed investigations into the actual process of rupture and   slip [12]. The study of injection veins of pseudotachylyte and fault gouge-bearing fault structures, however, can provide, by direct observation, an unique view of the process of earthquake rupturing at shallow crustal levels that cannot be obtained through indirect seismic studies.

This paper discusses the nature and origin of pseudotachylyte veins and some fault gouge veins associated with the injection and fluidization of fine-grained crushed-materials. The paper is based on the field and petrologic studies of the Iida-Matsukawa fault, southern Nagano Prefecture, central Japan (Fig.1).

FIELD OCCURRENCES OF PSEUDOTACHYLYTE AND FAULT GOUGE VEINS

The Iida-Matsukawa fault extends 12 km in a NW-SE direction, in the southern part of the Kiso range, Nagano Prefecture, central Japan (Fig.1). The wall rocks consist of granitic rocks of the Ryoke belt. The fracture zone of the Iida-Matsukawa fault generally ranges from a few meters to a few tens of meters in thickness and consists of variable fault rocks including cataclasite, protocataclasite, fault gouge and fault breccia (using the fault-rock terminology of Sibson [17], Figs.2, 3). Some cataclasites show foliated textures characterized by the orientation of biotite and aggregates of quartz and feldspar clasts.

Pseudotachylyte veins
Pseudotachylytes have been found at several locations along the Iida-Matsukawa fault, and the main locations are shown in Fig.1. The pseudotachylyte veins are dark-brown to black in color, locally show a vitreous luster similar to that of glassy pseudotachylyte as described by Lin [3]. The rocks are compact and aphanitic in appearance, consisting of fine-grained matrix and some fragments of the host rocks. In the matrix, no mineral particles can be recognized with the naked eye. They occur as simple veins along the fault plane (called here fault veins) and as complex network of veins (injection veins) injected along fractures in the cataclastic rocks (Figs.2,3a,b). Some margins of the pseudotachylyte veins have become new fault surfaces along which striations and fault gouge ranging from a few mm to a few cm in thickness are found (Fig.3a). The pseudotachylyte veins occur in fault breccias, and some pseudotachylyte fragments are present in these fault breccias. Locally, injection veins can be traced continuously back to the parent pseudotachylyte-generated fault plane. The contacts between the pseudotachylyte veins and the cataclastic rocks are generally sharp as shown in Fig.3a,b. Such veins are commonly a few mm to a few cm in thickness, and some individual veins can be followed up to a few meters along the fault zone.

Fault gouge veins
Fault gouge has been found at many locations along the Iida-Matsukawa fault, and can also be divided into two types of occurrence similar to that of the pseudotachylyte veins: gouge veins occurred along the main fault plane, and injection veins in the country granitic rock (Fig.3c,d). Injection veins have been found only at Locs.1 and 2 (Fig.1), one is sketched in Fig.2. Fault veins along the fault plane generally range from a few mm to a few tens of cm in thickness. In some, foliation characterized by grayish-green, and grayish-brown to gray color layers can be observed. The gouge veins injected in the country rocks occur as simple veins or injected complex networks in some fractured zones (Figs.3c,d,4). The contacts between the injected gouge veins and the country rock, as seen with the naked eye, are generally sharp (Fig.3c,d). Locally, the injection veins can be traced back to parent gouge veins along the fault plane. The injection gouge veins are variable from a few mm to a few tens cm but generally from 4~6 mm to 10 mm in thickness. The fault veins typically occur as thin layers ranging a few mm to a few cm in thickness, or locally as a lump on concave surfaces up to 10~20 cm in diameter.

MICROSCOPY OF PSEUDOTACHYLYTES AND FAULT GOUGES

Pseudotachylyte veins
In thin section, the Iida-Matsukawa pseudotachylytes are generally gray-brown to dark-brown in plane polarized light and dark in crossed polarized light. The injection

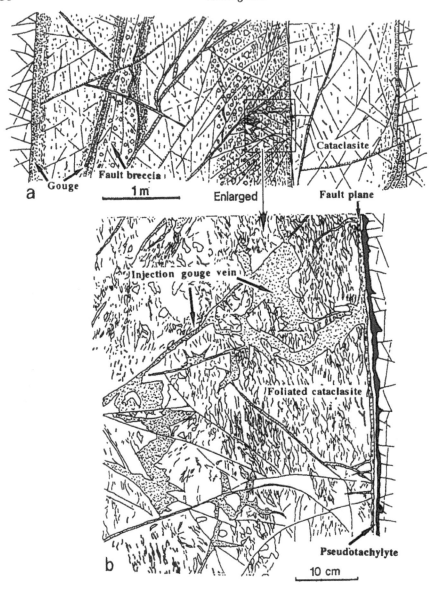

Figure 2.   Sketch of the fault outcrop at Loc. 2. Pseudotachylyte veins occurred along the fault plane (F).   Fault gouge occurred along the fault plane as simple vein and injected in the cataclasite as irregular   network veins.

relations of pseudotachylyte veins   can   be observed   under the microscope with the contacts between the veins and the host   cataclastic rocks generally being sharp (Fig. 5a, b). These veins consist of a fine-grained matrix that exhibits the optical character of glass and fragments of   quartz and   feldspar scattered   in the matrix.   Fragments   of   biotite present in the host granitic rocks have not been observed. Some fragments of saussuritized

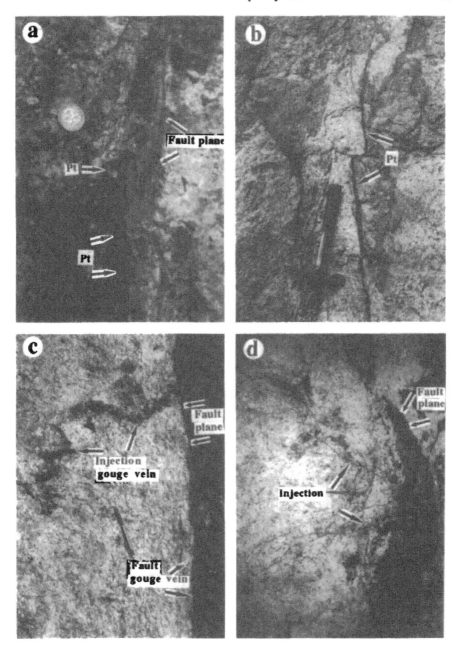

Figure 3. Photographs showing the occurrence of pseudotachylyte and fault gouge veins. (a): Pseudotachylyte vein (Pt) occurred along the fault plane. (b),(c): Pseudotachylyte veins (Pt) injected in the granitic cataclasite. The pseudotachylyte vein was displaced a few cm by later slipping along the cracks. (c), (d): Fault gouge veins occurred along the fault plane and injected into the foliated cataclasite in the fractured zone.

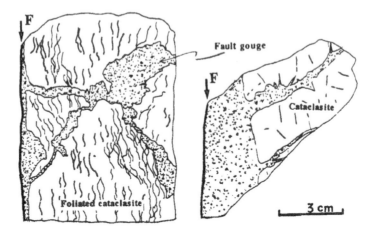

Figure.4. Sketches of hand specimens showing the occurrences of the fault veins and injection veins of fault gouge taken from Loc.2. F:Fault plane.

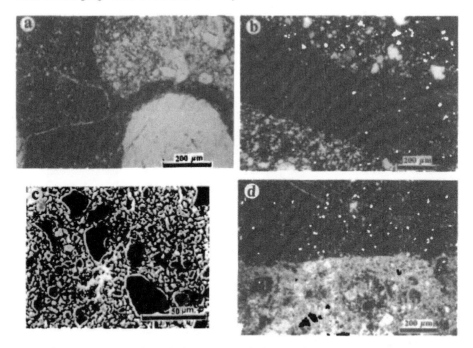

Figure 5. Photomicrographs of the pseudotachylyte and gouge veins (Pt). (a), (b): Pseudotachylyte veins injected in granitic cataclasite. (c):SEM photomicrograph showing the textures of the pseudotachylyte injection vein taken from Loc.1. Note that the fine-grained matrix of pseudotachylyte consists almost entirely of angular fine-grained clasts. (d):Fault gouge veins occurred along the fault plane. The boundary between the pseudotachylyte, fault gouge and granitic cataclasite is sharp. Note that the cataclasite was cemented by calcareous materials. (a):Plane-polarized light, (b),(d):Crossed-polarized light.

plagioclase, showing a plume-texture similar to that of microlites observed in melting-originated pseudotachylyte as described by Lin [3], have been noted. Under the electronic microscope (SEM), it is clear that the pseudotachylyte matrix consists dominantly of fine-grained angular fragments (Fig.5c). These fragments are generally larger than $2\sim3~\mu$m, and are smaller than that of cataclasite in size.

Fault gouge veins

Fault gouge veins, generally showing a character similar to that of pseudotachylyte, are yellowish-brown to brown in plane polarized light and dark in crossed polarized light. It is hard to recognize whether the fine-grained matrix materials consist of crystalline clasts or are aphanitic cryptocrystalline materials. The contacts between the injection veins and the host cataclastic rocks are generally sharp (Fig.5d), but contacts between the fault gouge generated on the fault plane and bordering cataclastic rocks are typically not so sharp.

DIFFRACTION PATTERNS OF PSEUDOTACHYLYTES AND FAULT GOUGES

The X-ray diffraction spectra of the pseudotachylyte veins and fault gouge veins are shown in Figs.6 and 7, respectively; one spectrum of quartz is also shown as a standard sample for comparison. All these X-ray diffraction spectra show diffraction patterns of crystalline materials. There is a similarity in diffraction patterns among the country granitic rock, pseudotachylyte, and fault gouge, in which the main crystalline peaks indicate the presence of quartz and feldspar in all of the samples (Figs.6,7). Some mica peaks such

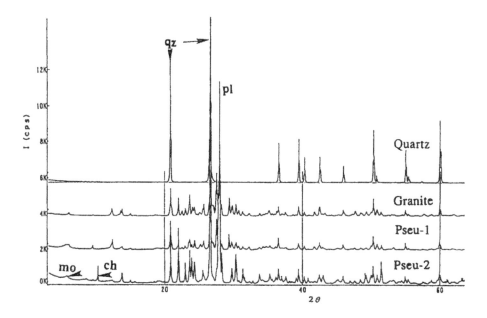

Figure 6. Powder X-ray diffraction spectra of the pseudotachylyte veins. Granite: country granite sample taken from Loc.1. Quartz: standard sample of quartz. qz:quartz, ch:chlorite, mo:montmorilinite, pl:plagioclase.

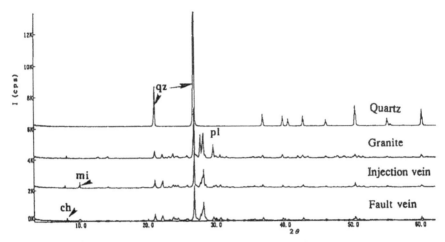

Figure 7. Powder X-ray diffraction spectra of the fault gouge veins. Granite: country granite sample taken from Loc.1. Quartz:standard sample. qz:quartz, ch:chlorite. pl:plagioclase.

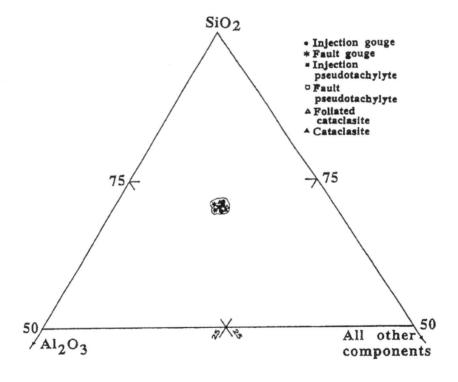

Figure 8. Si_2O-Al_2O_3-all other components diagram showing the relation among the pseudotachylyte, gouge and country granitic rock.

as biotite and muscovite can only be recognized in the granitic country rocks. Minor clay mineral peaks such as chlorite and montmorilonite have been recognized only in the pseudotachylytes and fault gouges. It is possible that the clay minerals formed from micas by hydrothermal alteration of feldspars and micas. These diffraction patterns suggest that the rock-forming minerals of the granitic country rocks are almost the same as those of the pseudotachylytes and fault gouges.

CHEMICAL COMPOSITIONS

The average bulk chemical compositions of the pseudotachylytes, gouges and their associated country rock as analyzed by XRF are shown in Fig.8. The average bulk compositions of the veins of pseudotachylyte and gouge are very similar to that of the country granitic rocks of the Iida-Matsukawa fault. Thus, based on this similarity between the chemical compositions of the veins of pseudotachylyte and gouge and the country granitic rock, the powder X-ray diffraction patterns, and the field occurrence relationships, the veins of the pseudotachylyte and gouge are interpreted to have formed from the rock in which they occur.

CLAST-SIZE DISTRIBUTION

The purpose of measurement of clast-size distribution is to determine whether the size distributions of fragments in the injected veins of pseudotachylyte and fault gouge are the same as that of fault veins of pseudotachylyte and fault gouge. If the injected veins have the same material source as the fault veins, they might be expected to have a similar or same size distribution pattern. The clast-size measurements were performed on the SEM photomicrographs because the matrix materials are too fine to recognize under the optical microscope using the measuring method described by Shimamoto & Nagahama [14].

The results of measurements were plotted as cumulative frequency diagram for the major diameter (r) along transverse axis and the total numbers of measured grains (N) in the vertical axis, using logarithmic scales on both axes (Fig.9). The results of grain-size analysis show that the matrices consist of 80-90 vol.% fragments larger than 2~3 μm, and only 10-20 vol.% fine-grained clasts smaller than 2~3 μm. It is clearly observed that there is a similar size-distribution pattern between that of the injection veins and fault veins, and a different pattern between the country granitic cataclasites and the veins of pseudotachylyte and fault gouge.

DISCUSSIONS

Origin of injection veins of the Iida-Matsukawa pseudotachylyte and fault gouge

Most of the melting-originated pseudotachylytes described in the literature have striking similarities: dense and aphanitic appearance, occurrence as irregular veins intruded into country rocks in both simple and complex networks, and generally, thicknesses of a few mm to a few cm. It was Shand[13] who first described and sketched the occurrence of irregular, branching pseudotachylyte veins in the Vredefort region, South Africa. Sibson [15] classified fault-generated pseudotachylyte veins into two fundamental classes of veins: (i) fault veins, laying along markedly planar shear fractures on which the pseudotachylyte has been generated, and (ii) injection veins, intruded into the country rocks and often appearing as dilational veins along which there has no lateral offset

Aiming Lin

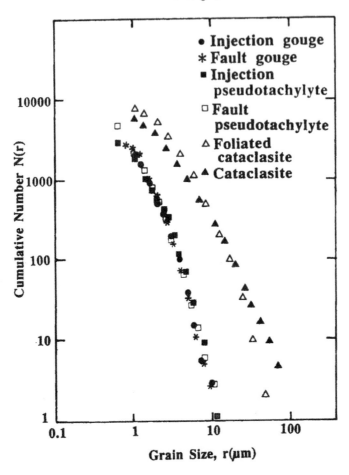

Figure 9. Size distributions of clasts contained in the fault veins and injection veins of pseudotachylyte and fault gouge as well as cataclasite occurred along the Iida-Matsukawa fault.

of markers.

As already stated above, the pseudotachylyte and fault gouge veins found in the Iida-Matsukawa fault zone also show a general character of intrusion occurring as simple vein and complex networks, have a dense and aphanitic appearance, and no observable minerals in the fine-grained matrix are microscopically visible. But, it is clear that the fine-grained matrix is composed of angular clasts observed under the SEM, showing little evidence of frictional attrition such as rounded, embayed and other irregular outlines as described by Lin [5]. This indicates clearly that this pseudotachylyte mainly formed by crushing rather than melting. The boundaries between these pseudotachylyte and fault gouge veins and the country granitic rock are sharp, both as seen in the field and under the microscope, where there is no lateral offset of markers and no striking shear textures. This shows that these veins formed at the shear zone and were intruded into the country granitic rock along the cracks.

Formation mechanisms of the injection veins of pseudotachylyte and fault gouge

Did the pseudotachylyte and fault gouge injection veins form by rapid intrusion of fine-grained clasts during seismic faulting or by the deposition of minerals transported by hydrothermal fluids along fractures. The powder X-ray diffraction patterns and the similarity of bulk chemical compositions between the veins of pseudotachylyte and gouge show that the minerals in these injection veins are similar to those of fault veins and the country granitic rock. This suggests that the injection veins have the same source materials as the fault veins and the country granitic rock. If the injection veins formed by the precipitation of minerals transported by the ground water along cracks, they would have a different composition from these veins and the country rock, and their texture would be microcrystalline, not fragmental. Finally, there is a similarity of size-distribution of clasts between the injection vein and fault vein, which indicates the genetic similarity of the two type veins. Field occurrences, powder X-ray diffraction patterns, chemical compositions, and the size-distribution patterns of clasts show that the injection veins formed by the rapid injection of fine-grained clasts during seismic faulting.

Therefore, it is suggested that the injection occurred by rapid fluidization of fine-grained clasts generated in the shear zone. Fluidization is defined as "the mixing process of gas and

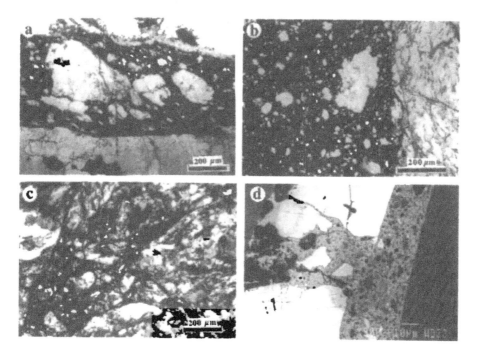

Figure. 10. Microstructural features of the melting-originated pseudotachylytes (**a~c**) and experimentally-generated pseudotachylyte (**d**). (**a**): Outer Hebrides pseudotachylyte, Scotland, described by Sibson (15). (**b**), (**c**): Fuyun pseudotachylyte, China, described by Lin [4~6]. (**d**): Experimentally-generated pseudotachylyte from gabbro described by Lin [4] and Lin et al. [9]. Note that there are more than 60~70 vol.% fine-grained crushed clasts included in all these pseudotachylytes.

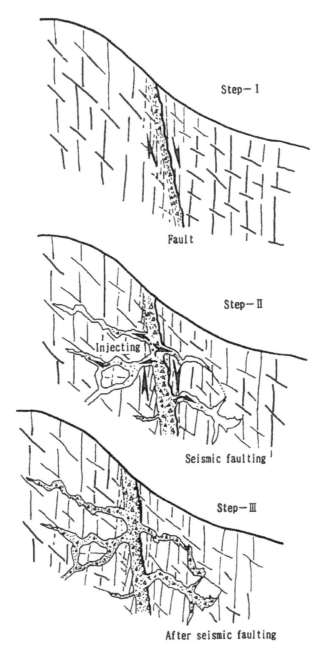

Figure 11.   Formation model of injection veins of pseudotachylyte and fault gouge.   Step-I: Fine-grained crushed materials formed in a fault zone. Step-II and Step-III: Fluidization and injection of fine-grained materials along fractures generated during coseismic faulting in fault zone.

loose fine-grained material so that the whole flows like a liquid, e.g. the formation of an ash flow or nuee ardente during a volcanic eruption " in the "Glossary of Geology" [1], and was first introduced to explain certain geologic phenomenon by Reynolds [11]. Although the term fluidization is applied specifically to gas-solid system, it is strikingly applicable to a suspension of solid particle in an upward flowing stream of liquid which has a lower density than that of the particles. The geological examples of intrusive fluidized systems are characterized by net veins of dikes, by breccias in which many of the fragments, even those derived from the adjacent wallrocks, are rounded as if by sandblast, and by mechanical hybrids [11]. Other geological examples of intrusive veins are associated with seismic faulting as melting-originated pseudotachylyte veins [4,5], and liquefaction veins formed during large earthquake, which also show the characteristics of simple and network veins injected in the wallrocks. It has been reobserved by the author that clasts may exceed over 60~70% volume in some injection veins of the typical melting-originated pseudotachylytes such as those described by Sibson [15] (Fig.10a), Lin [4,5] (Fig.10b,c), Toyoshima [19], and even in the experimentally-generated pseudotachylyte veins [7] (Fig.10d). This hints that the clasts mixed with the melt were injected into open space generated during seismic slip by rapid intrusive-like spraying in a gas-solid-liquid system like a pyroclastic flow, rather than by slow flowing of liquid. Substantial cavity development may accompany seismic slip in strong rocks at depths of several kilometers [17]. These cavities form transitory low-pressure channels are particularly favorable sites for the rapid passage of fluidized particles. The rapid injection of the fluidized particles may be formed by the sudden fluid-pressure differentials generated during seismic faulting as shown in Fig.11.

It is important to realize that the bubble phase of gas-solid system has no counterpart in liquid-solid systems and that the turbulent expanded bed is, in consequence, specific to gas-solid systems. This is of importance to the geologist because, from recognition of turbulent expanded and rock fragments which have not been appreciably transported away from their source rock, together with a lack of sorting of the fragments concerned - it can be inferred that the field agent was gas and not liquid. Closely allied in mechanism to the process of fluidization of solid particles by gas is the method of painting by spraying [11]. It is possible that the injection veins found in the Iida-Matsukawa fault formed by such fluidization of fine-grained materials during seismic faulting.

CONCLUSIONS

As stated above, the following conclusions can be obtained.

(1) The dark, dense and aphanitic pseudotachylyte found in the Iida-Matsukawa fault formed mainly by crushing rather than melting.

(2) The injection veins of pseudotachylyte and fault gouge formed by rapid injection of fine-grained crushed materials during seismic faulting.

(3) It is suggested that the injection formed by the fluidization of fine-grained crushed materials generated in fault zone in a gas-solid-liquid system.

Acknowledgments

I would like to express my sincere thinks to Professor T. Matsuda of the Kumamoto

University and Professor T. Shimamoto of the University of Tokyo for their advice and discussions in this study. Thanks are also due to the two referees, Professor G.A. Davis and Professor Y. Zheng whose suggestions and valuable critical reviews greatly improved the manuscript.

REFERENCES

1. R.L. Bates and J.A. Jackson (Eds). *Glossary of geology, third edition.*. American geological Institute, Alexandria, Virginia (1987).
2. P.E. Gretener. On the character of thrust sheets with particular reference to the basal tongues. *Bull. Can. Pet. Geol.* **2 5**, 110-122 (1977).
3. A. Lin. ESR and TL datings of active faults in the Iida area of the southern Ina valley (in Japanese). *Active fault research* **7**, 49-62 (1989).
4. A. Lin. Origin of fault-generated pseudotachylytes. *Ph.D. thesis, Tokyo University*, 108pp (1991).
5. A. Lin. Glassy pseudotachylytes from the Fuyun fault zone, northwest China. *J. Struct. Geol.* **1 6**, 71-83 (1994a).
6. A. Lin. Microlite morphology and chemistry in pseudotachylyte, from the Fuyun fault zone. *J. Geol.* **1 0 2**, 317-329 (1994b).
7. A. Lin and T. Shimamoto. Chemical composition of experimentally-generated pseudotachylytes (in Japanese, with English abstract). *Structural Geology* **3 9**, 85-101 (1994).
8. A. Lin. Injection veins of crushing-originated pseudotachylyte and fault gouge formed during seismic faulting. *Engineering Geology*, **4 3**, 213-224 (1996).
9. A. Lin, T. Matsuda and T. Shimamoto. Pseudotachylyte from the Iida-Matsukawa fault, Nagano Prefecture: Pseudotachylyte of crush origin? (in Japanese, with English abstract). *Structural Geology* **3 9**, 44-51(1994).
10. R.H. Maddock. Melt origin of fault-generated pseudotachylytes demonstrated by textures. *Geology* **1 1**, 105-108 (1983).
11. D.L. Reynolds. Fluidization as a geological process, and its bearing on the problem of intrusive granites. *Am. J. Sci.* **2 5 2**, 577-614 (1954).
12. C.H. Scholz. *The mechanics of earthquakes and faulting.* Cambridge University Press, Cambridge, 433pp (1990).
13. S.J. Shand. The pseudotachylyte of Parijs (orange Free State), and its relation to "trap-shotten gneiss" and "flinty crush-rock". *Quart. J. Geol. Soc. London* **7 2**, 198-221 (1916).
14. T. Shimamoto and H. Nagahama. An argument against the crush origin of pseudotachylytes based on the analysis of clast-size distribution. *J. Struct. Geol.* **1 4**, 999-1006 (1992).
15. R.H. Sibson. Generation of pseudotachylyte by ancient seismic faulting: *Geophys. J. Royal Astron. Soc.* **4 3**, 775-794 (1975).
16. R.H. Sibson. Fault rocks and fault mechanisms. *J. Geol. Soc. London.* **1 3 3**, 191-213 (1977).
17. R.H. Sibson. Brecciation processes in fault zones: Inferences from earthquake rupturing. *Pure Appl. Geophys.* **1 2 4**, 159-176.
18. R.H. Sibson. Earthquake rupturing as a mineralizing agent in hydrothermal systems. *Geology* **1 5**, 701-704 (1987).
19. T. Toyoshima. Pseudodachylyte from the Main Zone of the Hitaka metamorphic belt, Hokkaido, northern Japan. *J. Meta. Geol.* **8**, 507-523 (1990).

Proc. 30th Int'l. Geol. Congr., Vol.14, pp. 41-52
Zheng et al. (Eds)
© VSP 1997

Dynamic Fault Motion under Variable Normal Stress Condition with Rate and State Dependent Friction

CHANGRONG HE and SHENGLI MA

Institute of Geology, State Seismological Bureau, Beijing 100029, China

Abstract

In this study, we discuss the transient process, steady state property and dynamic motions of an inclined fault with constant minimum principal stress, based on a velocity and history dependent friction law with varying normal stress. Examination of the constitutive equations reveals that there exists an instantaneous response in friction stress to any normal stress change. The steady state value of friction stress in such an inclined geometry is not linearly related to logarithm of velocity as in the constant normal stress case, and the curve shape even depends on the coefficient of friction at the reference velocity. The dynamic stick-slip motion shows features similar to that in the constant normal stress case on the whole, but they are different from each other in the following aspects: 1) in the inclined system, velocity covers about 20 orders of magnitude in the deceleration phase, which is much wider than that in the constant normal stress case, 2) the value of stress drop in the inclined fault is greater than that in the constant normal stress case due to the effect of decrease in normal stress during the slip, 3) a relation between stress drop and load point velocity that is similar to the constant normal stress case may be employed for inferring the parameter B–A of velocity dependence from stick-slip motions in triaxial test, but overestimation may arise if the factor $1-\mu \cdot \cot\varphi$ is not taken into account.

Keywords: rock friction, inclined fault, steady state, dynamic motion, stress drop

INTRODUCTION

Stick-slip motion in rock friction is a problem of continuous interest for its possible relevance to the mechanism of earthquakes. Experimental works so far indicate that friction stress in steady state is velocity-weakening under certain conditions [3,4,15-17,10], where stick-slip occurs as the result of interaction of the slipping surface with an elastic system [4,14,17,7,8,19]. For the purpose of gaining better understanding of the physics responsible for this process, both stability criterion and dynamic motion of a single degree of freedom spring-slider system have been investigated extensively by a number of authors under the framework of velocity and history dependent friction [13-15, 17, 1,2, 5-9,19]. Most of these works were conducted under the condition of constant normal stress because of lack of data on normal stress dependence. However, a few experimental works and some analyses on the effect of normal stress were also conducted lately [11,5,18].

The effect of normal stress is important for two reasons: (a) variable normal stress is more realistic than constant normal stress for natural faults; (b) variable normal stress exists in some experimental procedures such as triaxial friction tests, for confining pressure (rather than normal stress on the slipping surface) is controlled in most triaxial tests. One may argue that normal stress can be controlled even in triaxial tests, but it is evidently not possible for the control system to follow dynamic motions with stick-slip. For these reasons, analysis based on the new experimental results under a certain variable normal stress condition is inevitable for practical purposes. In this work, we address the problem by considering one case, in which a uniform rock fault is loaded in a direction inclined to the fault strike, holding the minimum principal stress constant at the same time. Another condition in which minimum principal stress is negligibly small compared with the maximum principal stress may also be catalogued into this type. The linear stability of such a system has been already analyzed by Dieterich and Liner [5], so in the subsequent sections, we try to answer the following questions: (a) Does the steady state friction stress vary with velocity in a similar way as that in the constant normal stress case? (b) Is there a similarity in the dynamic motion between the two cases? (c) If any, what is the basic difference between the two cases? Asking these questions will help us to understand the nature of more realistic fault motions.

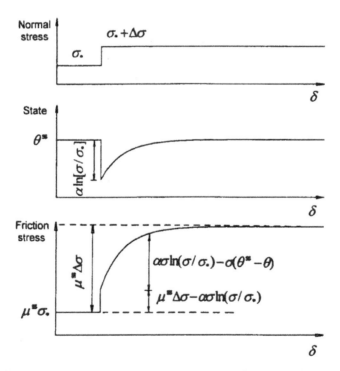

Figure 1. Effect of a step change in normal stress in the rate and state dependent friction law that incorporates varying normal stress. The change of normal stress causes instantaneous responses in both state and friction stress, followed by an exponential evolution to a new steady state. An increase in normal stress causes a negative instantaneous change in state and a positive instantaneous change in friction stress.

THE FORMULATION WITH NORMAL STRESS DEPENDENCE

Following Linker and Dieterich [11], the constitutive equation of rate and state dependent friction with variable normal stress are described as follows,

$$\tau = \mu_{\bullet}\sigma + A\sigma \ln\frac{V}{V_{\bullet}} + \sigma\theta \tag{1a}$$

$$\frac{d\theta}{dt} = -\frac{1}{L}\left(\theta + B\ln\frac{V}{V_{\bullet}}\right)\frac{d\delta}{dt} - \frac{\alpha}{\sigma}\frac{d\sigma}{dt} \tag{1b}$$

where $V_{\bullet}$ and $\mu_{\bullet}$ are the reference slip velocity and corresponding coefficient of friction respectively. When normal stress is constant, if the slipping surface is in a certain steady state and a step jump of velocity is imposed on it, state variable θ will evolve with sliding distance δ to a new steady state value. The evolution follows an exponential function in which L is the characteristic sliding distance. This form of state variable θ has been commonly used [13,6,17]. The second term on the right side of equation (1b) denotes instantaneous response of state to normal stress. This response means that for a change in normal stress, there is an instantaneous response with opposite sign in the state. α is a new parameter in the description of this effect. This formulation agrees fairly well with the experimental results of Linker and Dieterich [11].

To give a portrait of normal stress dependence of state and friction stress, consider a step jump of normal stress to σ from a steady state at velocity V and normal stress $\sigma_{\bullet}$. Both the state and friction stress experience instantaneous changes before the succeeding process that evolves with slip distance δ. For instantaneous changes, the increment of friction stress can be evaluated by

$$\Delta\tau_{\text{ins}} = [\mu_{\bullet} + A\ln(V/V_{\bullet}) + \theta'']\Delta\sigma - \sigma\cdot\Delta\theta_{\text{ins}}$$

with $\Delta\sigma=\sigma-\sigma_{\bullet}$ and

$$\Delta\theta_{\text{ins}} = \int_{\delta_j}^{\delta_j} -\frac{1}{L}[\theta + B\ln(V/V_{\bullet})]\,d\delta - \int_{\sigma_{\bullet}}^{\sigma}\frac{\alpha}{\sigma}\,d\sigma$$
$$= -\alpha\cdot\ln(\sigma/\sigma_{\bullet})$$

where δ_j denotes sliding distance at the instant when velocity jump is imposed. Because no slip-controlled evolution occurs in the instantaneous response, the first integral in the equation above is 0. So

$$\Delta\tau_{\text{ins}} = [\mu_{\bullet} + A\ln(V/V_{\bullet}) + \theta'']\Delta\sigma - \sigma\cdot\alpha\ln(\sigma/\sigma_{\bullet})$$

Note $\theta''=-B\ln(V/V_{\bullet})$, then

$$\Delta\tau_{\text{ins}} = \mu''\Delta\sigma - \alpha\sigma\ln(\sigma/\sigma_{\bullet}) \tag{2}$$

with $\mu'' = \mu_{\bullet} - (B-A)\ln(V/V_{\bullet})$.

Equation (2) is a result derived from the general equations (1a) and (1b), which means that there is an instantaneous response of friction stress to the step change in normal stress. This phenomenon was also found in the experiments of Linker and Dieterich [11], who referred to it as "elastic coupling". Note that equation (2) may be expressed approximately by $(\mu^{ss}-\alpha)\,\Delta\sigma$ for small $\Delta\sigma$. So in Linker and Dieterich's experiments at $V_0=1\mu m/s$, $\alpha=0.56$ and $\mu^{ss}-\alpha\approx0.2$, hence $\mu^{ss}\approx0.76$.

In the succeeding evolution that follows the instantaneous process, $d\sigma/dt=0$, and the process is governed by

$$\frac{d\theta}{dt} = -\frac{1}{L}\left(\theta + B\ln\frac{V}{V_*}\right)\frac{d\delta}{dt}$$

The value of θ recovers exponentially from $\theta^{ss}-\alpha\ln(\sigma/\sigma_*)$ to its original steady state value θ^{ss}. The corresponding friction stress at a certain value of θ during the evolution is given by an equation as follows,

$$\tau = \mu^{ss}(\sigma_* + \Delta\sigma) + (\theta - \theta^{ss})\sigma \tag{3}$$

For the value of θ evolves to its original value $\theta^{ss}=-B\ln(V/V_*)$, friction stress reaches to its steady state value $\tau^{ss}=\mu^{ss}(\sigma_*+\Delta\sigma)$ after sufficient slip distance. Expressions (2) and (3) are illustrated in Fig. 1.

INCLINED SPRING-SLIDER FAULT SYSTEM

A simplified spring-slider fault system is shown in Fig. 2. For the boundary condition under consideration, σ_3=Const, and friction stress τ and normal stress σ on the slipping surface are related by

$$\sigma = \sigma_3 + \tau\cot\varphi \tag{4a}$$

or

$$\frac{d\tau}{dt} = \frac{d\sigma}{dt}\tan\varphi \tag{4b}$$

where φ is the fault azimuth shown in Fig.2. For friction stress on the slipping surface varies with velocity, state and normal stress, equations (4a) and (4b) mean that normal stress is also a variable in this system.

Let m be mass of the slider and k the equivalent stiffness in the slipping direction, the equation of motion is

$$m\frac{d^2\delta}{dt^2} = k(\delta_0 - \delta) - \tau \tag{5a}$$

where δ_0 is the load point displacement from which initial spring contraction due to uniform pressure($\sigma_1 = \sigma_3$, or $\tau = 0$) is deducted, with $d\delta_0/dt = V_0' = $ Const. For quasi-static analysis, however, (5a) reduces to

$$\tau = k(\delta_0 - \delta) \tag{5b}$$

The equivalent stiffness k in the slipping direction can be estimated by

$$k = k_1 \sin^2\varphi\cos\varphi \tag{6}$$

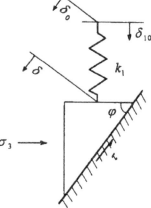

Figure 2. The spring-slider model of an inclined fault. The minimum principal stress σ_3 is held constant throughout this work.

Stability Criterion for Small Perturbation

Dieterich and Linker[5] has performed a linear stability analysis of the inclined fault system for small perturbation. The system will be unstable when

$$k < k_{cr} \tag{7a}$$

where k_{cr} is the critical stiffness. For notations adopted in Fig.2, it takes the form

$$k_{cr} = \frac{(B - A)\sigma^{ss}}{L\left[1 - \cot\varphi(\mu^{ss} - \alpha)\right]} \tag{7b}$$

where μ^{ss} and σ^{ss} are the coefficient of friction and normal stress respectively in a certain steady state, with $\mu^{ss} = \mu^* - (B-A)\ln(V/V_*)$.

No matter whether equation (7a) holds or not, the system will also be unstable when

$$\varphi < \tan^{-1}\mu^{ss} \tag{7c}$$

where $\tan^{-1}\mu^{ss} = \varphi_{cr}$ is the critical fault azimuth. The critical values of stiffness k_{cr} and critical fault azimuth φ_{cr} in the two criteria are no longer constants as in the constant normal stress case, but depend on slip velocity. This means that the stability criteria change dynamically with slip velocity.

Steady State Value of Friction Stress
With stiffness that is greater than the critical value, the spring-slider system evolves to a steady state after a sufficient long slipping distance when disturbed by a load point velocity jump. At the steady state, $d\theta/dt=0$, $d\sigma/dt=0$, equation (1b) reduces to

$$\theta^{ss} = -B\ln\frac{V}{V_\bullet} \tag{8}$$

Consider a load point velocity jump from an initial steady state with reference velocity $V_\bullet$ and reference normal stress $\sigma_\bullet$, friction stress in the ultimate steady state can be calculated from equation (1) combined with equation (4b), simply

$$\tau^{ss} = \tau_\bullet - \frac{1}{1-\mu_\bullet\cot\varphi}(B-A)\sigma^{ss}\ln\frac{V}{V_\bullet} \tag{9a}$$

where $\tau_\bullet = \mu_\bullet\sigma_\bullet$.

The relation between τ^{ss} and $\ln(V/V_\bullet)$ is not linear because σ^{ss} is coupled to friction stress τ^{ss}. It can be rewritten in an explicit form as follows,

$$\frac{\tau^{ss}}{\sigma_\bullet} = \mu_\bullet - \frac{(B-A)\ln\dfrac{V}{V_\bullet}}{1+\left[(B-A)\ln\dfrac{V}{V_\bullet}-\mu_\bullet\right]\cot\varphi} \tag{9b}$$

Fig.3 shows the relation between τ^{ss} and $\ln(V/V_\bullet)$ with $(B-A)/A=1$, $\mu_\bullet=0.76$. Moreover, the coefficient $1/(1-\mu_\bullet\cot\varphi)$ on the right side of equation (9a) depends on the coefficient of friction at the start point. Despite this fact, because μ^{ss} changes only slightly around $\mu_\bullet$ even the velocity varies within two orders of magnitude due to the small value of A and B, this factor may be approximately taken as a constant when the reference velocity is shifted within a certain range. These equations are also relevant to variation of stress drops in the dynamic motion, which will be discussed in the next section.

Transient Process in a Perfectly Rigid System
In a sufficiently stiff system, friction stress will evolves to its steady state value after an instantaneous response to a step change in load point velocity. The transient process in a perfectly rigid system is governed by equations (1a), (1b) and (4b). During the process, $d\sigma/dt<0$ because of the velocity weakening effect. This leads to a smaller $|d\theta/dt|$ than that in the constant normal stress case(see equation (1b)) which means slower evolution. Friction stress after the instantaneous change can be calculated from

equations (1a) and (1b), and its steady state value is given by equation (9b). So friction stress during the transient process varies between these two values. To see the complete feature of the evolution, we determine the relation between friction stress and displacement with a numerical procedure described in the next section, setting $B/A=2$, $\mu_*=0.76$, $\alpha=0.56$, $k=10^6 k_{cr}$ and $\varphi=60°$. We fitted the results by a decreasing exponential function to see the feature of evolution. The fitting shows that the relation is non-exponential, and, slower than that in the constant normal stress case.

DYNAMIC MOTIONS WITH INERTIA

The analysis in this section is conducted with a numerical procedure. The complete motion of the system is determined by solving equations (1a), (1b) and (5a) simultaneously. equation (5a) reduces to (5b) in quasi-static motion. It is convenient to

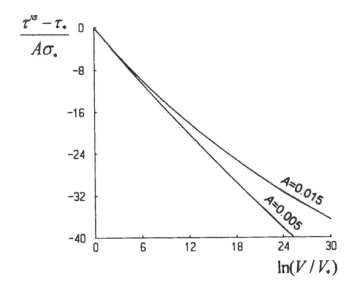

Figure 3. Steady state lines of friction stress vs. logarithm of velocity with $(B-A)/A=1$, $\mu_*=0.76$. The relations are curves rather than straight lines because normal stress is coupled to the friction stress.

replace m/k with $(T/2\pi)^2$, where T is the period of free vibration of the system without friction. In quasi-static motion, the state relaxation time scale is L/V, which may be in the order of a few years for natural faults [14]. On the other hand, the time scale for an instability is characterized by T, which is in the order of several seconds for natural earthquakes. Because of this contrast that exists between the two time scales [14], switching between two corresponding time steps is necessary at a certain point for the efficiency of calculation. The quasi-static calculation is switched over to dynamic motion when $2\pi L/(VT)<5\times10^{-4}$, following the procedure of Rice and Tse [14]. The time step is switched back to that for quasi-static motion when the difference between force in the spring and frictional resistance reduces to a value less than a tolerance level. Based on Euler's method, the numerical procedure employs a two-step scheme which compares the result of one step integration and the result by two half steps. The result is selected to be the solution when the difference between the two results comes to a value less than a tolerance level, otherwise it is discarded and the step Δt is cut to $\Delta t/2$ for further trial. The trial calculation is continued until it attains the desired accuracy. The tolerance level is set as follows,

$$\left| X_{\Delta t} - X_{\Delta t/2} \right| < \left| X_{\Delta t} \right| \times 10^{-8}$$

where $X_{\Delta t}$ and $X_{\Delta t/2}$ denote values of a variable calculated by one step and two half steps respectively.

The major parameters used for calculation are listed in Table 1, most of which are selected from the work of Linker and Dieterich [11]. Note that k_{cr} is the critical stiffness for the steady state prior to a jump of load point velocity.

Table 1. Major parameters used in the calculation

φ	μ_*	α	A	$\lambda{=}(B{-}A)/A$	k	T(sec.)
60°	0.76	0.56	0.0145	0.1, 1, 2	$0.8k_{cr}$	5

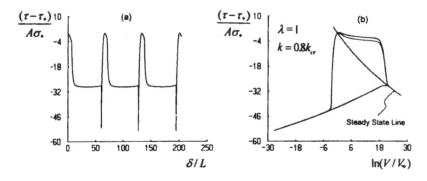

Figure 4. Dynamic stick-slip motion caused by applying a load point velocity jump from V_* to 1.5 V_*, with $\lambda{=}1$. (a) Plot of friction stress versus displacement, (b) phase plane plot of the solution. Stress drops are greater than the constant normal stress result due to the normal stress decrease during slips. Velocity covers more than 20 orders of magnitude during decelerating stage, which is much broader than the constant normal stress result.

The calculation is started from a steady state when $V = V_0 = V_* =$ Const, $\mu{=}\mu_*$, $\sigma{=}\sigma_*$, $\tau^* = \mu_* \, \sigma_*$, and the slip motion is then made unstable by giving a step jump of load point velocity from V_* to a new value βV_*.

Feature of the Dynamic Motion
One of the results is shown in Fig. 4(a), (b), with $(B{-}A)/A{=}1$. In this case, V_0 is changed from V_* to 1.5V_* at $t{=}0$. In Fig.4(a), friction stress is plotted against slip

distance, which shows an evident feature of stick slip, or mathematically, limit cycle. In Fig.4(b), orbit of the solution on the phase plane of friction stress and velocity is shown. Line in the middle of the closed locus is the steady state line to which the orbit converges around maximum velocity. Comparison with the constant normal stress case shows significant difference between magnitudes of stress drops in the two cases. Stress drop in the inclined fault is greater because of the decrease in normal stress during dynamic slip. The phase plane portrait in Fig. 4(b) shows that velocity covers about 20 orders of magnitude in the deceleration phase, which is much broader than that in the constant normal stress condition(which is about 10 orders of magnitude, see Rice and Tse [14]). On the whole, however, the dynamic motion is similar to that in constant normal stress condition as shown in the works of Rice and Tse [14] and Gu and Wong [7].

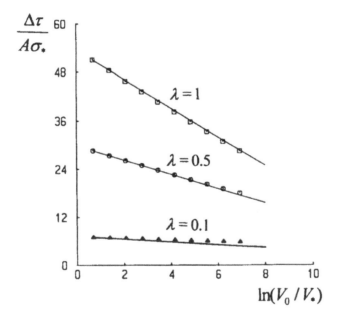

Figure 5. Stress drops plotted against load point velocity. Straight lines are drawn according to equation (10) with slopes determined by $2\lambda/(1-\mu\cdot\cot\varphi)$. Deviation from the straight lines is evident at high velocities.

Load Point Velocity Dependence of Stress Drop
To investigate the trend of stress drop with load point velocity, the value of velocity jump at the load point is systematically changed in the simulation. The system is initially slipping in steady state before the velocity jump. Stress drop is defined here as the difference between the initial and final values of friction stress during the calculation with inertia. The calculation shows a negative dependence of stress drop on the load point velocity, as in the constant normal stress case [7]. Fig. 5 shows three sets of stress drops plotted against load point velocity, and each set has a value of λ in common. Solid lines show the slope $-2\lambda/(1-\mu\cdot\cot\varphi)$, which is twice the initial slope for the steady state line as in equation (9b). Though the points deviate slightly from the

straight lines at high velocity in cases with $\lambda=0.1$ and $\lambda=1$, the lines may serve as simple approximation to the numerical result. Based on this point, stress drop may be approximated by

$$\frac{\Delta\tau}{\sigma_*} = C - \frac{2(B-A)}{1-\mu_*\cot\varphi}\ln(V_0/V_*) \tag{10}$$

where C=Const. This result is analogous to that of Gu and Wong [7], though attention must be paid to the difference that stems from the variable normal stress.

DISCUSSION

Implications to Experimental Work
Triaxial compression is a popular method for friction test for its convenience in applying high confining pressure, hence it is worthwhile to examine its compatibility with other test methods such as double direct shear test, rotary shear test, etc., in which normal stress is held constant [3,17]. For steady state friction test, result of triaxial test with constant confining pressure will be influenced by the angle of slipping surface, coefficient of friction at the start point and the final normal stress. According to (9a), B–A should be calculated with

$$B - A = -\frac{\tau'' - \tau_*}{\sigma''\ln(V/V_*)}(1 - \mu_*\cot\varphi) \tag{11}$$

The factor $(1-\mu_*\cot\varphi)$ in this equation is less than unity for φ is usually around 60° in most experimental works. Accordingly, if not multiplied by this factor, value of B–A (velocity dependence of steady state friction) will be overestimated. An example of calculation with μ_*=0.75, φ=60° shows that the factor turns out to be 0.567, hence B–A may likely be overestimated by a factor of 1.76.

For triaxial tests in which control unit regulates the normal stress to keep it constant, as adopted by Lockner et al. [12], the result will be equivalent to that obtained by other types of test with constant normal stress. This is immediate from the constitutive equations (1a) and (1b).

As suggested by Gu and Wong [7], the parameter of velocity dependence may also be estimated by the relation between the stress drop and load point velocity. This is convenient under circumstances where stick slip happens. For triaxial compression test, however, this must be done carefully for the reasons mentioned above. Equation (10) may serve as a good approximation for such purpose. Note again that the factor $(1-\mu_*\cot\varphi)$ must be taken into account, otherwise an overestimation may arise for the same reason as in the calculation about steady state. In the work of Wong and Zhao [20], the coefficient of friction varied in a range of 0.73–0.78 at a reference velocity about 1μm/s under confining pressure of 50 MPa. In this case, the parameter B–A may be overestimated by a factor of 1.7–1.8 if the effect of variable normal stress is not taken into account.

About The Parameter α

From the analysis given above, it is clear that α is not relevant to steady state values of friction stress, normal stress and state variable, whereas it affects the instantaneous response and the transient process. From the work of Linker and Dieterich [11], instantaneous response of friction stress to the change in normal stress is positive. For a small change in normal stress $\Delta\sigma$, equation (2) can be approximated by $\Delta\tau_{ins} \approx (\mu^{ss} - \alpha)\Delta\sigma$. If α is a constant, then the positive response can only be an assumption because it corresponds to an expression $\mu^{ss} - \alpha \geq 0$, where μ^{ss} decreases with velocity and the sign of $\mu^{ss} - \alpha$ may turn negative at higher velocity. This seems to contradict the common knowledge about friction. Accordingly, the nature of this parameter needs further examination with experimental work.

CONCLUSIONS

In the foregoing sections, we examined the elements of the modified constitutive equation, the variations of steady state value of friction stress, the dynamic motions with inertia in an inclined spring-slider fault system.

Examination of the constitutive equation indicates an existence of instantaneous response of friction to any change in normal stress, which is in accordance with the experimental result of Linker and Dieterich [11].

For an inclined fault with variable normal stress, friction stress in steady state is no longer linearly related to the logarithm of velocity. However, the relation is a well-defined expression, and can be used for estimation of velocity dependence in steady state by experimental work.

Dynamic motion of an inclined fault has two significant differences compared with the constant normal stress case. Namely, (a) velocity covers about 20 orders of magnitude in the deceleration phase, which is much wider than that in the constant normal stress condition; (b) in an inclined fault, the value of stress drop is greater than that in the constant normal stress case due to the decrease of normal stress during the slip.

The magnitude of stress drops decreases with the increase of load point velocity. This can be approximately expressed by

$$\frac{\Delta\tau}{\sigma_{*}} = C - \frac{2(B-A)}{1 - \mu_{*}\cot\varphi}\ln(V_{0}/V_{*})$$

within three orders of magnitude of load point velocity range. This equation may be used to estimate the parameter $B-A$ in triaxial tests.

On the whole, variations of friction stress in both steady state and dynamic motions of an inclined fault with constant minimum principal stress are similar to that in the constant normal stress case. This is important because we have known a lot about the fault motion under constant normal stress condition. As an answer to one of the questions set in the introduction, critical stiffness in the inclined spring-slider fault system is "basically" different from that in the constant normal stress case, for the latter does not change with velocity.

Acknowledgments

We are grateful to Ren Wang and Yadong Zheng whose reviews improved the presentation.

REFERENCES

1. M. L. Blanpied and T. E. Tullis. Stability and behavior of a frictional system with a two state variable constitutive law, *Pure and Appl. Geophys.* 124, 415-444(1986).
2. T. Cao and K. Aki. Effect of slip rate on stress drop. *Pure and Appl. Geophys.* 124, 515-529(1986).
3. J. H. Dieterich. Modeling of rock friction: 1. Experimental results and constitutive equations. *J. Geophys. Res.* 84, 2161-2168(1979).
4. J. H. Dieterich. Constitutive properties of faults with simulated gouge. In: *Mechnical Behavior of Crustal Rocks, Geophys. Monogr. Ser.* 24, N. L. Carter, M. Friedman, J. M. Logan, and D. W. Stearns(Ed.), pp.103-120, AGU, Washington D. C. (1981).
5. J. H. Dieterich and M. F. Linker. Fault stability under conditions of variable normal stress. *Geophys. Res. Lett.* 19, 1691-1694(1992).
6. J.-C. Gu, J. R. Rice, A. L. Ruina and S. T. Tse. Slip motion and stability of a single degree of freedom elastic system with rate and state dependent friction. *J. Mech. Phys. Solids* 32, 167-196(1984).
7. Y. Gu and T.-F. Wong. Effects of loading velocity, stiffness, and inertia on the dynamics of a single degree of freedom spring-slider system. *J. Geophys. Res.* 96, 21677-21691(1991).
8. Y. Gu and T.-F. Wong. Nonlinear dynamics of the transition from stable sliding to cyclic stick-slip in rock. In: Nonlinear Dynamics and Predictability of Geophysical Phenomena, *Geophysical Monograph* 83, IUUG Volume 18, pp.15-35(1994).
9. F. G. Horowitz. Mixed state variable friction laws: some implications for experiments and a stability analysis. *Geophys. Res. Lett.* 15, 1243-1246(1988).
10. B. D. Kilgore, M. L. Blanpied and J. H. Dieterich. Velocity dependent friction of granite over a wide range of conditions. *Geophys. Res. Lett.* 20, 903-906(1993).
11. M. F. Linker and J. H. Dieterich. Effects of variable normal stress on rock friction: observations and constitutive equations. *J. Geophys. Res.* 97, 4923-4940(1992).
12. D. A. Lockner, R. Summers and J. D. Byerlee. Effects of temperature and sliding rate on frictional strength of granite. *Pure and Appl. Geophys.* 124, 445-469(1986).
13. J. R. Rice and A. L. Ruina. Stability of steady state frictional slipping. *J. Appl. Mech.* 50, 343-349(1983).
14. J. R. Rice and S. T. Tse. Dynamic motion of a single degree of freedom system following a rate and state dependent friction law. *J. Geophys. Res.* 91, 521-530(1986).
15. A. L. Ruina. Slip instability and state variable friction laws. *J. Geophys. Res.* 88, 10359-10370(1983).
16. T. Shimamoto. Transition between frictional slip and ductile flow for halite shear zones at room temperature. *Science* 231, 711-714(1986).
17. T. E. Tullis and J. D. Weeks. Constitutive behavior and stability of frictional sliding of granite. *Pure and Appl. Geophys.* 124, 383-414(1986).
18. W. Wang and C. H. Scholz. Micromechanics of the velocity and normal stress dependence of rock friction. *Pure and Appl. Geophys.* 143, 303-315(1994).
19. J. D. Weeks. Constitutive laws for high-velocity frictional sliding and their influence on stress drop during unstable slip. *J. Geophys. Res.* 98, 17,637-17,648(1993).
20. T.-F. Wong and Y. Zhao. Effects of load point velocity on frictional instability behavior. *Tectonophysics* 175, 177-195(1990).

Proc. 30th Int'l. Geol. Congr., Vol.14, pp.53-65
Zheng et al. (Eds)
© VSP 1997

The Tectonic Significance of Active Middle Crust

ZHU'EN YANG [1], ZONGXU WU [1], WANCHENG BAI [2] AND MIN QING [2]
[1] Institute of Geology, State Seismological Bureau, Beijing 100029, China
[2] Gold Geological Institute of Ministry of metallurgical Industry, Beijing 102800, China

Abstract

A complicated giant ductile shear system exposed in Wutaishan area of North China. It consists of three mega-ductile shear belts and each belt is composed of a number of small-scale parallel or low-angle intersecting shear zones. This system looks like a tensile network-type structure with about several kilometers to more than ten kilometers in width and tens of kilometers in length. Its scale is comparable to the layer or inclusion with high conductivity and lower wave velocity of middle crust from geophysical detection. Its texture is analogous to the number of reflector in "Crocodile" texture from vertical seismic reflection in the middle crust. It could be inferenced that the process of formation of giant ductile shear system would explain the phenomenon rationally from geophysical detection in active middle crust.

Keywords: middle crust, ductile shear system, mylonite, Wutaishan area

INTRODUCTION

The knowledge of middle crust has been changed in recent years, because more and more results from geophysical sounding in the crust show that the middle crust is a more nonhomogeneous layer. The relative simple structure of middle crust exist in some stable area, such as in Ordos in North China [8, 14]. It shows the simple layered structures. The more complex structure of middle crust always appear in active area, especially in Mesozoic and Cenozoic rift faulting area or Cenozoic graben series, and seismic active area, such as Shanxi Graben series in North China, and Yinchuan Graben [8, 12, 14]. This kind of complex structure in middle crust show more distinctly from deep seismic reflection detection [15, 16]. Fig.1 show the feature from Tibetan and Xintai earthquake area. This kind of middle crust consists of near horizontal multi-interfaces that look like "Crocodiles", and in seismic refraction, they show the anomalous layers or inclusions with lower seismic wave-velocity and relative high conductivity [8, 12, 14]. What geological significance to this kind of middle crust has attracted more and more geologists and geophysicists, but has reached no consensus views.

The temperature and pressure condition of middle crust with depth of 10 km to more than 20 km belong to those of green-schist and amphibolite facies. Where is the transform position from the brittle deformation regime in upper crust to ductile deformation regime in lower crust. It is a gradual change process under the stable area, and it would experience active deformation under the active area. The difference of lithology, geothermal gradient, stress condition and properties of minerals would cause the complication of deformation in the middle crust, and result in complex nature from geophysical detection. No definite explanation for the complex structure of middle crust was concluded.

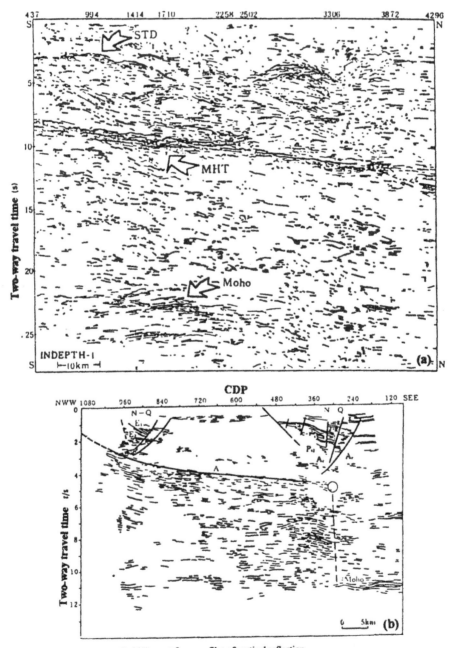

Figure 1. The structure of middle crust from profiles of vertical reflection

A: INDEPTH-1 profile of vertical reflection in Tibetan(from Teng Jiwen , 1995)

B: Profile of vertical reflection in Xingtai earthquake area(from Wang Chunyong et al, 1994)

Studies on the area exposed of metamorphic rocks with green-schist facies and upper amphibolite facies could give a lot of information about deformation process of middle crust. The research of giant network-type ductile shear system that formed in middle crust exposed at Wutaishan area, North China would provide some beneficial inspirations.

OUTLINE OF GEOLOGY

Wutaishan area is located at northern segment of Mt. Taihangshan, central part of North China. Most rocks exposed on surface are metamorphic series from late Archean to early Proterozoic Era. There are Fuping system, Longquanguan system, Wutai system, Hutou system respectively from low to high in a stratigraphic sense [4, 6] (Fig. 2).

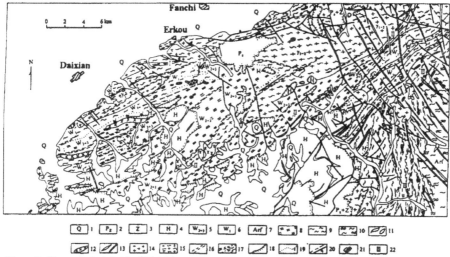

Figure 2. The geological sketch of Wutaishan region

1 Quaternary system; 2 Paleozoic group; 3 Sinian system; 4 Hutou system; 5 Middle and upper of Wutai system; 6 Lower of Wutai system; 7 Fuping system; 8 Strata with greenshist facies; 9 Strata with amphibolite facies; 10 Strata with high amphibolite facies; 11 Ultrabasic intrusive; 12 Basic intrusive; 13 Diabase dyke; 14 Intermidiate-acid intrusive; 15 Gray gneiss; 16 Ductile shear zone; 17 Complicated ductile shear belt and augen gneiss; 18 Nonconformity; 19 Boundary of strata; 20 Fault and infered fault; 21 Residential area; 22 No. of mega-ductile shear belt

Fuping system is exposed at the middle and northen segments of Mt. Taihangshan. It consists of volcano-sediments with metamorphic phase from high amphibolite facies to granulite facies. The formation age is 2700-2900 Ma [4, 6]. The main rock types are plagiognesis, diopside-plagio-amphibolite, hornblende-pyroxene-ganulite. These rocks experienced deeper migmatization. The tectonic orientation is near the direction of E-W.

Longquanguan system is distributed along a belt with width of 1.2-5 km between Fuping system and Wutai system. It consists of augen gneiss with metamorphic phase of lower amphibolite facies. Its metamorphic age is 2000-2186Ma [4, 7]. It used to be taken as a stratigraphic unit, due to the exposed location. The results from our research show that it is not a stratigraphic unit, but a tectonite belt formed by ductile shearing, and is a part of complicated ductile shear system in this area.

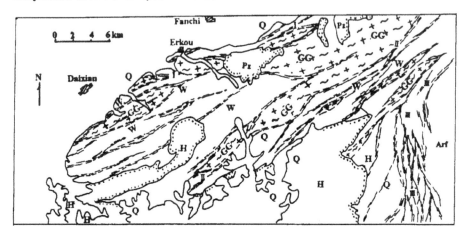

Figure 3. The sketch of structure of ductile shear system and distribution of gray gneiss (GG) (Legend is same as Fig.2)

Wutai system is popular in Wutaishan area which with metamorphic phase from amphibolite facies in its lower part to green-schist facies of its upper part. It constitutes of a formation of metamorphic volcano-sediment and exposed over the Fuping system. The rock association in its lower part with amphibolite facies consists of plagio-amphibolite, magnetite-quartzite, garnet-staurolite-biotite-leucogranulitite, kynaite-leucogranulitite. The migmatization is unpopular. The rocks of middle and upper part with metamorphism of green-schist facies are of chlorite-quartz-schist, metasandstone, magnetite-quartzite, marble and quartzite. The formation age is 2300-2500 Ma.

Hutuo system is distributed widely on the south of Wutaishan Mt. and covers on Wutai system with unconformity. Some boundary experienced ductile deformation. It consists of volcano-sediment series with light metamorphism and deformation. Main type of rock are phyllite, slate, quartzite, leucogranulitite, marble, dolomitite etc. The metamorphic age in phyllite is 2047 Ma and the formation age of basic volcanic rock is 2366 Ma. Therefore, the lower limit age of sedimentation is estimated about 2400 Ma.

The strata of Sinian system and Paleozoic group are a clastic-carbonate formation with unmetamorphism which covers all metamorphic series with unconformity. The distribution area of cover is relatively limited. The main tectonic orientation of this area is in the direction of northeast, except the Fuping system.

TYPES OF TECTONITE FORMED BY DUCTILE SHEARING

Mylonite rocks are the main tectonite rocks formed by ductile shearing. It could be divided into three main types, although its composition are relatively complicated.

Meta-mylonite: It shows intense ductile deformation in hand specimen. The quartz experienced plastic deformation like filamentous. The feldspar deformed as banded. The mica shows strong orientation and the hornblende shows both intense ductile deformation and orientation. The flow structure is popular in meta-mylonite. Under the polarizing microscopy, the steady recrystallization was well developed in meta-mylonite. It is the main difference from mylonite that means its deformation temperature is relatively high, or its formation depth is relatively deeper. The microstructures of ductile shearing are distinct, such as kinked band in feldspar (Fig.4a), divergent texture due to subgrainization, rolling of porphyroblast (Fig.4b, Fig.5a) etc. The differential stress from quartz porphyroblast is about 10-14 Mpa (Table 1).

Mylonite: The ductile deformation of mylonite is evident in hand specimen. Under the polarizing microscopy, The quartz porphyroblast shows obvious plastic deformation. The deformed feature in feldspar porphyroblast is complicated. The phenomenon of plastic deformation and kinks are less than those of meta-mylonite. Some rolling structure can be observed. such as "S" type structure formed by rolling of hornblende or garnet porphyroblast. Some "tectonic fish" are developed (Fig.4c). The dynamic recrystallizatiom of quartz is relative popular (Fig.4d). The differential stress from dynamic recrystallized grain size of quartz is about 28-33 Mpa.

S-C mylonite: This is a typical product from ductile shearing. Augen structure is well developed (Fig.5a). S-C fabrics is the representative fabrics (Fig.5b). The "tectonic fish" is popular, including "feldspar fish", "hornblende fish" (Fig.5c) and "mica fish". They are formed by the difference of property of ductile and brittle in different mineral porphyroblast. The recrystallization of quartz is distinct. Brittle figures are developed in some brittle mineral porphyroblasts, for instance feldspar, hornblende and garnet (Fig.4b, Fig.5a,5d). The differential stress from subgrain size of quartz porphyroblast is about 16-28 Mpa.

Although the characteristics of three type of mylonite are difference, but they are all originated from ductile shearing. Based on their deformed features, it shows that the difference of their formation temperature or depth, and virtually, the difference in formation period.

THE DIFFERENTIAL STRESS, STRAIN RATE AND DEFORMED PERIOD OF DUCTILE SHEARING

Differential stress: A number of methods for calculating the differential stress of tectonite were put forward in 1970's and 1980's based on the study of laboratory and statistics. In general, the recrystallized grain size of quartz with dynamic equilibrium or subgrain size of quartz porphyroblast is taken to calculate. Most formulae are similar in form but the constant. In this paper, The differential stress ($\sigma_1 - \sigma_3$) is calculated from formula (1) of Mercier (1977) as follows:

$$\sigma_1 - \sigma_3 = 381D^{-0.71} \tag{1}$$

Table 1 The differential stresses in ductile shear zones of Wutai area

No	Samples	D.Stress MPa	Relief MPa	S.S. deviation	Strain rate /sec	Belt
1	WT-04-1	11.33	6.36-21.33	0.33	1.11E-14	I
2	WT-70-2	13.07	7.33-21.33	0.40	1.46E-14	I
3	WT-72-6	11.88	6.36-15.99	0.34	1.21E-14	I
4	WT-73-3	9.76	5.36-15.99	0.31	8.35E-15	I
5	WT-73-3"	28.54	19.15-41.79	0.81	6.42E-14	I
6	WT-74-2	11.22	5.36-15.99	0.33	1.09E-14	I
7	WT-75-1'	11.81	7.97-15.99	0.29	1.20E-14	I
8	WT-75-1"	30.38	20.82-41.77	0.73	7.23E-14	I
9	WT-76-1'	11.86	7.33-21.33	0.34	1.21E-14	I
10	WT-76-1"	29.45	20.82-41.77	0.65	6.81E-14	I
11	WT-78-3	10.95	7.33-15.99	0.26	1.04E-14	I
12	WT-78-3"	30.14	19.15-41.77	0.63	7.12E-14	I
13	WT-84-4	26.32	17.77-41.77	0.58	5.50E-14	I
14	WT-09-4'	10.96	6.36-21.33	0.33	1.04E-14	II
15	WT-09-4"	31.5	20.82-41.77	0.64	7.74E-14	II
16	WT-11-1	12.28	6.36-21.33	0.36	1.29E-14	II
17	WT-15-2	27.9	20.82-41.77	0.64	6.15E-14	II
18	WT-15-3	12.17	7.97-21.33	0.28	1.27E-14	II
19	WT-17-1'	12.9	8.76-21.33	0.28	1.42E-14	II
20	WT-17-1"	26.82	20.82-41.77	0.51	5.70E-14	II
21	WT-18-2	14.17	8.76-21.33	0.42	1.70E-14	II
22	WT-18-2"	33.45	22.89-41.77	0.61	8.68E-14	II
23	WT-19-1	33.12	22.89-55.70	0.64	8.51E-14	II
24	WT-20-1'	14.58	8.76-21.33	0.33	1.79E-14	II
25	WT-20-1"	34.8	25.54-41.77	0.62	9.35E-14	II
26	WT-23-1	13.97	9.78-21.33	0.37	1.65E-14	II
27	WT-24-2	12.52	7.97-15.99	0.30	1.34E-14	II
28	WT-24-2"	33.45	22.89-41.77	0.69	8.69E-14	II
29	WT-66-3	12.6	7.97-21.33	0.31	1.36E-14	II
30	WT-66-7	11.38	6.36-15.99	0.32	1.12E-14	II
31	WT-80-1	31.54	19.15-41.77	0.84	7.76E-14	II
32	WT-81-2	13.32	7.97-21.33	0.35	1.51E-14	II
33	WT-86-4'	12.89	7.97-21.33	0.34	1.42E-14	II
34	WT-86-4"	33.65	22.89-41.77	0.64	8.77E-14	II
35	WT-93-8	14.31	9.78-21.33	0.33	1.73E-14	II
36	WT-93-8"	31.4	20.82-41.77	0.58	7.69E-14	II
37	WT-28-1	28.78	16.61-41.77	0.63	6.52E-14	III
38	WT-29-1	28.88	19.15-41.77	0.60	6.56E-14	III
39	WT-32-1	15.14	7.33-26.13	0.43	1.92E-14	III
40	WT-34-1	14.23	7.97-21.33	0.39	1.71E-14	III
41	WT-34-2	21.55	14.33-34.88	0.45	3.76E-14	III
42	WT-34-2"	10.84	6.36-13.04	0.24	1.02E-14	III
43	WT-35-3	10.55	5.97-15.99	0.29	9.69E-15	III
44	WT-38-1'	17.05	10.49-25.34	0.50	2.41E-14	III
45	WT-38-1"	6.95	3.77-9.79	0.16	4.38E-15	III
46	WT-28-2	17.01	13.04-21.33	0.35	2.40E-14	III
47	WT-40-2	16.62	11.13-21.33	0.34	2.30E-14	III
48	WT-41-1	17.58	11.13-21.33	0.38	2.56E-14	III
49	WT-42-1	17.73	13.04-21.33	0.34	2.60E-14	III

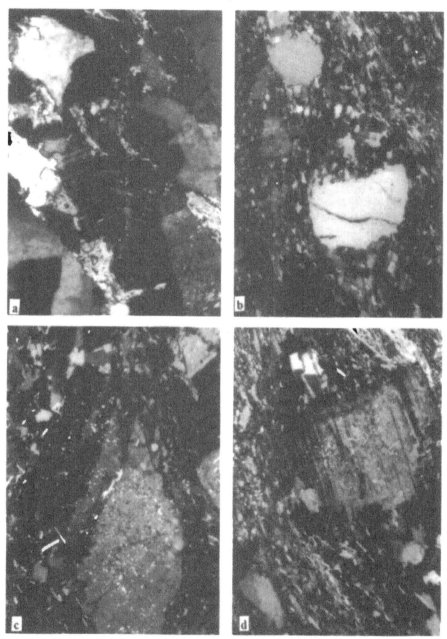

Figure 4. a: Kinked plagioclase in meta-mylonite and its brittle fractured in late period. X100, Mt-29-1, crossed nicols; b: "S" structure and brittle fractured in feldspar porphyroblast, the divergence by subgrainization of feldspar porphyroblast in pressure shadow, X25, Mt-84-4, crossed nicols; c: The dynamic recrystallized quartz distribute surrounding the "feldspar fish", X25, Mt-29-4, crossed nicols; d: The augen structure, augen of feldspar porphyroblast subgrainized in its pressure shadows, the dynamic recrystallized quartz distribute surrounding the augen, X25, Mt-24-2, crossed nicols

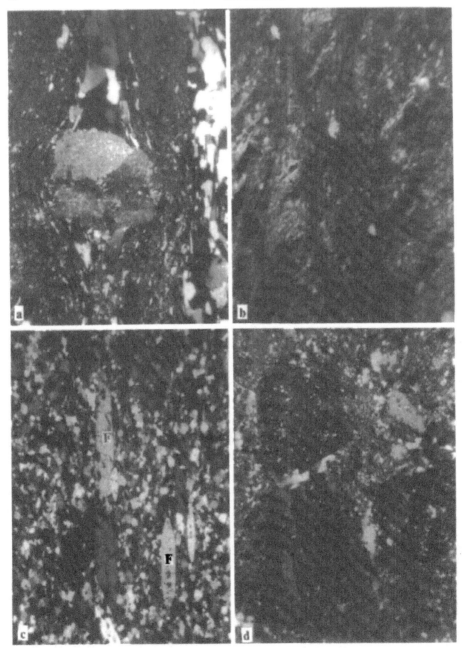

Figure 5. a: The augen structure, feldspar porphyroblast subgrainized and fractured, quartz porphyroblast deformed as tensile band and subgrainized, X25, Mt-18-4, crossed nicols; b: S-C structure, deformed minerals oriented as "S" plane, rheomorphic plane of shearing is "C", X25, Mt-17-2, crossed nicols; c: The oriented "hornblende fish"(F), X25, Mt-93-7, crossed nicols; d: The brittle fractured and rolling structure of plagioclase porphyroblast, X25, Mt-23-1, crossed nicols

Here, D is the recrystallized grain size of quartz or subgrain size of quartz porphyroblast. The average differential stresses, their variation and statistical standard errors are shown in Table 1. The statistical standard errors from more than 65 percent of samples are below 0.5. Therefore, these results are reliable statistically.

Figure 6 is the projection of average differential stress. It shows that these results display concentrations. They are divided into two groups. The regression equations using the method of least squares are respectively as follows:

$$y = 30.4714 + 0.0103x \tag{2}$$
and
$$y = 12.6565 + 0.027x \tag{3}$$

The equations (2) and (3) actually present two near horizontal lines and represent two groups of average values of 30 Mpa and 12 Mpa. They could be the differential stresses of ductile shearing at different period.

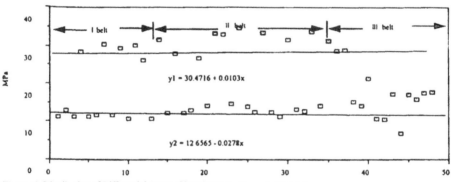

Figure 6. Distribution of Differential Stress of Ductile Shear Zones in Wutai Area

Strain rate during ductile shearing: In recent years, a number of equations of strain rate calculation during ductile shearing were put forward by many scholars based on the study of strain experiments of quartzite and theoretical inference. Hacker et al (1990) pointed out that there is no obvious difference among these equations, but differ in parameter. The deviation from results of these equations are undistinctive. In this paper, the equation of strain rate used is from Heasen (1982) as belows:

$$e = A \cdot^n \exp(-H/PT) \tag{4}$$

Here, e is strain rate; A is constant, equal 3.14×10^{-1}; $\cdot$ is differential stress; H is mobilized entropy; R is gas constant; T is absolute temperature of deformation. Take the average temperature of 273–315 °C of ductile shearing in this area [7, 13]. The results are listed in Table1.

Figure 7 shows the distribution of strain rate. There are two concentrated belt with $0.8 \times 10^{-14} \sim 2.0 \times 10^{-14}$ /s and $5.5 \times 10^{-14} - 9 \times 10^{-14}$ /s respectively. The average values are

1.36×10^{-14}/s and 7.36×10^{-14}/s. This concentrated distribution is consistent to distribution of differential stress.

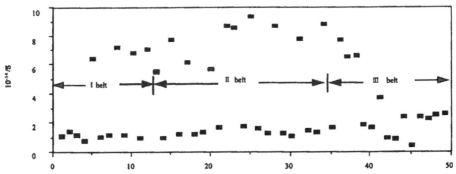

Figure 7. Distribution of Strain Rates of Ductile Shear Zones in Wutai Area

The periods of ductile shearing: The ductile shearing in different period would result in different products due the corresponding differential stress of their formation. There are a lot of ductile shear zones exposed in field of this area. More evidence is necessary for distinguishing these products which were formed in the same period or not. Among the mylonite rocks by ductile shearing, it could be divided into two periods of products from ductile shearing based on the research under polarizing microscopy. Most of dynamic recrystallized quartz from early period of ductile shearing had experienced recovery. Only the dynamic recrystallized quartz of later period could be preserved clearly. The differential stress from subgrain size of quartz are about 30 Mpa and 12 Mpa respectively (Fig. 6, 8). Therefore, the strain rate shows the similar concentrated distribution (Fig.7). This consistent distribution from different ductile shear zones means that they had experienced the relevant tectonic condition. More complexities appear in ductile shear belt III (Fig. 8). It could be interrelated to its relative high metamorphism.

STRUCTURE OF DUCTILE SHEAR ZONES

A lot of ductile shear zones developed in Wutaishan area, especially in Wutai system and gray gneiss (TTG). The scale of these ductile shear zones ranges from several centimeter to tens of meter in width. A series of shear zones with same orientation concentrated as a large ductile shear belt (Fig. 3). Most of TTG which were formed in early period have an outline of vast augen or olive shape with the same orientation of ductile shear belt (Fig. 2, 3). The boundary of TTG are shear zones and there are a lot of shear zones in different scale developed within gray gneiss. Three mega–ductile shear belts oriented in Northeast direction. Each belt is composed of a number of small scale ductile shear zones, these small ductile shear zones are parallel or intersected in small angle with each other. All the ductile shear zones constitute a giant ductile shear system which look like a vast tensile network.

Belt I is located at Northwest part of Wutaishan area and oriented in north-east direction. It contains tens of ductile shear zones with total width of 2~8 km. Several TTG, such as

Wangjiahui rock body and Erkou rock body show obvious ductile deformation. Their long axis orient in the same direction of ductile shear belt.

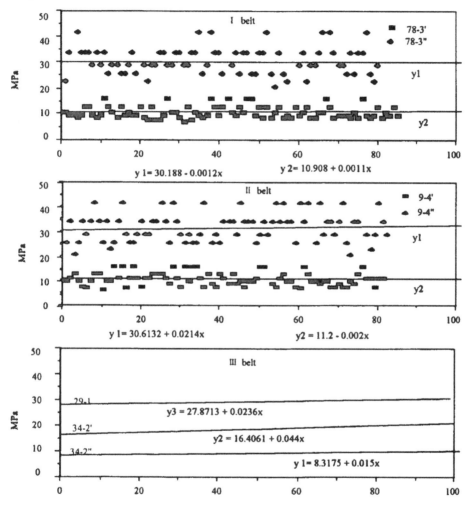

Belt II with width of 2~5 km distributes along the central axis of Mt. Wutaishan in north-east direction. Several of gray gneiss body arrange in the same orientation. Their boundary of them with Wutai system are ductile shear zones.

Belt III basically consists of Longquanguan system, and orients in North-south direction at its southern segment and north-east direction at its northern segmet. It is composed of augen gneiss and mylonite banding. The contact boundary zone with Fuping system is transitional type and distinct ductile shear zone from Wutai system. Most of metamorphic age from Lonquanguan system are about 2000-2100 Ma [4]. This result is younger both than Fuping system of its basement and Wutai system of its cover. It is evident that this relationship is unreasonable on stratigraphy. Therefore, it would be readily receptible to take "Longquanguan system" as a ductile shear zone but strata unite [7].

DISCUSSION: THE GIANT DUCTILE SHEAR SYSTEM AND ACTIVE MIDDLE CRUST

The information from geophysical detection provides us with a more complex structure in the middle crust. It is the comprehensive result from the complicated geological process in active middle. The "Crocodiles" texture in seismic reflection means that a number of reflectors exist in the middle crust. The layers or inclusions with lower seismic wave- velocity in seismic refraction and relative high conductivity in magnetotelluric sounding show the difference of geophysical field. The genesis could be originated from the change of physical property of rock due to the complicated deformation and metamorphism.

Depth of the middle crust is about ten to more than twenty kilometers. Its temperature and pressure condition belong to these of green-schist facies and amphibolite facies. Therefore the exposed rocks with metamorphic facies of green-schist and lower amphibolite would represent the deformation condition in middle crust. The rocks in middle crust are easy to ductile shearing. This flow process by ductile shearing would be complicated due to the non-homogeneous of middle crust. Some part of the middle crust could be thickened by material flow in, and some could thined due to material loss. Therefore, the process of deformation in middle crust becomes quite complicated.

The giant ductile shear system in Wutaishan area shows considerably complex structure, although lack of knowledge for its 3-D structure. The formation temperature of these ductile shear zones is about 260-370 °C [7, 13]. This condition is analogous to middle crust. About the scale, the giant network-type ductile shear system is about several kilometers to more than ten kilometers in width; and tens of kilometers in length. Its scale is comparable to the layer or body with high conductivity and lower wave velocity of the middle crust from geophysical detection. Based on the style of texture from this giant network-type ductile shear system, the well-oriented each ductile shear zone could form good reflective interface, and the "crocodiles" textures in reflection phenomenon could be resulted from a giant network-type ductile shear system. On the effect of ductile shearing process, the decomposition of dark minerals causes hydroxyl free and favorite for the formation of high conductivity. The heavy ions liberated from the mineral's separation would cause decrease of rock density and result in formation of lower wave velocity in the middle crust. The texture of the giant network-type ductile shear system show the deformation state with both ductile and brittle deformation, and reflect the more nonhomogeneous existance in the middle crust.

The complex net-works ductile shear system in Wutaishan area was formed in early geology age. It is necessary to do more research in detail to use this structure and formation process to explain the geophysically sounding phenomena, including construction of 3-d texture of ductile shear system based on mapping and simulating the result by geological model computation etc. Moreover, only the research of structure and petrography in detail from the exposed middle crust can provide us the reasonable and important information for the tectonic process in the active middle crust.

Acknowledgments

The authors thank professor Wang Ren, Zheng Yadong, Deng Jifu and Lin Chuanyong for their helpful reviews and suggestions of an early manuscript. Thanks also to Miss Luo Shulan and Guo Fang for their helps in field and laboratory work. This research is supported by National Natural Science Foundation of China, project number are 49372099 and 48970100.

REFERENCES

1.R. J. Twiss, Theory and applicability of a recrystallized grain size paleopiezometer, *Pure Appl. Geophys.*, 115, 227-244 (1977)

2.J-C. C. Mercier et al, Stress in the lithosphere: Inference from steady-state flow of rocks, *Pure, Appl. Geophys.*, 115.199-226 (1977)

3.M. A. Etheridge and Wilkie J. C. An assessment of dynamically recrystallized grain size as a paleopiezometer in quartz-bearing mylonite zones, *Tectonophysics*, 78 (1-4), 475-508 (1981)

4.Bai Jin(Chief editor), *Early Precambrian geology of Wutaishan Mt.*, Science and Technology Press of Tianjin, China (1986)

5.Yuan Guiping, A discussion on the classify of "Banyukou group" of Wutai system in Wutaishan Mt., *Geology in Shanxi Province*, (2), 176-185 (1986)

6.Wang Qichao, A study on stratigraphy of early Precambrian in Wutaishan area, Yinshan area and eastern segment of Yinshan Mt., *Acta Geologica Sinica*, (1), 16-29 (1988)

7.Li Jianghai, A study on regulation of "longquanguan system" in Northern Taihang Mt. and its ductile deformation in late Archean, *A unpublished doctoral thesis*, Department of Geology, Peking University (1989)

8.Ma Xingyuan(Chief editor), *Lithospheric dynamics atlas of China*, China Cartographic Publishing House (1989)

9.Wu Wenlai, Li Naihuang, Studies on ductile shear zones of western segment of Wutaishan Mt., *Metallurgical Geology in Shanxi Province*,(1), 33-41 (1989)

10.B. R. Hacker, A. Yin and J. M. Christie, Differential stress, stress rate and temperatures of mylonitization in the Ruby Mountain, Nevada: Implication for the rate and duration of uplift, *JGR 95 (B6)*, 8569-8580 (1990)

11.Li Jiliang, Wang Kaiyi, Wang Qingchen et al, Early Proterozoic collision orogenic belt in Wutaishan area, China, *Scientia Geologica Sinica*, (1), 1-11, (1990)

12.Ma Xingyuan, Liu Changquan and Liu Guodong(Principal Compilers), *Xiangshui to Mandal Transect, North China (GGT2)*, Copublished by The Inter-union Commission on the Lithosphere and American Geophysical Union. (1991)

13.Luo Shulan, A study on granitoid rocks of Wutaishan Mt. and the evolution of continental crust, *A unpublished doctoral thesis*, China University of Geoscience, Beijing (1992)

14.Sun Wucheng, Xu Jie, Yang Zhu'en Zhang Xiankang (Principal Compilers), *Fengxian to Alxazouqi Transect, China*, (GGT17), Seismological Press, China (1993)

15.Wang Chunyong, Zhang Xiankang, Lin Zhongyang and Li Xueqing, The characteristics of crust structure in Shulu fault basin and its neighbor region, *Acta Seismologica Sinica*, 16, (4), 472-479 (1994)

16.Teng Jiwen, An introduction of geophysical research of crust and upper mantle in Tibetan, *Acta Geophysica Sinica*, 14, Supplement 1, 1-15 (1995)

Proc. 30ᵗʰ Int'l. Geol. Congr., Vol. 14, pp. 66−73
Zheng et al. (Eds)
©VSP 1997

Study on the Mylonite of the Shallow Structure Level in Southeastern China

Y. SUN, L. WAN and L. Zh. GUO

(*Department of Earth Sciences, Nanjing University, Nanjing 210093, China*)

Abstract

Based on systematic observation and synthetic analysis of mylonites formed at different depths three categories of mylonites have been classified, they are shallow level mylonite with cataclastic rheological deformation (type B), medium-deep level mylonite with ductile-slip rheological deformation (type D), and deep level mylonite with viscous rheological deformation (type V). The type B has been called cooling mylonite not only because of its shallow structure level, but also its low temperature of formation ($<250°C$-$300°C$). Type B mylonites have obvious characteristics compared with types D and V, such as semipenetrative lineation, foliation and fabric, special substructure and dislocation, light colour stress minerals and the semi-reduction condition forming. Finally it is shown that an ordered texture of the cooling mylonite can be caused through the dynamometamorphism, and this mylonite is closely corresponding with the stratabound ore deposit in southeastern China

Keywords: Cooling mylonite, shallow structure level, semipenetrative lineation, ductile deformation domain, stress mineral, substructure, dynamo-metamorphism, southeastern China

INTRODUCTION

Much research in the mylonite field has been carried out since the last century, after C. Lapworth (1885) first coined the mylonite terms from study the Moine Fault in United Kingdom[2]. A Penrose Symposium on mylonitic rocks and its geneses was held in Califonia in 1981, therefore this learning conference provided an important summary for mylonite researches[7]. Most of researchers at home and abroad agree with the idea suggested by R. H. Sibson (1977), i. e. the mylonite is formed during ductile deformation level in crustal depths more than 10 km[4].

On the basis of systematic observation and analysis of fault rocks including more than 100 fault zones and 1500 thin sections as well as 500 chemical specimens in southeastern China, three categories of mylonites have been classified as follows [5]:

First type (B): mylonites of shallow level and lower temperature can be caused by cataclastic rheological deformation, and are often seen as calcareous mylonites.

Second type (D): mylonites of medium-deep level and medium temperature can be formed by ductile-slip rheological deformation, and are commonly found as felsic mylonites.

Third type (V): mylonites of deep level and higher temperature are produced by viscous rheological deformation, and it is generally discovered as a ferromagnesian mylonite (Tabe 1).

One of the authors has discussed the shallow level mylonites with Prof. R. H. Sibson in Berkeley, California, (1988) and in Canberra, Australia (1991). Some characteristics of the shallow level (or cooling mylonite) are briefly described below.

DUCTILE DEFORMATION DOMAIN IN A BACKGROUND OF THE BRITTLE DEFORMATION

On the basis of geological and geophysical exploration in southeastern China especially in the Middle-Lower Yangtze area, phenomena of the regional layering are very obvious not only the layerslip faulting but also the ductile or ductile-brittle deformation. Parameters of the rock mechanics and rock physics of the incompetent bed in the southern Jiangsu Province were individually determined: the uniaxial pressure strength (R_P) 52. 33MPa, the elasticity module (E) 0. 71 $\times$ 10^4MPa, the Poission's ratio (υ) 0. 406, the angle of interfriction (φ) 28°, the density of rock (D) 2. 60g/cm^3, the magnetic susceptibility (k) 12. 0 $\times$ 10^{-6} CGSM, the apparent electric resistivity (ρ) 50-150Ωm, and the velocity of elasticity wave (Vp) 5. 29km/s, (Vs) 2. 59km/s[5]

These parameter values mentioned above when compared with the competent bed, are all lower except for the Poission's ratio. Generally the ductile deformation microdomains within multiple bands could be formed in the footwall of layerslip fault. It is quite obvious that there are the given ductile deformation domain under the brittle deformation background, and the shallow level mylonite may be produced [6]. Practically determined examples taken from Suzhou, Jiangsu Province; Dayu, Jiangxi Province; Anhua, Hunan Province etc. (Fig. 1) will be progressively explained below.

SEMIPENETRATIVE LINEATION FOLIATION AND FABRIC

Some mylonitic samples of the ductile deformation microdomains were collected from Devonian rocks in Suzhou Dongshan, Jiangsu Province. Quartz crystal orientation parameters including the columnar plane ($10\bar{1}0$), the pole lines of rhombic plane r($10\bar{1}1$), z($01\bar{1}1$) and the crystal axis a (100) had been carried out . Among these parameters the angle between poles of r and z planes are strictly 38°. The fabric diagrams of cooling mylonite are shown in Fig. 2.

In a general way, both S and R tectonites present a monoclinic symmetry type in the type B mylonite (Fig. 3), and the B tectonite and rhombus symmetry type commonly occur in the mylonite type D (Fig. 4) (Table-1). As compared with the mylonite type D. The development of lineation, foliation and fabric and the

Table 1　The Faulted Rock, Rheology and Dislocation of Different Deformational Structural Levels in the Yangtze Area

Deformational Structural Level				Deformation and Rock	Rheology and Slip		Fabric and Dislocation	
Level	Failure	Depth (km)	Similar Body	Faulted Rock	Rheology	Slip	Fabric	Density of Dislocation(ρ) Flow stress(σ)
Brittle Deformational Level	brittle failure	<3 or <5	solid (Hooke)	cataclasite	cataclastic flow	not obvious	homogeneous	not obvious
	parabrittle failure	3—10 or 5—10	elasticity (Kelvin) or viscosity—elasticity (Biulges)	Lower temperature, cataclastic flow mylonite (B type), calcareous mylonite	granulated flow, difference flow	stick slip fast slip a creep	S tectonite as a dominant, found the R and SF tectonites	slipping movement, disruptive wall, $\rho \geqslant 10^4 - 10^5/cm^2$ $\sigma \geqslant 80 - 100MPa$
Ductile Deformational Level	paraplastic yield	10—15	elasticity —viscosity (Maxwell)	medium temperature, ductile slip mylonite (D type), felsic mylonite	plastic ductile flow	stick — creep slip, α — β creep	B tectonite as a dominant, in the characteristic of rhombic symmetry	climb movement, thick wall, $\rho = 10^7 - 10^8/cm^2$ $\sigma = 10 - 80MPa$
	plastic yield	15—20	plasticity (Saint—Venant)					
Viscous Deformation Level	viscous flow	>20	viscosity —plasticity (Bingamowu) or viscosity (Newton)	high temperature, viscous mylonite (V type), ferromagnesian mylonite	viscous flow Ode flow	Creep slip β creep, slowly slip	tectonite of compounding type flow fabrics	$\rho \leqslant 10^7 - 10^8/cm^2$. $\sigma \leqslant 10 - 80MPa$ or dislocation disappear

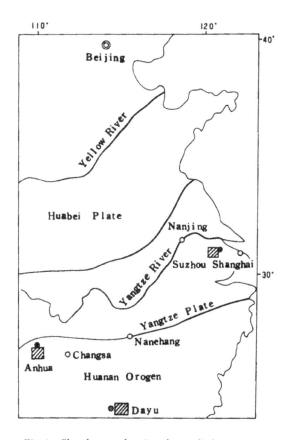

Fig. 1   Sketch map showing the studied areas

degree of movement symmetry within the mylonite type B, which was formed in the semi-penetrative and lower temperature realm(<250℃-300℃ determined using inclusions), are distinctly lower, so it is called a slip mylonite, or cooling mylonite or cooling type structure.

SPECIAL SUBSTRUCTURE AND DISLOCATION OF PRIMARY STAGE IN THE PLASTIC DEFORMATION

Broadly speaking the subgrain manifests a narrow and disconnected feature (Fig. 5), and the deformation bands and lamellae are also rarely consummate in the cooling mylonite. It should be shown that an obvious subgrain within the dynamic recrystallization can be found in a wide type B mylonite band (Fig. 6)

The dislocation state of a cold processing-quench stage, in which there is a larger dislocation density (d), $d \geqslant 10^8$-$10^9 g/cm^2$ (Fig. 7) and a higher differentiation flow stress (σ),$\sigma \geqslant 80$-100MPa, has been mainly obtained in the light of observation statistics of more than 200 micro-photographs taken with TEM from quartz

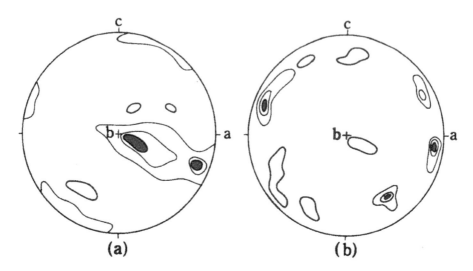

Fig. 2. a. b Fabric diagram of the pole-line of m plane of quartz preferred orientation in both the outer band (a) and inner band (b) in the type B mylonite domain at Suzhou Dongshan, in Jiangsu province (the fabric diagrams of the pole-lines of r plane and z plane are similar to the m plane, to be omitted).
Determined 200 times, Density grade: 7-6-4-2%

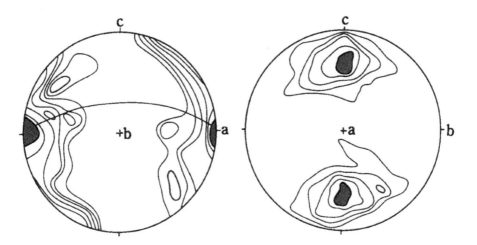

Fig. 3 Illustration of orientation developed for the c-axis of quartz in the mylonite type B in the shearing zone at Nanjing.
Determinied 141 times, ⊥ b, Density grade: 12-10-8-4-2-1-0%

Fig. 4 Illustration of orientation developed for the c-axis of quartz in the mylonite type D in the compressive zone at Suzhou.
Determined 127 times, ⊥ a, Density grade: 9-7-3-1-0%

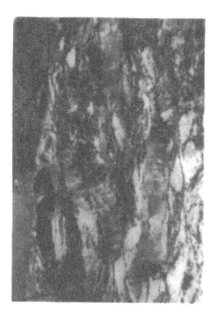

Fig. 5 Cooling mylonite, the preferred orientation and the disconnected subgrains.
Orthogonally polarized × 100, Anhua, Hunan Province (No. 88-11-31)

Fig. 6 Cooling mylonite, the obvious connected subgrain zones,
Orthogonally polarized × 100, Anhua, Hunan Province (No. 88-11-39)

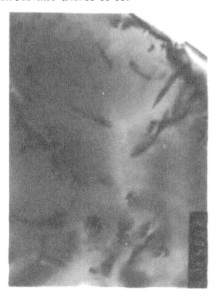

Fig. 7 Free dislocations and appearing state × 20000 (No. 84101)

Fig. 8 Discontinuous dislocation wall and trending into a subgrain × 2000 (No. 84100)

grains mineral of type B mylonite. and most of them as mentioned above are free dislocations (Fig. 7) and discontinuous dislocation wall (Fig. 8)

LIGHT COLOUR STRESS MINERALS AND THE SEMI-REDUCING CONDITION OF SHALLOW MYLONITIZATION

In thin section observation it is often seen that there were pressure sensitive mineral calcite and mobile mineral quartz in the mylonite type B. As far as the authors are aware that the dynamic twinning $(01\bar{1}2)$ is easily seen in the calcite, and the segregate band or silicified zone is widely formed with the quartz. In addition shallow minerals such as muscovite, sericite, potash feldspar, the layer silicated mineral with light colour mineral and the acidic plagioclase as well as foliate chlorite can be also found[3]. As result of chemical analyses, in general, $Fe^{3+}/Fe^{2+}=0.$ 2-2 indicates a sem-reducing condition. Meanwhile, the chemical composition of type B mylonite shows a high alkaline (K_2O, Na_2O), rich Si and much more Fe_2O_3, H_2O compared with type D mylonite. In addition some SiO_2 might result from a hypergene action.

Therefore there is a reaction formula for the sericite forming

$3KAlSi_3O_8 + 2H^+ \rightarrow KAl_3SiO_3O_{10}(OH)_2 + 2K^+ + 6SiO_2$
potash feldspar   sericite

CONCLUSION

According to explanation cited above it is deduced that the mylonite forming does not entirely depend the depth of crust as the sole condition. Under a given temperature ($<250°C - 300°C$) and pressure, the cooling mylonite (dynamometamorphic tectonite) might be produced in the shallow structural level[1].

It is considered that a ductile deformation domain of the cooling mylonite with a relatively thick-bedded superimposition could occur in brittle fault zones of regional layerslip systems in the shallow crust in southeastern China, the stratabound ore deposit of a medium-lower temperature concerning the dynamometamorphism or a flow shearing can be often formed.

Acknowledgements

Thanks are due to Prof. T. Suzuki of Department of Geology, Kochi University, Japan and Prof. M. Faure of Department of Earth Sciences, Orleans University, France for the thorough guiding and discussion and the writing of this paper.

Project supported by the National Natural Science Foundation of China

REFERENCES

1. J. V. Godard and J. P. Evans. Chemical change and fluid-rock interaction in faults of crystalline thrust sheets, northwestern Wyoming, U. S. A. , J. Struct. Geol. , 17, 533-547 (1995).

2. C. Lapworth. The Highland Controversy in Brittle Geology: Its Cause, Course and Consequences, Nature, 32, 558-559 (1885).

3. G. E. Lioyd and B. Freeman. Dynamic recrystallization of quartz under greenschist facies conditions, J. Struct. Geol. , 16, 867-882 (1994).

4. R. H. Sibson. Fault rocks and fault mechanism, J. Geol. Soc. , 33, 191-213 (1977).

5. Y. Sun and X. Z. Shen. Determination of the parameters of Rock Mechanic and physical properties from slip systems of the Nappes in the Lower Yangtze Area, South China, Geotectonica et Metallogenia, 16, 161-170 (1992).

6. Y. Sun, X. Z. Shen and T. Suzuki. Study on the Ductile Deformation Domain of the Simple Shear in Rocks-Taking Brittle Faults of the Covering Strata in the Southern Jiangsu Area as an Example, Science in China(Series B), 35, 1512-1520 (1992).

7. J. Tullis. Penrose conference report: significance petrogenesis of mylonite rock, Geology, 10, 227-230 (1982).

PART 2

TECTONOPHYSICAL SIMULATION AND DIGITAL MODELING

Proc. 30ᵗʰ Int'l. Geol. Congr., Vol.14, pp 77-88
Zheng *et al.* (Eds)
© VSP 1997

Simulation of Some Geological Structures*

ZUOXUN ZENG, LILIN LIU, SITIAN LI, YONGTAO FU AND SHIGONG YANG

Faculty of Earth Sciences, China University of Geosciences, Wuhan, 430074, China

Abstract

This paper presents the results from simulation of seven types of geological structures by use of mathematical and physical methods. Curvilinear equations for plume and rib marks are derived. From mathematical simulation, the theoretical values of plume fringe angle for monoplume structure and half-plume structure are 72° and 81°, respectively. Different plume and rib structures are achieved from physical models. Measurement values of plume fringe angle from physical models and natural rocks are statistically identical with the theoretical values. Experiments are performed for ductile shear zones by use of Chen's grid method. The finite shear strain in the model is similar to that in natural shear zones. In simulation of chess-board structures (or conjugate fractures), the clay models are adopted to examine the effect of material behavior, mean stress and sustained loading on conjugate shear angle, respectively. The results from simulation of epsilon-type structure by use of superposition principle, finite element method, photoelastic method and in-plane moiré method indicate that a large plate sustained by two nuclear columns and loaded by unidirectional body forces is an ideal model for this type of structure. It is concluded from the simulation of uninucleus-type torsional structures by use of the stress function method, finite element method, brittle coating and bubble method that the nearer to the nuclear column, the greater the values of the principal stresses, maximum shear stress and differential ture strain, and that these values decrease rapidly with the increase in distance away from the center of the nuclear columns. In particular, the experiment by the bubble method indicates that the distortion mainly takes place near the nuclear column and that nearly only rigid rotation takes place far from it. At the same time, the Weissenberg effect is observed in the bubble method experiment. The binucleus-type torsional structure is a new kind of structural type presented by the first author. It is characterized by two nuclear columns which distinguish it from uninucleus-type torsional structures. Analytic method in viscoelasticity, tissue paper method, clay model, shadow moiré method are adopted in the simulation of this type of structure. It is shown by both mathematical and physical simulation that torsional movement around two nuclear columns can produce s-shape, reversed s-shape, elliptic arc, hyperbolic and turbine-like structures. The simulation of India-Eurasia collision come to the conclusion that the inversion of Tan-Lu faults and Red River fault, the clockwise rotation of Southeast Asia may be the result of the collision.

Keywords: simulation, plume and rib structure, ductile shear zone, chess-board structure, epsilon-type structure, uninucleus-type torsional structure, binucleus-type torsional structure, India-Eurasia collision

INTRODUCTION

Simulations have been carried out throughout the whole history of geology. The earlier researchers made a lot of contributions to this domain [e.g. 1, 4-13, 15-26]. Simulation of

---

* The paper is supported by the State Education Commission of China, the open laboratory fundation and research project of the Ministry of Geology and Mineral Resources of China

some geological structures are done by use of mathematical and physical methods, including analytic methods, the bubble method, the shadow moiré method, the in-plane moiré method, Chen's grid method, and the brittle coating method. The structures are plume and rib structures, ductile shear zones, chess-board structures, epsilon-type structures, uninucleus-type torsional structures, binucleus-type torsional structures and the collision tectonics of India into Eurasia.

PLUME STRUCTURES AND RIB STRUCTURES

Plume and rib structures are classified into eight patterns. They are monoplume, biplume, furcate plume, bend plume, half-plume, counter plume, rib and plume-rib compound structures.

Curvilinear equations for plume and rib marks are derived, respectively, as

$$y = 2ae^{\frac{\pm i}{2L}(x+nb)} \tag{1}$$

and

$$y^2 = 4a(x+nb) \tag{2}$$

where x coincides with plume axis, y is perpendicular to it, n is an arbitrary constant and b is the unit along the x-axis. From mathematical simulation, the theoretical values of plume fringe angle for monoplume structure and half-plume structure are 72° and 81°, respectively. Different plume and rib structures are achieved from scaled models. The statistical results of physical simulation show that the mean values of plume fringe angle for monoplume structure and half-plume structure are 72° and 83°, respectively. From observation on some natural structures, they are 75° and 80°, respectively.

DUCTILE SHEAR ZONE

Card stack model was used in the simulation of inhomogeneous finite strain fields in ductile shear zones [17]. Chen's grid method is adopted in this paper. It is a grid method in co-moving system developed based on the strain-rotation decomposition theorem of Chen [3]. As long as the displacement differences $\Delta u, \Delta v$ and the distances $\Delta l_x, \Delta l_y$ between two neighbouring points along the co-moving coordinate axes determined, the angle of whole rotation (Ω) and strain components can be computed as follows:

$$\Omega = arcsin\left[\tfrac{1}{2}\left(\tfrac{\Delta v}{\Delta l_x} - \tfrac{\Delta u}{\Delta l_y}\right)\right] \tag{3}$$

$$\varepsilon_x = \tfrac{\Delta u}{\Delta l_x} + (1 - cos\,\Omega) \tag{4}$$

$$\varepsilon_y = \tfrac{\Delta v}{\Delta l_y} + (1 - cos\,\Omega) \tag{5}$$

$$\varepsilon_{xy} = \tfrac{1}{2}\left(\tfrac{\Delta u}{\Delta l_y} + \tfrac{\Delta v}{\Delta l_x}\right) \tag{6}$$

The undeformed and deformed grids of model 1 of a ductile shear zone are shown in Fig.1(a). The finite rotational field and the finite shear strain field for model 1 are given in

Figs.1(b) and (c). Showing of the rotational field and lateral effect of simple shear are two advantages of this result.

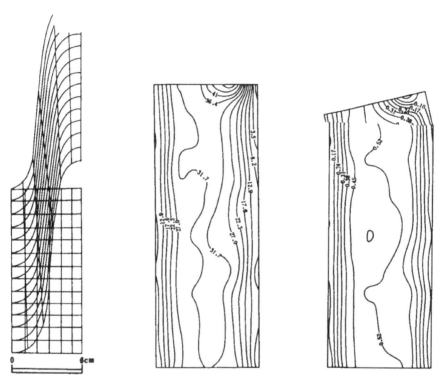

Figure 1. Undeformed and deformed states of grid (a), finite rotation field (b) and shear strain field (c) in model I of a ductile shear zone.

CHESS-BOARD STRUCTURES

In simulation of chess-board structures (or conjugate fractures), the clay models are adopted to examine the effect of material behavior, mean stress and sustained loading on the conjugate shear angle, respectively. Based on the simulation, the following conclusion can be made: (1) initial conjugate shear angles decrease with the increase of internal friction angle of the model material (Fig.2), (2) conjugate shear angles increase with the decrease of the limit mean stress (Table 1), and (3) conjugate shear

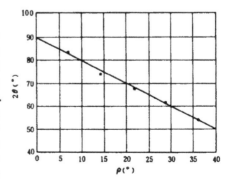

Figure 2. Relationship between conjugate shear angle and the internal frictional angle of the model material.

angles can increase from smaller than to greater than 90° under sustained loading. It is obvious that here is a quantitative result instead of qualitative simulation of chess-board structures in earlier clay model [12].

Table 1. Experiment result showing the effect of mean stress on conjugate angle.

No.	limiting stress state	limiting mean stress	average conjugate angle (°)
1	$\sigma_b \leftarrow\square\rightarrow \sigma_b$	$\sigma = \sigma_v / 2 > 0$	32.2
2	$\tau_v\ \square\ \tau_v$	$\sigma = 0$	55.7
3	$\sigma_b \rightarrow\square\leftarrow \sigma_b$	$\sigma = -\sigma_v / 2 < 0$	61.3
4	$\sigma_b\ \square\ \sigma_b$	$\sigma = -(\sigma_v + \sigma_v') \ 2 < 0$	85.0

EPSILON-TYPE STRUCTURE

The results from simulation of epsilon-type structure [11, 13] by use of superposition principle, finite-element method, photoelastic method and in-plane moiré method indicate that

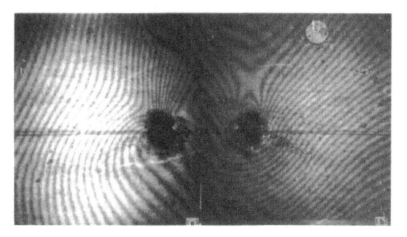

Figure 3. Photographs of moiré pattern of u-displacement (a) and v-displacement (b) fields in the model of epsilon-type structure.

a large plate sustained by two nuclear columns and loaded by unidirectional body forces is an ideal model for this type of structure. The in-plane moiré experiment was performed with a centrifuge. Figs.3(a) and (b) show that the moiré patterns of u-displacement and v-displacement fields, respectively. From the in-plane moiré analysis, the

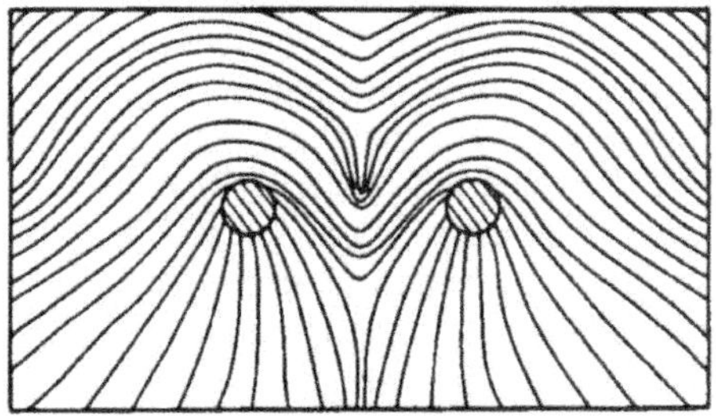

Figure 4. Trajectories of principal strain (ε_1) from in-plane moiré method for epsilon-type structure.

trajectories of the principal strain (ε_1) (Fig.4), contours of the maximum shear strain, the displacement field and contours of the rigid rotational angle of the model of epsilon-type structures are obtained.

UNINUCLEUS-TYPE TORSIONAL STRUCTURES

In the infinitesimal deformation theory of elasticity, the displacements of uninucleus-type torsional structures are given as

$$u = 0, \quad v = \frac{Mr}{4\pi Gh}\left(\frac{1}{a^2} - \frac{1}{r^2}\right) \qquad r \geq a \qquad (7)$$

where u and v are the radial and circumferetial displacement, respectively, r the polar dis-

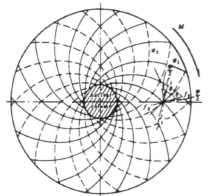

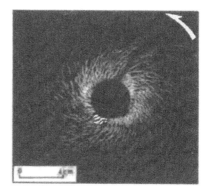

Figure 5. Trajectories of principal stress of uninucleus-type torsional structures from elasticity solution. f_1-f_1' and f_2-f_2' are the directions of conjugate fractures, 2θ the conjugate angle and φ the internal friction angle. Half arrows show the directions of relative movements.

Figure 6. Brittle-coating crack pattern in the model of uninucleus-type torsional structures. Half arrow shows the direction of relative movements.

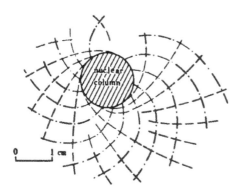

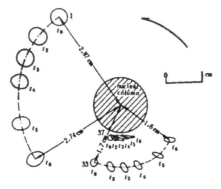

Figure 7. Trajectories of principal stretch in the model of uninucleus-type torsional structures. Half arrow shows the direction of relative rotation.

Figure 8. Configurations of three bubbles at different deformation stages. Half arrow shows the direction of relative rotation.

tance, a the radius of the nucleus column, M the moment of torsion, and G the shear modulus of elasticity. The trajectories of principal stress from elasticity solution is shown in Fig.5. By use of a new brittle coating [24], the turbine-like fracture patterns which initiated from the nuclear columns are obtained (Fig.6). The inhomogeneous finite strain field of the model for uninucleus-type torsional structures is obtained in the bubble method proposed by the first two authors of this paper [23]. As deformation markers, bubbles are put into transparent model materials and the finite strains can be determined from the deformation of the bubbles. The trajectories of the principal stretches are developed as a turbine-like pattern (Fig.7).

The configurations of bubbles at different stages (Fig.8) indicate that distortion mainly takes place near the nuclear column and that the nearly only rigid rotations take place far from it. The particle movement paths are not exact in circles as predicted by the infinitesimal deformation theory of elasticity. They show characteristics of finite deformation with the Weissenberg effect.

BINUCLEUS-TYPE TORSIONAL STRUCTURES

Binucleus-type torsional structures are characterized by two nuclear columns which distinguish them from uninucleus-type torsional structures. Six examples are developed in Hunan, Fujian, Guangxi, Xinjiang of China, respectively [22, 25].

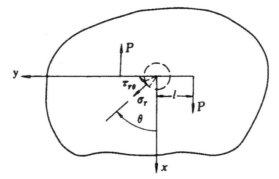

Figure 9. Infinite plate model of binucleus-type torsional structures. The two P are concentrated forces representing the reactions of the nuclear columns.

We assume that a round plate is cut off from the Earth's crust, in which there are two fixed nuclear columns, and let M be the moment of a force couple acting on the plate by the surrounding landmass. This is the geometry and displacement and loading conditions of the mechanical model. For convenience of calculation, in the light of de Saint-Venant's principle, the original model may be reduced to a problem of a Burgers viscoelastic infinite plate subject to the action of two concentrated forces P that are equal in magnitude, opposite in direction and $2l$ apart (Fig.9).

The stress components can be obtained as

$$\left.\begin{aligned}
\sigma_r &= M_1\xi_1(-M_2 + M_3\lambda_1\xi_2) - M_1\eta_1(-M_2 + M_3\lambda_1\eta_1) \\
\sigma_y &= M_1\xi_1(M_4 - M_3\lambda_1\xi_2) - M_1\eta_1(M_4 - M_3\lambda_1\eta_1) \\
\tau_{by} &= -M_1\xi_2(M_4 + M_3 X\xi_1) + M_1\eta_1(M_4 + M_3 X\eta_1)
\end{aligned}\right\} \tag{8}$$

where

$$M_1 = \frac{P_0}{4\pi}, \quad M_2 = 3 + f_0(t), \quad M_3 = 2\left[1 + f_0(t)\right] \text{ and } M_4 = 1 - f_0(t)$$

while

$$f_0(t) = \frac{1}{W}\left(\frac{W_1}{J} + \frac{-E_6E_5 - 2JR + 2W_1}{IE_5 - 4J}exp\frac{E_2}{2}t + \frac{-E_6E_4 - 2JR + 2W_1}{IE_4 - 4J}exp\frac{E_3}{2}t\right) \quad t > 0 \tag{9}$$

where

$$W = 2a_1a_5 + 1 \quad \text{and} \quad E_5 = I - E_1 \tag{10}$$

The displacement components and strain components are also obtained.

As shown in Fig.10, between the two nuclear columns, the trajectories of the principal compressive stresses are distributed in an s-shape while those of the principal tensile stresses in a reversed s-shape. Fig.11 shows a brittle-coating crack pattern on the model. Fig.12 shows a shadow moiré pattern of the model, in which, two depressions and two upwards are developed. The relation between the angle of regional rotation (θ) and deflection (W) is given in Fig.13. The trajectories of principal stretch are in s-shape and reversed s-shape determined from the bubble method (Fig.14).

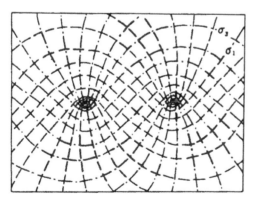

Figure 10. Trajectories of principal stress in the model of binucleus-type torsional structures

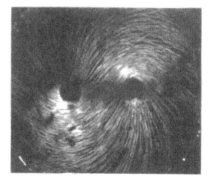

Figure 11. Brittle-coating crack pattern on the model simulating binucleus-type torsional structures after the outside of the model undergo a clockwise rotation of 10°.

It is shown by means of both analytical and scaled-model simulation that torsional movement around two nuclear columns can produce s-shaped, reversed s-shaped, elliptic arc, hyperbolic and turbine-like structures. And uninucleus-type torsional structures can be thought of as a special case of binucleus-type torsional structures.

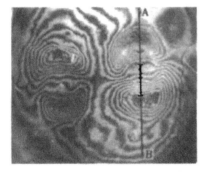

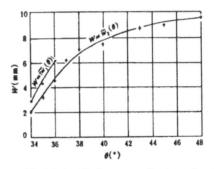

Figure 12. Shadow moiré pattern of model II of binucleus-type torsional structures.

Figure 13. Relationship between regional rotation (θ) and deflection (W) in the binucleus-type model.

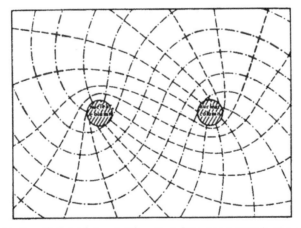

Figure 14. Trajectories of principal stretch from bubble method of binucleus-type torsional structures.

INDIA-EURASIA COLLISION

Tapponnier's continental extrusion model [15, 18] has played an important role in the explanation of the forming mechanism of the Cenozoic tectonics and sedimentary basin in eastern and southeastern Asia. Further physical and mathematical simulations based on Tapponnier's model have made new progress [e.g. 5, 26]. However, these simulation works are mainly done for southwestern China. Much information indicate that right-lateral shear has been taking place along NNE-trending preexisting faults in eastern China since the beginning of the Cenozoic. It is considered that the opening of the Bohaiwan basin (or North China basin) is relative to the right-lateral shear of these NNE-trending faults [2, 9, 14]. In Tapponnier's model, the formation of Cenozoic fault basin in east China was explained as resulting from left-lateral slip of NE-trending and NEE-trending faults. It is clear that there is difference between this consideration and observed practical data. Jolivet *et al.* [9] suggested that the Japan Sea and the Bohaiwan basin opened as pull-apart basin along two N-S right-lateral systems confined between the Baikal-stanovoy system and the left-lateral Qing-Lin fault. However, more and more geological information indicate that the NNE-trending faults such as Tan-lu faults extend southward beyond QinLin. Based on geological practice, Li *et al.* [14] suggested that the India-Euarsia collision casused the inversion of preexisting NNE-trending faults from left-lateral compresso-shear to right-lateral tenso-shear, triggered the deep processes and formed the largescale fault basins in east China. This viewpoint is supported by recent researches. Our aim is to apply simulation method to examine the possibility that India-Eurasia collision cause the right-lateral tenso-shear along the NNE-trending faults in east China, and trigger the opening of fault basin. The inversion of Red River faults and the opening of Yinggehai basin and the clockwise rotation of Southeast Asia are also discussed by means of model analysis.

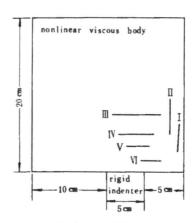

Figure 15. Mechanical model for the India-Eurasia collision. The western and northern boundaries are fixed. The Southern, eastern and top boundaries are free. Faults I to VI represent, respectively, Zhenhai, Nan'au, Tan-Lu, Aerjin-Haiyuan, Kunlun, Xianshuihe and Red River before the collision.

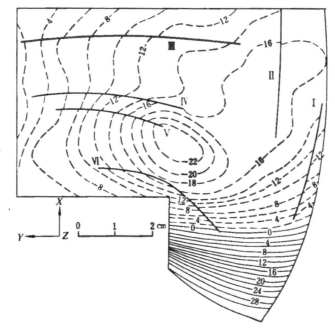

Figure 16. *v*-displacement field for the intender pressing forward from 0 to 5 cm. The values of displacement are all had an enlargement factor of 2.85 (in mm).

Taking a similar approach to the experimental model used by Peltzer and Tapponnier [15], we let a rigid indenter represent the relative rigid Indian subcontinent. The width of the indenter is 5 cm corresponding to 2000 km in practical distance. The Eurasian continent is represented by a rectangular parallelepiped. The plane dimension is 20 cm × 20 cm and the height is, 0.6, 0.9, 1.3, 1.6, and 3.0 cm (physical models) and 1.0 cm (numerical model), respectively, instead of 11 cm, with the western and northern boundaries fixed, the bottom boundary varies in different experiments: fixed, slideable with friction, slideable without friction, or free, respectively, and the top boundary is free, so the mechanical model is a three-dimensional deformation model instead of two-dimensional deformation one. In Fig.15, faults I to VI represent, respectively, the Zhenhai-Nan'ao, Tan-Lu, Aerjin-Haiyuan, Kunlun, Xianshuihe and Red River before the India-Eurasia collision.

From the physical and numerical model analyses, we acquired the following insights into the India-Eurasia collision:

1. As long as a regional basement décollement layer or an asthenosphere exists, India-Eurasia collision can influence east China and even farther, reactivating the preexisting NNE-trending faults in that region to have right-lateral slide and triggering the opening of the basins producing oil and gas as the Bohaiwan basin in east China.

2. With Indian subcontinent pressing into Eurasia, the Red River fault led to left-lateral slide first and then right-lateral later. The inversion from left-lateral slip to right-lateral slip is attributed result of progressive deformation. The latter slip may be the major controlling factor for the opening of the Yinggehai basin producing oil and gas.

3. According to the model result, the maximum extrusion is found in the area between the Xianshuihe and Red River faults (Fig.16), and the maximum extrusion is about 300 km,

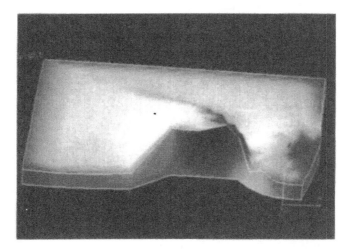

Figure 17. One of the results of a numerical model of three-dimension for the India-Eurasia collision.

but it is not an extrusion of the whole block. It is a displacement field of inhomogeneous extrusion. Much energy is absorbed by the uplift in the front of the indenter (Fig.17).

4. In the process of the pressing of India into Eurasia, Southeast Asia rotated clockwise in different degrees. The maximum rotation is greater than 60°. According to the model result, this kind of rotation is not regional block rotation but a finite rotation field (Fig.18).

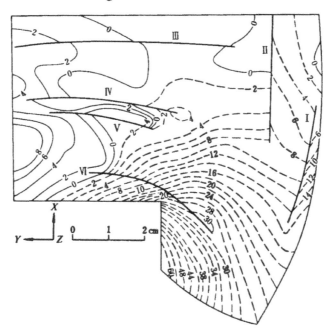

Figure 18. Finite rotation field around Z-axis for the indenter pressing forward from 0 to 5 cm (in degree). The solid lines represent counter clockwise rotation and the dotted lines represent clockwise rotation.

SUMMARY

Some quantitative experiment methods, especially large deformation methods, are applied in the simulation of geological structures. Looking forward to the future, tectonic simulation will play even more important part in geomechanics and modern tectonics. It is considered that three dimensional modeling, taking rheological behavior, finite deformation and structural association in different tectonic setting into simultaneously account, represents one of the main direction of attack in tectonic simulation for some time to come.

Acknowledgements

We thank An Yin and Yadong Zheng for their constructive criticism and helpful reviews on the manuscript.

REFERENCES

1. V.V. Belousov and M.V. Gzovsky. Experimental tectonics, *Physi. and Chem. Earth.* 6, Pergamon Press (1965).
2. W. Chen and J. Nabelek. Seismogenic strike-slip faulting and the development of the North China basin. *Tectonics* 7, 975-989 (1988).
3. Z. Chen. *Rational mechanics*, China Institute of Mining & Technology Press , Xuzhou (1988) (in Chinese).
4. H. Cloos. Experimente zur inneren Tektonik, *Z. Miner. Geol. Pal.* 12, 609-621 (1928).
5. Ph. Davy and P.R. Cobbold. Indentation tectonics in nature and experiments, 1.Experiments scaled for gravity, *Bull. Geol. Ins. Univ. Uppsals.* 14, 129-141 (1988).
6. J.M. Dixon. Recent developments in centrifuge modelling of tectonics processes:equipment, model construction techniques and rheology of model materials. *J. Struct. Geol.* 7, 83-102 (1985).
7. S.K. Ghosh, N. Mandal, D. Khan and S.K. Deb, Models of superposed bulking in single layers controlled by initial tightness of early folds, *J. Struct. Geol.* 14, 381-394 (1992).
8. M.K. Hubbert. Theory of scale models as applied to the study of geologic structures, *Geol. Soc. Am. Bull.* 48, 1459-1520 (1937).
9. L. Jolivet, P. Davy and P. Cobbold. Right-lateral shear along the Northwest Pacific margin and the India-Eurasia collision, *Tectonics* 9, 1409-1419 (1990).
10. L. Lan and P. Hudleston. Rock rheology and sharpness of folds in single layers, *J. Struct. Geol.* 18, 925-931 (1996).
11. J.S. Lee. Experimental and theoretical study on the ε structure, *Sci. Rec. Acad. Sin.* 1, 461-470 (1945).
12. J.S. Lee, C.H. Cheng and M.T. Lee. Experiments with clay on shear fractures, *Bull. Geol. China.* 28,25-32 (1948).
13. J.S. Lee. *Introduction to geomechanics*, Science Press, Beijing, China; Gordon and Breach, Science Publishers, Inc., New York (1984).
14. S. Li, S. Yang, C. Wu and S. Cheng. Geotectonic background of the Mesozoic and Cenozoic rifting in East China and adjacent areas. In: *Tectonopalaeogeography and Palaeobiogeography of China and adjacent regions.* H. Wang, S. Yang and B. Liu (Eds). pp.109-126. Press of China University of Geosciences, Wuhan (1990) (in Chinese with English abstract).
15. G. Peltzer and P. Tapponnier. Formation and evolution of strike-slip faults, rifts and basins during the India-Asia collision: an experimental approach, *J. Geophys. Res.* 93 (B12), 15085-15117 (1988).
16. H. Ramberg. *Gravity, deformation and the earth's crust: in theory, experiments and geological application, 2nd. ed.,* Academic Press Inc. London (1981).
17. J.G. Ramsay and M.J Huber. *The Technique of modern structural geology.* 1, Strain analysis, Academic Press, London (1983).
18. P. Tapponnier, G. Peltzer, A.Y. LeDain, R. Armijo and P. Cobbold. Propagating extrusion tectonics in Asia: insights from simple experiments with plasticine, *Geology* 10, 611-616 (1982).
19. R. Wang. Some problems of mechanics in tectonic analysis, *Advances in Mechanics* 19, 145-157 (1989) (in Chinese with English abstract).
20. W. Wang. Mechanical research on typical fracture system, *Phys. Chem. Earth.* 17,149-158 Pergamon Press (1990).
21. C. Wang. *Theory and application of global tectonics stress field*, Changchun Publishing House, Changchun (1994) (in Chinese with English abstract).
22. Z. Zeng. Mechanical research on binucleus-type vortex structures, *Acta Geologica Sinica* 3,343-362 (1990).
23. Z. Zeng and L. Liu. Bubble method in experimental analysis for finite strain, *Exploration of Geosciences* 3, 43-49. China University of Geosciences, Wuhan (1990) (in Chinese with English abstract).
24. Z. Zeng and L. Liu. Experimental research on typical fracture systems by use of a new brittle coating, *Exploration of Geosciences*, 6, 40-45. Press of China University of Geosciences, Wuhan (1992).
25. Z. Zeng and L. Liu. Experimental research on binucleus-type vortex structure, *Bull. Insti. Geome.CAGS* 15, 111-119. Geological Publishing House, Beijing (1993) (in Chinese with English abstract).
26. J. Zhong. Tectonic feature of Qinghai-Xizhang plateau and its stressed state, *Seismology and Geology* 10, 67-87 (1988).

Proc. 30 Int'l. Geol. Congr., Vol. 14,* pp. 89-105
Zheng *et al.* (Eds)
© VSP 1997

Mechanical Behavior of Halite and Calcite Shear Zones from Brittle to Fully-Plastic Deformation and A Revised Fault Model

EIKO KAWAMOTO and TOSHIHIKO SHIMAMOTO

Earthquake Research Institute, University of Tokyo, Tokyo 113, Japan

Abstract

Shearing experiments on halite, calcite and mixed halite-calcite layers of about 0.7 mm-thick have been performed to study mechanical properties of fault, using a high temperature biaxial testing machine. Ambient temperature was increased in linear proportion to the normal stress on the simulated fault to study the effect of geothermal gradient. Results from halite shear zones provide an experimental strength profile of the lithosphere, which can be divided into brittle, intermediate, and fully-plastic regimes, with the peak strength slightly below the middle of the intermediate regime. Stick-slip (unstable fault motion) was recognized down approximately to the strength peak at which the velocity dependency of friction changes from the potentially unstable velocity weakening to the velocity strengthening behavior. Because both power and exponential laws can fit to the data of halite shear zones in the fully-plastic regime, the transitional mechanisms between high and low temperature plastic deformations may operate in this regime. Results from similar experiments on mixed halite-calcite shear zones indicate that the strength profile and the behavior of the shear zone are determined nearly solely by halite, the weaker mineral, even when halite content is as small as 5 vol. % at large displacements near the residual frictional strength. Existing fault models all contradict with these results and a new fault-zone model is proposed.

Keywords: fault model, rock friction, strength profile of lithosphere, halite shear zones, intermediate deformation, fault rock

INTRODUCTION

Mechanical properties of rocks under large shearing deformation from brittle to fully plastic deformation are required for modeling the mechanical behavior of large-scale faults and plate boundaries, including the generation of large earthquakes. Fault zone models have been proposed [12, 17-19], but the supporting data are poor owing to the lack of experimental data at high temperatures and under large shearing deformation. The latter point is significant because the fault property changes substantially after the strain localization at large strains [5, 15, 16]. Recently, Kawamoto [8] provided an experimental strength profile of the lithosphere ranging from brittle to fully-plastic regimes by shearing experiments on halite, under the condition that the test temperature is increased in proportion to the normal stress, trying to simulate the geothermal gradient in the earth. A complete set of revised mechanical data, together with representative deformation textures, for halite shear zones are presented herein to show the overall features of the complete transition of monomineralic shear zone.

Another critical issue is what mineral controls the properties of natural multimineralic faults zones in nature. Previous work [3, 7, 20] attempts to model deformation and properties of two phase aggregates. Applicability of existing models to faults is hard to be evaluated due to the lack of systematic experimental data for multi-phase aggregates under large shearing deformation. We thus examined the behavior of mixed halite-calcite shear zones at temperatures to 700 °C. Calcite is still in the brittle regime, in which the friction law holds, at this temperature range and at low normal stresses (up to 32 MPa), whereas halite undergoes complete transition under the same conditions and is much weaker than calcite. The only representative data are presented herein (see [9] for complete data). The results clearly demonstrate that the mechanical behavior of the aggregate is primarily controlled by halite, the weaker member, when its content exceeds about 5% in volume at large shear stains. Data for the ultimate frictional strength (peak strength) and the residual frictional strength are consistent with Tharp's framework [20] and with Jordan's two block model [7], respectively. The results brings out the significance of phyllosilicates in fault zones in determining the properties of natural faults.

The data contradict all previous fault models and a new fault-zone model is proposed. Future tasks are also suggested based upon the present results.

EXPERIMENTAL PROCEDURES

The high-temperature biaxial testing machine (Fig. 1) described by Kawamoto [8] was used for this study. The machine was designed to study the frictional property at elevated temperatures. The machine consists of a gear-train loading system equipped with a servo-motor, a ball screw that converts rotary motion to axial displacement, axial and normal force gauges, specimen assembly and a horizontal hydraulic press to apply the normal force to the sliding surface . With this type of direct shear arrangement, the vertical loading system and horizontal hydraulic ram independently control the shear and normal stresses, respectively, on the sliding planes. The axial and normal force gages have rated capacity of 500 kN and 200 kN, respectively. By combining the speed change of the servo-motor with four different sets of gear arrangements, the machine is capable of producing nine-orders of magnitude speed-change, the slip rate along simulated faults ranging from 1.5 mm/s down to less than 0.1 mm/yr. The displacement rate can be varied almost instantly by any amount within these 9 orders of magnitude either by changing the motor speed or by turning electromagnetic clutches on and off to select a gear assembly. Thus, the machine is suitable for the velocity stepping experiments to study frictional properties of faults.

Iodized cooking salt (0.2 - 0.3 mm in grain size) is used for 0.7 mm-thick shear zones between three blocks of granite, and the same salt with crushed Yule marble (0.1 - 0.25 mm in grain size) mixed in various proportion for the 0.7 mm-thick bimineralic shear zones are used between three blocks of gabbro (Fig. 2). First, the three-block specimens were pressed for 30 minutes at room temperature and normal stress of 40 MPa to compact the gouge layers, before the specimens were set in the biaxial machine. The test temperature was increased in proportion to the normal stress, trying to simulate the geothermal gradient in the earth. As the uniaxial strength of the specimens substantially decreases with increasing temperature due to thermal cracks [*e.g.* 1], the temperature versus pressure gradient we could set in our experiments is much greater than the actual geotherm.

Figure 1. A photograph of the high-temperature biaxial testing machine. This machine consists of a gear-train loading system (A) equipped with an AC servo-motor (B), a ball screw (C), axial and normal force gauges (D and E), specimen assembly (F), a horizontal hydraulic press (G) and displacement transducers (H). A split furnace (I) is set around the specimen assembly.

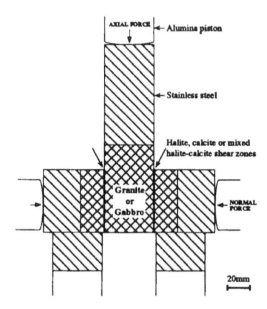

Figure 2. A schematic diagram showing the specimen assembly in the furnace. Three blocks of granite and gabbro have been used, respectively, for experiments on halite shear zones and for those on calcite or mixed halite-calcite shear zones.

Experiments have been performed on the halite shear zones at temperatures to 500 °C, at the normal stresses to 4.4 MPa, and the slip rate ranging from 0.0003 to 30 µm/s, and on the mixed halite-calcite shear zones at temperatures to 700 °C, at the normal stresses to 32 MPa, and under the slip rate of 0.3 µm/s. For some experiments with pure halite or pure calcite shear zones, a test was continued under the same conditions until a steady-state or nearly steady-state in the shear stress is reached, and then the conditions were changed for the next test at different conditions. The other experiments were performed until the total displacement reached the limit of 20 mm.

EXPERIMENTAL RESULTS

Mechanical behavior of halite shear zones
Figure 3 shows the shear stress versus normal stress and temperature, during steady-state or nearly steady-state fault motion from several early runs on the halite shear zones. Temperatures and normal stresses are given, respectively, on the right and left vertical axes. At lower temperatures and normal stresses, the shear stress increases in proportion to the normal stress with the frictional coefficient of 0.6, which is slightly smaller than that for many rocks [*cf*. 2]. At higher temperatures and normal stresses, the data deviate from this linear relationship and reach the strength peak. At further higher temperatures and normal stresses, the shear strength decreases with increasing temperature irrespective of increasing normal stress. In this strength profile, the boundary between the brittle and the intermediate regimes was taken at the point where the friction law breaks down (P in Fig. 3).

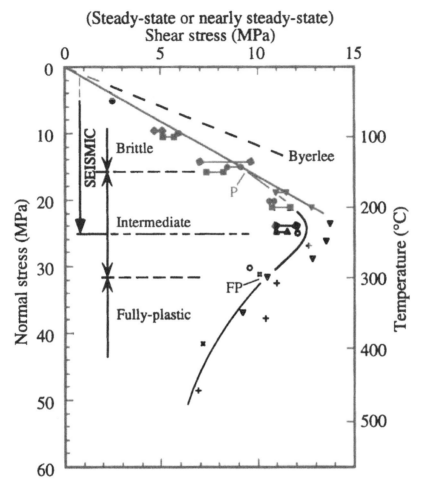

Figure 3. Strength profile for halite shear zones, deformed at elevated temperatures increasing in linear proportion to the normal stress. The dashed line indicates Byerlee's equation [2] for the frictional strength for many rocks. Each symbol denotes a set of data collected from a single experiment. Stick-slip is denoted by closed symbols, tied with a horizontal line showing the maximum and minimum shear stresses during stick-slip.

Figure 4 shows the shear stress versus normal stress on the simulated halite shear zones at steady-state or nearly steady-state shearing deformation at elevated temperatures. At 200 °C and 250 °C, the shear stress increases in proportion to the normal stress until about 15 MPa and 10 MPa, respectively, and then deviates from the linear relationship. At 350 °C and 500 °C, the shear stress becomes nearly insensitive to the normal stress above about 20 MPa and 5 MPa, respectively. The arrows in Fig. 4 point to the normal stress values for the experiments in Fig. 3. Open and closed arrows indicate that the shear stress is sensitive and insensitive to the normal stress, respectively. Fully-plastic flow is characterized by the shear stress nearly independent of the normal stress (*i.e.* Bridgman effect). It is obvious that the halite shear zones are in fully-plastic regime at 350 °C and 500 °C, whereas the shear zones are pressure-dependent at 200 °C and 250 °C in the intermediate regime. FP around 300 °C in Fig. 3 roughly coincides with the pressure-sensitive to pressure-insensitive transition, that is the boundary between the intermediate and the fully-plastic regimes.

Based on these definitions, the intermediate regime is found to be as broad as the brittle and the fully-plastic regimes (Fig. 3). Stick-slip was clearly recognized in the brittle regime and the upper half of the intermediate regime (closed symbols in Fig. 3). However, the unstable slip changed to the stable slip towards the fully-plastic regime.

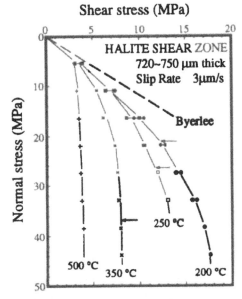

Figure 4. Normal stress-dependence of friction (a) at several temperatures [8]. The dashed line indicates Byerlee's equation [2] for the frictional strength for many rocks. Stick-slip is denoted by closed symbols tied with horizontal bars indicating the maximum and minimum shear stresses during stick-slip.

The shear stress versus normal stress on the simulated halite shear zones have been studied with various slip rates at room temperature [14, 15]. The fully-plastic flow takes place at much higher normal stress at room temperature. The halite shear zone seems to be in fully-plastic regime at the normal stresses of about 350 MPa at fast slip rates, about 250 MPa at intermediate slip rates, and about 150 MPa at slow slip rates. At confining pressures above about 100 MPa, the shear stress is lowest at slow slip rates and highest at fast slip rates. Such an effect of slip rates on the shear resistance at room temperature is also found in Fig. 4 at a certain slip rate by increasing temperature.

Flow laws in the fully-plastic regime
The effect of strain rate on the flow stress is studied with the data in the fully-plastic regime to determine the appropriate flow laws. In general, the flow law is expressed as:

$$\dot{\gamma} = f(\tau) \exp(-Q/RT) \tag{1}$$

where $\dot{\gamma}$ is the shear strain rate, τ is the shear stress, Q is the activation energy, R is the gas constant, and T is the absolute temperature [*e.g.* 11]. In the case of high temperature and relatively low temperature plastic deformations, the following equations describe well many experimental results, respectively.

$$f(\tau) = A\sigma^n \qquad (2)$$

$$f(\tau) = A' \sinh (B\sigma) \quad \text{or} \quad f(\tau) = A'' \exp (B'\sigma) \qquad (3)$$

When (2) or (3) is used as $f(\tau)$, the flow law is called power law or exponential law, respectively. Least-squares fit to the data yields $A = 3.55 \times 10^5$ (MPa^{-n}/s), $n = 7.6$ and $Q = 178.7$ (kJ/mol) for the power law (Fig. 5a), and $A'' = 8.95 \times 10^7$ (s^{-1}), $B' = 1.54$ (MPa^{-1}) and $Q = 188.3$ (kJ/mol) for the exponential law (Fig. 5b).

Data of the present experiments are fairly well described by both flow laws, which may indicate that the transitional deformation mechanisms between the high and low temperature plastic deformations are operating. The above values of Q appear to be greater than Q of the creep rate of halite [10; at T<550 °C]. Such high Q-value may be due to the impurities contained in our iodized cooking salt as nutriments.

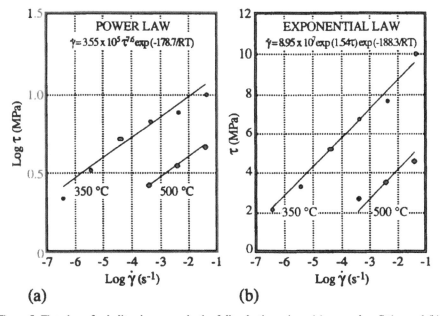

Figure 5. Flow laws for halite shear zone in the fully-plastic regime: (a) power law fitting and (b) exponential law fitting. $\dot{\gamma}$ is the average shear strain rate as determined from the displacement rate and the shear-zone thickness (0.7 mm), and τ is the nearly steady-state shear stress.

Velocity-dependence of halite shear zones

Figure 6 shows the effect of slip rate on the shear stress at elevated temperatures. In the brittle regime (Fig. 6a), the velocity weakening with stick-slip is recognized at each temperature. At 200 °C in the intermediate regime (Fig. 6b), the velocity weakening with stick-slip changes through the transitional regime to velocity strengthening due to decreasing slip rate. At higher temperatures in the intermediate regime (Fig. 6b), the shear resistance increases with increasing slip rate (velocity strengthening), and the velocity weakening is no more observed in this slip rate range.

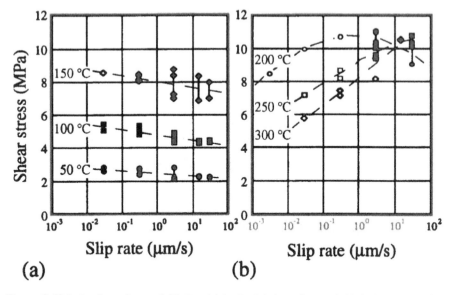

Figure 6. Velocity-dependence of friction (a) in the brittle regime and (b) in the intermediate regime. The normal stress is the same as in Fig. 3. Closed symbols denote stick-slip, with the maximum and minimum shear stresses during stick-slip tied with vertical bars.

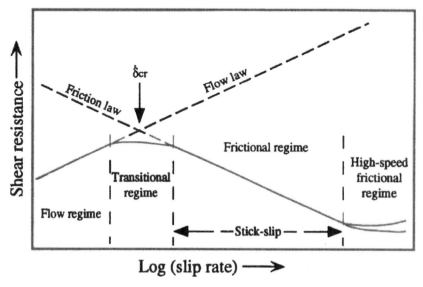

Figure 7. A schematic diagram showing the transition from frictional slip to ductile flow with decreasing slip rate [14]. The velocity, δ_{cr} at the intersecting point of the flow law and the friction law gives the approximate critical velocity at which the logarithmic friction law breaks down.

Figure 7 is a schematic diagram showing the effect of slip rate on the shear resistance at room temperature by Shimamoto [14]. Flow stress is lower than the expected friction at slow slip rates, and friction is lower than the expected flow stress at the intermediate and fast slip rates. Stick-slip occurs only in the frictional regime with a negative velocity dependence of friction (velocity weakening). At room temperature, frictional slip changes to ductile flow due to decreasing slip rate.

In the present experiments at elevated temperature, the change from frictional slip to ductile flow is recognized within the same range of slip rate due to increasing temperature. Small stick-slip occurs with positive velocity dependence of friction (velocity strengthening) at 250 °C and 300 °C in the intermediate regime, which contradicts well accepted idea. This may be due to the temperature heterogeneity along the sliding surface.

Figure 8 shows the stress drop during stick-slip. The stick-slip amplitude increases with increasing normal stress in the brittle regime and reaches its maximum near the boundary between the brittle and the intermediate regimes, and subsequently decreases towards the base of the intermediate regime. The results indicate that the fault is most unstable near the boundary between the brittle and the intermediate regimes.

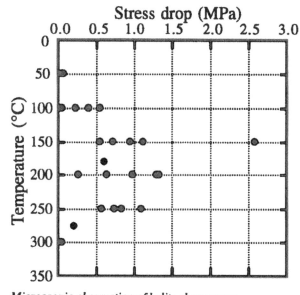

Figure 8. Stress drop during stick-slip of halite shear zones. Maximum stress drop is observed at 150 °C, where is the boundary between brittle and intermediate regimes.

Microscopic observation of halite shear zones

Observations were made with a polarized microscope under reflected light. Specimens were first polished dry with sandpaper and emery paper, and the final polish was made with carborundum (#6000) using benzine as a lubricant. Grain boundaries were etched using a small amount of saliva and were enhanced by vacuum-coating with carbon.

The starting material, precompacted under a normal stress of 26 MPa and at a temperature of 300 °C for an hour, is a structureless aggregate of halite clasts. Some recrystallized subgrains with clear surfaces are recognized along the original grain boundaries.

In the brittle regime, the major part of halite remained nearly undeformed and deformation was sharply concentrated at halite-rock interface since halite grains were nearly perfectly compacted prior to the shearing experiments. Figure 9 shows a halite shear zone deformed at 100 °C with normal stress of 9.6 MPa. Although the mechanical behavior is brittle type, some grains begin to elongate to form incipient foliation, which suggests partial operation of plastic deformation. Foliation is markedly dragged towards the zone of concentrated deformation at a rock-halite interface (Fig. 9a). Foliation is often cut by shear surfaces, the most predominant being R_1 Riedel shears (Fig. 9b).

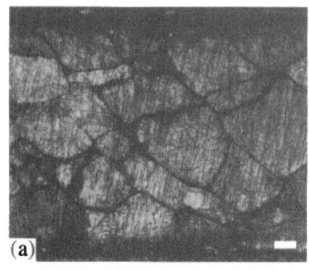

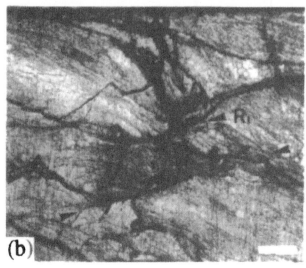

Figure 9. Photomicrographs of halite shear zone deformed in the brittle regime: test conditions are temperature T = 100 °C, normal stress σ = 9.6 MPa and shear strain γ = 30 with a slip rate of 3 μm/s. Sinistral shear sense. (a) Overall view of deformed shear zone. Deformation is concentrated along the upper rock-halite interface. (b) Different portion of the same specimen shown in (a). The solid triangles indicate a discontinuity cutting the foliation with the orientation of a R_1 Riedel shear. All photomicrographs in this and the next figures were taken with a polarizing microscope under reflected light. Many sub-vertical parallel lines in this and the next figures were created during the etching procedures when saliva was wiped off from the polished surface. Scale bars are 0.1 mm.

Before the peak strength in the intermediate regime, the texture of halite shear zone resemble those in Fig. 9a, in which the original grain boundaries are visible. After the peak strength is reached at 250 °C with normal stress of 25 MPa, the original grain boundaries disappear and two layers composed of fine and very fine recrystallized grains formed. Foliation defined by straight and continuous grain boundaries develops almost uniformly in the fine-grained region. On the other hand, foliation in the very-fine-grained region is distorted throughout the shear zone along some discontinuities exhibiting S-C like structure (Fig. 10a).

(a)

(b)

Figure 10. Photomicrographs of halite shear zone deformed (a) in the intermediate regime; test conditions are temperature T = 250 °C, normal stress σ = 25 MPa and shear strain γ = 27 with a slip rate of 3 μm/s; dextral shear sense: and (b) in the fully-plastic regime; test conditions are T = 350 °C, σ = 35 MPa and γ = 29; dextral shear sense. (a) Foliation defined by the continuous recrystallized grain boundaries is distorted throughout the shear zone along some discontinuities (triangles). (b) Foliation is almost uniformly and pervasively developed although it is partially folded when γ is large enough. Scale bars are 0.1 mm.

In the fully-plastic regime, the texture of halite shear zone deformed at 350 °C with normal stress of 35 MPa is characterized by the well developed foliation. Although foliation is partially folded by some discontinuities when shear strain, γ, is large enough (Fig. 10b), it is uniformly and pervasively developed when γ is small.

Mechanical behavior of mixed halite-calcite shear zones

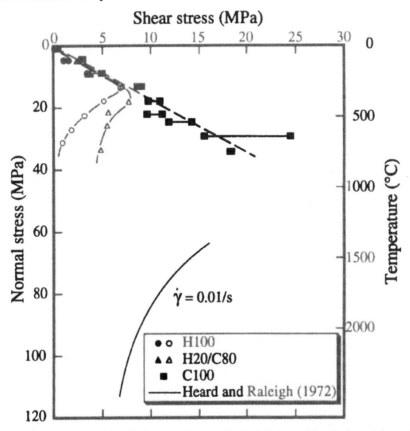

Figure 11. Strength profiles for halite, calcite and mixed halite-calcite (halite/calcite = 20/80 in volume) shear zones, deformed at elevated temperatures increasing linearly with an increase in the normal stress (22 °C/MPa). The solid line indicates flow stress of calcite predicted from the flow law after Heard and Raleigh (1972) for the average strain rate of our shearing experiment. Closed symbols tied with horizontal lines denote stick-slip showing the ranges of shear stress.

Figure 11 shows the shear stress and normal stress during steady-state or nearly steady-state shear resistance on the halite, calcite and mixed calcite-halite shear zones. Temperatures and pressure are given, respectively, on the right and left vertical axes. Calcite shear zones exhibits stick-slip in all tests at temperatures to 700 °C as shown by closed squares in Fig. 11, the horizontal bars indicate the range of shear stress during nearly regular stick-slip. The shear stress increases almost linearly with increasing normal stress for temperatures up to 700 °C with the frictional coefficient of 0.6, and stick-slip amplitude is greatest at 600 °C. For 100% halite shear zones, at lower temperatures and normal stresses, the shear stress increases in proportion to the normal stress with a frictional coefficient of 0.5. At higher temperatures and normal stresses, the shear stress reaches the strength peak at about 300 °C. At further higher temperatures and normal stresses, the shear strength decreases with increasing temperature irrespective of increasing normal stress. Stick-slip vanishes at around the strength peak. For mixed halite-calcite (halite/calcite = 20/80 in volume %) shear zones, the strength profile is close to that for 100% halite shear zones. Stick-slip disappears around the point where the stick-slip of pure halite shear zones vanishes. Figure 11 clearly indicates that when a fault zone consists of a mixture of minerals, the strength and behavior of the fault is determined primarily by the weaker mineral.

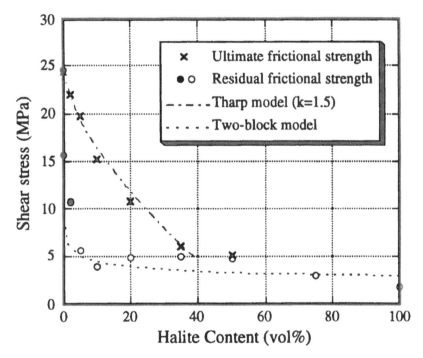

Figure 12. Ultimate frictional strength (**x**) and the residual frictional strength in nearly steady state (**●, o**) plotted agaist the composition of halite, the weaker mineral, of mixed halite-calcite shear zones. All experiments were performed at 600 °C in temperature and at ca.27 MPa in the normal stress. The maximum and minimum shear stresses during stick-slip are denoted by closed circles.

In order to evaluate the content of weaker mineral controlling the bulk strength of mixture, experiments were performed on bimineralic shear zones with various portions of halite. Figure 12 shows the bulk strength of mixed shear zones as a function of the halite content, at the strength peak and at 20 mm of displacement (nearly steady state). Tharp's formula [20] is expressed as $\sigma = \sigma_0 (1-ka^{2/3})$; where σ is the bulk strength, σ_0 is the strength of the pure stronger phase, k is a geometrical value and a is the fraction of the weaker phase. In two-block model [7], the stronger phase forms a rigid block, and the thickness of blocks is proportional to volume fraction of stronger phase. The ultimate frictional strength decreases rather gradually with increasing halite content, and the data correspond with Tharp model [20] with k-value of 1.5. When the bulk strength attains the nearly steady-state at 20 mm of displacement, a dramatic reduction of the residual frictional strength is observed with only 5% of halite, and the data are well fit by the two-block model of Jordan [7]. The weaker mineral, whose content is as small as 5% in volume, primarily determines the properties of the mixed shear zones at large strains.

A REVISED FAULT MODEL

Based on these experimental results and microscopic observation, we propose a revised fault model (Fig. 13). This model consists of brittle, intermediate and fully-plastic regimes of about the same extent, as in the previous model [15]. Shape of the strength profile in the intermediate regime is determined experimentally. It must be emphasized that mylonitic textures form not only in the fully-plastic regime, but also from the lower end of the brittle regime, being consistent with previous work [5, 13, 15, 16]. Stick slip is recognized from the brittle regime down to the middle of the intermediate regime near the peak strength. All previous fault models are not consistent with our experimental data as elaborated below.

Sibson's fault model [17, 18] consists of a seismogenic frictional slip regime (brittle regime) and aseismic plastic regime with a small transitional regime. In his model, mylonite is assumed to form in the aseismic plastic regime below the strength peak. This model contradicts with our experimental results with respect to the existence of a wide intermediate regime and the genesis of mylonites.

Strehlau's model [19] comprises three structural and rheological regimes; the upper frictional regime, the middle transitional regime and the lower ductile (plastic) regime. He assumed seismogenic behavior down to the bottom of the intermediate regime, or even deeper. Although Strehlau postulated three regimes, his model is a model for bimineralic fault zone, each mineral showing brittle and plastic behavior as in Sibson's model (17, 18). Thus his model is entirely different from our model and contradicts with our data in Fig. 11.

Scholz's model [12] is also intended for quartzo-feldspathic rocks and consists of brittle, semi-brittle and plastic regimes. He assumed 300 °C (T_1) and 450 °C (T_2) as the onset of quartz and feldspar plasticity, respectively, and the semi-brittle regime is bounded by T_1 at the top and by T_2 at the bottom. In his model, the lower limit of the seismogenic layer is approximately at T_1, below which mylonite begins to form. However, as mentioned above, earthquakes do occur within the regime where mylonites form, and T_1 is actually less than the temperature at the cutoff depth of seismicity [6]. In addition, he concluded that large earthquakes would propagate below T_1 to some greater depth (T_3) and that the shear zone strength should increase approximately linearly with normal stress, and hence depth, from the surface to T_3 as long as the behavior is essentially frictional. It is not clear, however, which mineral controls the strength profile of the semi-brittle regime in his model.

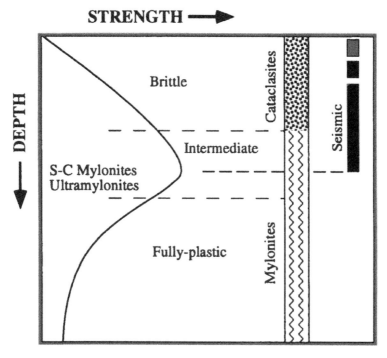

Figure 13. A revised fault model [8] consisting of the brittle, intermediate and fully plastic regimes. Mylonites begin to form from the intermediate regime, and the seismogenic fault motion is extended well into the intermediate regime down approximately to the strength peak.

Shimamoto's model [15] is based on the shearing experiments at room temperature, and the present revised model is on the whole very similar to his model. The major differences are the shape of strength profile in the intermediate regime and the lower limit of the seismic behavior of the fault.

CONCLUSIONS

Results of this study that warrant emphasis are as follows:

(1) Shearing experiments on halite layers provide an experimental strength profile of the lithosphere, which can be divided into brittle, intermediate and fully-plastic regimes. In the brittle regime the shear stress increases in linear proportion to the normal stress, and it becomes insensitive to the normal stress in the fully-plastic regime. The intermediate regime bounded by the two regimes is as broad as the brittle and fully-plastic regimes, and the peak of shear strength is located roughly at the middle of this regime. Stick-slip is recognized from the brittle regime down to the middle of the intermediate regime near the peak strength, and the stick-slip amplitude is at its maximum near the boundary between the brittle and intermediate regimes. The most unstable fault motion near the brittle-intermediate transition probably corresponds to the peak of microseismicity, very often located at about one-third of the seismogenic zone from its bottom. The transition from

the velocity weakening to the velocity strengthening is observed in the intermediate regime near the strength peak. This probably corresponds to the bottom of the seismogenic zone, although our data are inconclusive owing to somewhat large temperature heterogeneity. Effect of strain rate on the flow stress is studied with the data in the fully-plastic regime to determine the appropriate flow law. Our data can be fairly well described by both power law and exponential law, which implies the transitional deformation mechanisms between the high and low temperature plastic deformations are operating.

(2) Microstructural observation revealed that deformation in the halite shear zone is highly concentrated at the gouge-rock interface at the beginning of the brittle regime. In addition, the plastic deformation occurs even in the mechanically brittle regime with elongate grains to form foliation dragged towards the concentrated deformation zone. The distortion of the foliation with R_1, Y and P Riedel shears is characteristic of intermediate deformation below the peak strength. Fully-plastic deformation is characterized by uniformly and pervasively developed foliation, but the deformation partly becomes heterogeneous at large strains. Based on the present experiments, a new fault-zone model consisting of brittle, intermediate and fully-plastic regimes is proposed. Plastic deformation is prominent not only in the fully-plastic regime but also in the intermediate regime. The lower limit of the unstable slip behavior reaches around peak strength in the intermediate regime. Mylonite-forming regime and significance of lower limit of seismicity have been made clear by the new fault-zone model.

(3) When a fault zone consists of weak halite and strong and brittle calcite, the strength profile and behavior of the mixed shear zone are controlled primarily by halite. The ultimate and residual frictional strengths can be described, respectively, by the Tharp's framework model [20] and by the two-block model of Jordan [7]. Halite, the weaker member, essentially determines the strength and behavior of the mixed shear zone when its content is more than only 5% in volume at large strains. This result brings out the significance of phyllosilicates in fault zones in determining their frictional properties. It is highly likely that the strength and constitutive properties of faults are determined by the weakest constituent mineral if its content is about 5%.

Acknowledgments

We would like to express our sincere thanks to T. Nakazono and A. Yamamura of Marui Co. Ltd. for building the biaxial testing machine and to Takata Co., Ltd. for making many gabbro blocks we used in this study.

REFERENCES

1. S.J. Bauer and B. Johnson. Effects of slow uniform heating on the physical properties of the Westerly and Charcoal granites, *Proc. 20th U.S. Symp. Rock Mechanics, Univ. Texas, Austin,* 7-18 (1979).

2. J. Byerlee. Friction of rocks, *Pure Apple. Geophys.* 116, 615-626 (1978).

3. M.R. Handy. Flow laws for rocks containing two-linear viscous phases: a phenomenological approach, *Jour. Struct. Geology* 16, 287-301 (1994).

4. H.C. Heard and C.B. Raleigh. Steady-state flow in marble at 500° to 800°C, *Bull. Geol. Soc. Am.* 83, 935-956 (1972).

5. H. Hiraga and T. Shimamoto. Textures of sheared halite and their implications for the seismogenic slip of deep faults, *Tectonophysics* 144, 69-86 (1987).

6. K. Ito. Regional variations of the cutoff depth of seismicity in the crust and their relation to heat flow and large inland-earthquakes, *J. Phys. Earth* 38, 223-250 (1990).

7. P. Jordan. The rheology of polymineralic rocks - an approach, *Geol. Rundsch.* 77, 285-294 (1988).

8. E. Kawamoto. The first experimental determination of the strength profile of the lithosphere : preliminary results using halite shear zones, *Jour. Geol. Soc. Japan* 102, 249-257 (1996).

9. E. Kawamoto. Strength profile for bimineralic shear zones: and insight from high temperature shearing experiments for calcite-halite mixture, *Tectonophysics* (in preparation).

10. A. Nicolas and J.P. Poirier. *Crystalline plasticity and solid state flow in metamorphic rocks*, Wiley Interscience, London, 444 p. (1976)

11. J.P. Poirier. *Creep of crystals*, Cambridge Univ. Press, London, 260 p. (1985).

12. C.H. Scholz. The brittle-plastic transition and the depth of seismic faulting, *Geol. Rundsch.* 77, 319-328 (1988).

13. T. Shimamoto. The origin of large or great thrust-type earthquakes along subducting plate boundaries, *Tectonophysics* 119, 37-65 (1985).

14. T. Shimamoto. Transition between frictional slip and ductile flow for halite shear zones at room temperature, *Science* 231, 711-714 (1986).

15. T. Shimamoto. The origin of S-C mylonites and a new fault-zone model, *Jour. Struct. Geol.* 11, 51-64 (1989).

16. T. Shimamoto and J.M. Logan. Velocity-dependent behavior of simulated halite shear zones: an analog for silicates, *Am. Geophys. Un. Geophys. Monogr.* 37, 49-63 (1986).

17. R.H. Sibson. Fault rocks and fault mechanisms, *J. Geol. Soc. Lond.* 133, 191-213 (1977).

18. R.H. Sibson. Continental fault structure and the shallow earthquake source, *J. Geol. Soc. Lond.* 140, 741-767 (1983).

19. J. Strehlau. A discussion of the depth extent of rupture in large continental earthquakes, *Am. Geophys. Un. Geophys. Monogr.* 37, 131-145 (1986).

20. T.M. Tharp. Analogies between the high-temperature deformation of polyphase rocks and the mechanical behavior of porous powder metal, *Tectonophysics* 96, T1-T11 (1983).

Proc. 30th Int'l. Geol. Congr., Vol. 14, pp. 106-118
Zheng et al. (Eds)
© VSP 1997

The Effect of Anisotropy on the Shape of Fault-bend Folds

LABAO LAN AND PETER HUDLESTON
Department of Geology and Geophysics, University of Minnesota, Minneapolis, MN
55455, U.S.A.

Abstract

Fault-bend folds form as a geometric consequence of the mismatch in shape of hangingwall and
footwall during displacement at a ramp in a thrust fault. Geometric and kinematic models of
such folds are straight-limbed and sharp hinged. Most existing physical analog models and
numerical models of fault-bend folds produce structures lacking angularity and with broad hinge
zones. Natural fault-bend folds show much variation in shape, from angular to broad-hinged.
Mechanical anisotropy is an important property of layered rocks that strongly influences rock
deformation and likely is an important factor in determining the shape of fault-bend folds. We
investigate this by simulating the deformation of layered rocks involved in thrust faulting at a
ramp by the flow of layered, anisotropic viscous materials. In our study, the constitutive
equations for orthotropic anisotropy are incorporated into a two-dimensional finite element code
in incompressible viscous fluids. The degree of anisotropy is given by normal viscosity/shear
viscosity ($A = N/Q > 1$). We take the footwall of the thrust as being nearly rigid, and represent
the hangingwall as a layered anisotropic viscous medium. The fault is represented by a very
'soft' isotropic viscous layer. Displacement on the fault is produced by a constant velocity
boundary condition at the ends of the hangingwall block.

Fault-bend folds of similar amplitude and dip are produced near the thrust in both isotropic ($A =$
1) and anisotropic ($A = 20$) cases. The ramp anticline becomes broader as displacement is
increased. The fold dies out into the hangingwall rapidly for an isotropic layer, but propagates
well into the hangingwall in the anisotropic case. In addition, the ramp anticline in the isotropic
layer is broad and lacks a sharp hinge point, whereas in the anisotropic models it is sharp-
hinged, becoming flat-topped and double-hinged as fault displacement increases. Anisotropy
results in folds very similar to those of the geometric kinematic models and some natural folds.

Keywords: anisotropy, fault-bend folds, finite-element modeling, rheology, flat-topped shape

INTRODUCTION

Folding and faulting are intimately related in the fold-and-thrust belts that occur
throughout the world at the borders of nearly all orogenic belts. The nature of the
association between the two types of structure has been the subject of many studies
since the classical work of Rich [28], who showed that folding can be the geometrical
consequence of movement of a thrust sheet over a ramp in the footwall of the thrust.
Most subsequent studies have focused on the geometrical association of folds and
faults [6, 3, 17], on the kinematics of movement on thrust faults [6, 31, 33, 24, 11,
34], and on the construction of balanced cross sections based on the principle of
restorability of the sections [5, 8, 26].

Although the style of folds in fold-and-thrust belts is highly variable, certain characteristic features of the association between folds and thrusts have been established. This has led to the recognition of three basic types of fault-related folds in the literature. 1) The folds that form as a geometric consequence of movement of the hangingwall over the footwall are termed fault-bend folds [31]; 2) those that occur at a fault ramp and accommodate the loss in displacement near the tip of a fault are called fault-propagation folds [33, 25; and 3) those that occur on a flat fault to accommodate loss in displacement near the fault tip are called detachment folds [6, 17]. A number of modifications to these basic fold types have also been described [22]. More recently, several workers have considered the ramifications of the three basic geometric models of fault-related folds in terms of the strain and fabric development that accompanies slip on the fault and the formation of folds. The pattern of strain and fabric development places constraints on the kinematic evolution of natural fault-related folds, and provides for the possibility of testing the geometric models [29, 11, 23, 12, 15-16]. The focus of this paper is on fault-bend folds.

The kinematic models developed to explain fault-related folds [31, 33, 17, 25] idealize the folds as being straight-limbed, of constant bed thickness, and sharp hinged. A typical shape is the flat-rootless anticline or syncline, formed by applying these geometric rules and keeping the footwall rigid (Fig. 1). This style of structure is indeed seen in natural fold-thrust belts (Fig. 2). Although some natural folds appear to match the geometric model predictions of shape quite well, not all folds do, and there is considerable deviation of fold shape from the straight-limbed ideal. Variation in natural fold style has in the past led some observers to characterize most folds in fold-and-thrust belts as being sinusoidal or broad-hinged [27, 13] and others to characterize them as being straight-limbed and kink like [10]. It is clear that the geometric models cannot explain all the variation seen in natural folds, and that other factors besides the presence and inclination of ramps and the location of fault tips must play a role in determining fold shape. To explore the topic further, we must turn to mechanical aspects of faulting and folding. This has been done by a number of workers using physical analog models, simple analytical treatments, and numerical models.

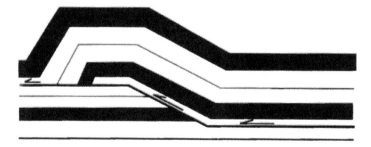

Figure 1. Geometric-kinematic model of a fault-bend fold (after [31-32]), with straight limb segments, flat-topped anticline, and sharp hinges.

All modeling is made difficult by the need to somehow treat simultaneously both brittle behavior (fracture and frictional slip) and ductile behavior (by a variety of deformation mechanisms). Physical model analog experiments of fold-and-thrust

Figure 2. Natural fault-related folds showing angular shapes and straight limbs in the hangingwall of a small thrust, Carboniferous sandstones and shales, Broadhaven, Pembrokeshire, Wales. Note the flat-topped shape of the main fold.

structures perhaps come closest to achieving this [7, 21], but such experiments do not allow all important physical properties to be controlled. Thus it is difficult to evaluate the effects of varying model parameters. These models do, however, result in structures similar to those seen in nature and provide valuable insight into understanding the general conditions that lead to fault-related folding.

Analytical treatments have treated the problem of fault-bend folding as the flow of a viscous hangingwall over a rigid [2] or deformable [18] footwall. Folding is then produced in marker horizons in the hangingwall in the first instance and in both hangingwall and footwall in the other by a combination of the geometric consequence of slip over the ramp and the mechanical effect of drag against the fault. Fold shape is rounded, with tightness depending on the amounts of slip and drag resistance.

A number of numerical models of fault-bend folding have now been carried out using finite element or finite difference methods [1, 9, 30]. These models employ homogeneous and isotropic media with elastic-plastic or elastic-plastic-viscous constitutive relations, and involve moving the hangingwall over a preexisting discontinuity that represents the thrust fault and ramp. Frictional slip is allowed on the fault. The hangingwall in these models becomes folded with increasing displacement on the thrust. In all these models fold shape is rounded and sharp hinges are lacking, quite unlike the straight-limbed flat-topped folds predicted by the geometric models (Fig. 1) and seen in many natural examples (Fig. 2). It seems likely that the difference in shape between the numerical fault-bend folds and many natural folds may be accounted for by the difference in constitution of the hangingwall: in the models this consists of just one or a few layers, whereas in nature it consists typically of many layers of individual thickness very much less than the height of the thrust ramp. This difference can be represented in models by formally introducing anisotropy.

Anisotropy is a prominent property of the sedimentary strata in fold-and-thrust belts, and we believe that mechanical anisotropy exerts a strong influence on the development of linked folding and faulting in fold-and-thrust belts. The purpose of the present study is to investigate, using a finite element analysis, how material anisotropy influences the shape of fault-bend folds as the fault hangingwall moves over a non-planar fault surface.

FINITE-ELEMENT MODEL SET UP

In our finite element models, we simplify the problem by taking the footwall of the thrust as being effectively 'rigid', and we represent the hangingwall as a layered viscous medium. The problem can be treated as dimensionless because all materials are Newtonian and we are concerned here with quasi-static flow (i.e. inertial forces may be neglected), in the absence of gravity. Overall, the model consists of materials of four different viscosities, defined with reference to the top layer in the model (stratigraphically above the level of the upper thrust flat), which is assigned unit viscosity (Fig. 3). The lower part of the hangingwall (stratigraphically between the levels of the upper and lower thrust flats) is a competent layer, marked by bold lines in Fig. 3. It has a viscosity of 10 units. The discontinuity that is the fault in nature is represented by a very 'soft' viscous medium (Fig. 3) with a viscosity of 0.01 units. Representing the fault as a thin viscous layer is a convenient approximation in our model, which is not designed to handle discontinuities. The footwall is assigned a viscosity of 100 units. Displacement on the 'fault' is produced by applying either a force or a velocity boundary condition to the vertical edges of the entire hangingwall block.

In the initial set of models, all the layers were isotropic. In a second set, the top and bottom layers of the hangingwall were given a planar (orthotropic) anisotropy, such that the normal viscosity is greater than the shear viscosity, as would be likely the case for layered rocks [4]. The basic equations of flow for an anisotropic

incompressible viscous fluid are, in two dimensions: $\sigma'_{xx} = N\dot{\varepsilon}_{xx}$, $\sigma'_{yy} = N\dot{\varepsilon}_{yy}$, and

$\sigma_{xy} = Q\dot{\varepsilon}_{xy}$, where N is the viscosity for normal stress and Q that for shear stress, and

σ'_{ij} and $\dot{\varepsilon}_{ij}$ are components of deviatoric stress and strain rate respectively. These equations are for the case in which the plane of anisotropy is parallel to x or y. If the plane of anisotropy is not parallel to x or y, the coefficients of viscosity can be found from N and Q using the rules of transformation of coordinates for fourth rank tensors [4]. The degree of anisotropy is given by the parameter $A = N/Q$.

For the boundary conditions, all the nodes along the edges of the footwall were fixed in both x and y directions; and the nodes along the two vertical walls of the hangingwall were given the same fixed horizontal displacement rate: no constraints

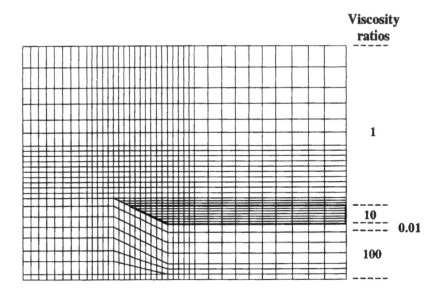

Figure 3. Initial finite element grid for a fault-bend fold model. The model consists of materials of four different viscosities, shown on the right and expressed in dimensionless form. The hangingwall is either isotropic or anisotropic.

were made on the vertical velocity components at these nodes. The nodes along the top boundary of the model were free. The dip angle of the ramp could be varied from 0°-90°. To isolate the effect of anisotropy on the resulting fold geometry, the ramp slope was kept fixed at 26.5° in this study. The plane of anisotropy is kept parallel to the layering during deformation, and thus the angle between the plane of anisotropy and the x coordinate direction changes as folding occurs.

The finite element code used in this work is adapted from one developed by Hansen [14]. It treats the case of two-dimensional, incompressible, quasi-static viscous flow. Modifications for application to the study of rock folding are given in Lan and Hudleston [19]. More detailed theoretical description of the code formulation and its modification to treat the case of orthotropic anisotropy is addressed elsewhere (Lan and Hudleston, in prep). All calculations in this study were carried out using vector-processing Cray-series Supercomputers.

RESULTS

Displacement of the hangingwall over the footwall results in a geometric change in the hangingwall to accommodate the motion over the ramp. In addition, in the case of anisotropy, there is the possibility of a mechanical instability as a result of stresses applied parallel to the anisotropy [4].

The initial state for all our models is shown in Fig. 3. A sequence of deformed states for increasing increments of hangingwall displacement for anisotropic behavior is shown in Fig. 4. One increment of displacement is equal to one tenth of ramp height of the model. Total displacement is given by summing the time increments. The viscosity ratios, defined by $N_1/N_2 = Q_1/Q_2$, for this model are shown to the right of the initial element grid (Fig. 3), and the anisotropy of the hangingwall in Fig. 4 is $N/Q = 20$. It can be seen that the layering in the hangingwall becomes increasingly folded with increase of displacement along the thrust. Maximum amplitude of the fold is attained once the lower cut-off point of the stiff layer in the hangingwall arrives on the upper thrust flat. It should be noted that the elements in the very 'soft' fault zone become highly distorted at fairly modest values of fault displacement (Fig. 4b). To avoid extreme element distortion in the fault zone, we updated the element grid periodically as appropriate.

As mentioned above, the footwall in the models was given a high viscosity so that it would stay effectively rigid. It can be seen from Fig. 4 that this is indeed the case. Thus in presenting data for other models, only the hangingwall is shown.

To study the effect of anisotropy on shape development in fault-bend folds, two models were made employing the same starting configuration and rheological parameters, and differing only in value of anisotropy, A. A fairly strong degree of anisotropy, given by $A = N/Q = 20$, was selected for this purpose. The results are shown in Figs. 5 and 6, in each figure structures being compared at equal values of total displacement.

In both the isotropic (Fig. 5a) and anisotropic (Fig. 5b) cases, the principal characteristics of the fold are determined by the ramp angle and ramp height. The cut-off angle of the bed truncated by the ramp is very nearly equal to the dip of the forelimb of the fold, and the backlimb dips at the same angle as the ramp. Differences in the two cases become pronounced, however, at increasing stratigraphic level above the thrust. In the isotropic case, fold amplitude and forelimb dip diminish with increasing height above the thrust: their values at the top of the model are small fractions of their values at the ramp level. In the anisotropic case, fold amplitude and forelimb dip diminish much less with height above the thrust. A second obvious difference between the two models is in the sharpness of the fold hinges. In the isotropic case (Fig. 5a) the anticline is quite rounded, with no well-defined hinge point. In the anisotropic case (Fig. 5b), by contrast, the hinge is sharp and well-defined.

Fault displacement in Fig. 6 is twice that in Fig. 5. As expected, the differences in fold shape between the two models become more distinct as fault displacement increases. The general roundness of the folds in the isotropic layer is maintained in Fig. 6a, whereas the angularity in the anisotropic case is more pronounced in Fig. 6b. Moreover, the hangingwall anticline in the latter case is 'flat-topped' or box shaped (Fig. 6b). The anticline has two distinct hinges, one corresponding to the base of the ramp in the hangingwall, and the other corresponding to the top of the ramp in the footwall. In the isotropic case, such a clear correspondence cannot be made although the principal elements of the fold are determined by the same cut-off points at the top and base of the ramp. Comparing Figs. 5 and 6, we can see that the ramp anticline becomes broader with increasing fault displacement, as the truncated strata in the

L. Lan & P. Hudleston

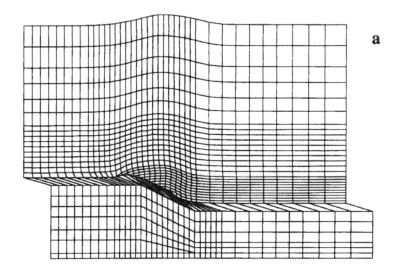

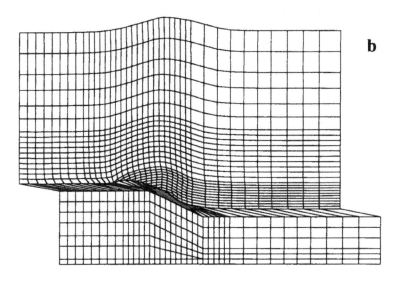

Figure 4. Deformed finite-element grid for an anisotropic hangingwall (N/Q=20) after 10 (a), and 15 increments (b) of time or displacement.

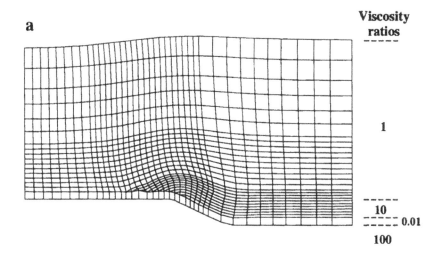

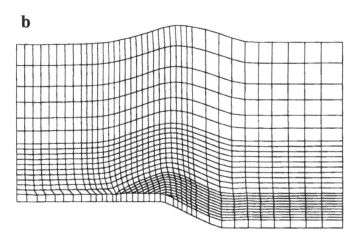

Figure 5. Shapes of fault-bend folds produced in (a) an isotropic hangingwall (N/Q=1), and (b) an anisotropic hangingwall (N/Q=20). Both after 20 increments of time or displacement.

hangingwall, defining the forelimb of the ramp anticline, are carried forward. The backlimb of the anticline is fixed above the ramp in the footwall. The distance between the two anticlinal hinges in Fig. 6b will increase as displacement increases. The same general phenomenon also occurs for the isotropic case (Fig. 6a), except that the hinges are not well defined - the deformation caused by the adjustment to the geometry of the footwall is dissipated with distance away from the thrust. If displacement in the isotropic case continues to increase, eventually the anticline would also become flat-topped, but in this case the hinges would be broad, not sharp.

If the degree of anisotropy is increased to 50, the flat-topped shape and sharp hinges of the folds become even more striking (not illustrated) although the geometric characteristics of the fault-bend folds are similar to those from the model with $A = N/Q = 20$.

a

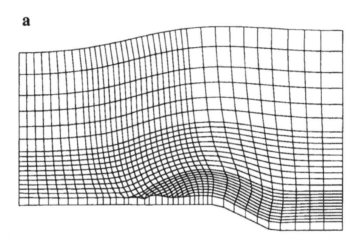

b

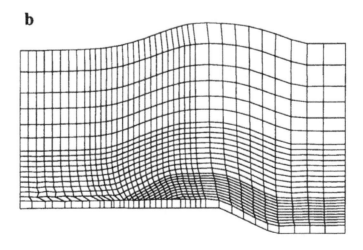

Figure 6. The same models as in Fig. 5 after 40 time increments. (a) Isotropic hangingwall. (b) Anisotropic hangingwall.

DISCUSSION AND CONCLUSIONS

Our numerical results are broadly consistent with the results of all other models of fault-bend folding, in that an open, rootless anticline is formed in the hangingwall above the zone between the hangingwall and footwall cut-offs. As the distance between the two cut-offs becomes greater, with increasing fault displacement, the anticline becomes broader, although maximum amplitude and limb dip remain fairly constant once the lowest point of hangingwall cut-off arrives at the top of the ramp.

Although the broad features of the folds are similar in both of our models and in other physical models and analog experiments of fault-bend folding, the shape of the folds clearly depends on whether or not the hangingwall is isotropic or anisotropic. We can infer from our results that the angularity of the folds in the hangingwall produced by movement over a 'ramp' increases as N/Q increases (cf. Fig. 5a and 5b or Fig. 6a and 6b). Moreover, as displacement on the fault increases, the broadening of the ramp anticline in an anisotropic hangingwall results in the development of the characteristic flat-topped shape (Fig. 6b). This is distinctly different from the rounded shape developed in the ramp anticline for an isotropic hangingwall (Fig. 6a). The effect of anisotropy on the shape of a fault-bend fold is consistent with its effect on the shapes of folds in single isolated layers, which tend also to become angular as anisotropy is increased [20].

The fold geometry and its evolution that result from a highly anisotropic hangingwall are very similar to the idealized straight-limbed, angular-hinged shapes used to develop geometric/kinematic models of fault bend folds (Fig. 1). In their angularity, they are also similar to many natural folds in fold-and-thrust belts (Fig. 2, also see [10]). By contrast, the broad folds with rounded hinges developed in our isotropic numerical models are similar to those developed in other numerical models employing a single competent hangingwall layer cut by the thrust ramp, or a hangingwall consisting of several layers. The shape of the ramp anticline of our anisotropic model is compared in Fig. 7 with the shapes of ramp anticlines produced by the numerical models of Erickson and Jamison [9] and Strayer and Hudleston [30].

We have only addressed one aspect of fault-bend folding in this paper. There are other problems of understanding the mechanisms of fold development in fold-and-thrust belts that may be tackled using an approach similar to that taken here or by using numerical models that allow faults to be represented as true discontinuities [1, 9, 30]. When the controls on folding associated with a single fault have been established, various combinations of fault geometry, fault displacement, and degree of anisotropy of both hangingwall and footwall can be used to examine how key parameters affect the resulting fold geometry, and more complex situations can be made, including attempting to match fold and thrust development in specific natural examples.

Acknowledgments

We acknowledge support provided by the National Science Foundation (EAR-9219702 and EAR-9526945) and the University of Minnesota Supercomputer Institute. We thank Professors R. Wang and Y. Zheng for their helpful review.

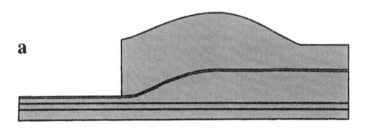

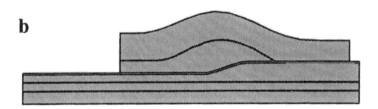

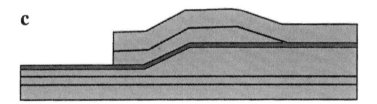

Figure 7. Comparison of the shapes of fault-bend folds produced by various numerical models. (a) Broad-hinged fault-bend fold produced in an isotropic elastic-viscous-plastic finite-element model [9]. (b) Broad-hinged fault-bend fold produced in an isotropic elastic-plastic finite-difference model [30]. (c) Angular, straight-limbed fault-bend fold developed in an anisotropic viscous finite-element model (this paper).

REFERENCES

1. K.D. Apperson and D.F. Goff. Deformation of thrust ramps and footwalls observed in numerical models, *Eos* **72**, 514-515 (1989).
2. P. Berger and A.M. Johnson. First-order analysis of deformation of a thrust sheet moving over ramp, *Tectonophysics*, **70**, T9-T24 (1980).
3. S.E. Boyer and D. Elliott. Thrust system, *Bull. Am. Ass. Petrol. Geol.* **66**, 1196-1230 (1982).
4. P.R. Cobbold. Mechanical effect of anisotropy during large finite deformation, *Bull. Soc. Geol. France*, 7 Ser. **18**, 1497-1510 (1976).
5. C.D.A. Dahlstrom. Balanced cross sections, *Can. J. Earth Sci.* **6**, 743-757 (1969).
6. C.D.A. Dahlstrom. Structural geology in the eastern margin of the Canada Rocky Mountains, *Bull. Can. Petrol. Geol.* **18**, 332-406 (1970).
7. J.M. Dixon and R. Tirrul. Centrifuge modeling of fold-thrust structures in a tripartite stratigraphic sequence, *J. Struct. Geol.* **13**, 3-20 (1991).
8. D. Elliott and M.R.W. Johnson. Structural evolution in the northern part of the Moine thrust belt, N.W. Scotland, *Trans. Roy. Soc. Edinburgh*, **71**, 69-96 (1980).
9. S.G. Erickson and W.R. Jamison. Viscous-plastic finite-element models of fault-bend folds, *J. Struct. Geol.* **17**, 561-573 (1995).
10. R.T. Faill. Kink-band folding, Valley and Ridge Province, Pennsylvania, *Bull. Geol. Soc. Am.* **84**, 1289-1314 (1973).
11. M.P. Fischer, N.B. Woodward and M.M. Mitchell. The kinematics of break-thrust folds, *J. Struct. Geol.* **14**, 451-460 (1992).
12. D.M. Fisher and D.J. Anastasio. Kinematic analysis of a large-scale leading edge fold, Lost River Range, Idaho. *J. Struct. Geol.* **16**, 337-354 (1994).
13. V.E. Gwinn. Thin-skinned tectonics in the Plateau and northwestern Valley and Ridge provinces of the central Appalachians, *Bull. Geol. Soc. Am*, **75**, 863-900 (1964).
14. B.H. Hanson. Thermal response of a small ice cap to climatic forcing. *J. Glaciology*, **36**, 49-56 (1990).
15. S. Hardy and J. Poblet. Geometric and numerical model of progressive limb rotation in detachment folds, *Geology*, **22**, 371-374 (1994).
16. C.A. Hedlund, D.J. Anastasio and D.M. Fisher. Kinematics of fault-related folding in a duplex, Lost River Range, Idaho, USA, *J. Struct Geol.* **16**, 571-584 (1994).
17. W.R. Jamison. Geometric analysis of fold development in overthrust terranes, *J. Struct. Geol.* **9**, 207-219 (1987).
18. B. Kilsdonk and R.C. Fletcher. An analytical model of hangingwall and footwall deformation at ramps on normal and thrust faults, *Tectonophysics*, **163**, 153-168 (1989).
19. L. Lan and P.J. Hudleston. Finite element models of buckle folds in non-linear materials. *Tectonophysics* **199**, 1-12 (1991).
20. L. Lan and P.J. Hudleston. Rock rheology and sharpness of folds in single layers, *J. Struct. Geol.* **18**, 925-931 (1996).
21. S. Liu and J.M. Dixon. Localization of duplex thrust-ramps by buckling: analog and numerical modeling, *J. Struct. Geol.* **17**, 875-886 (1995).
22. K.R. McClay (ed). *Thrust tectonics*. London, Chapman and Hall (1992).
23. M.A. McNaught and S. Mitra. A kinematic model for the origin of footwall synclines, *J. Struct. Geol.*, **15**, 805-808 (1993).
24. S. Mitra. Three-dimensional geometry and kinematic evolution of the Pine Mountain thrust system, southern Appalachians, *Bull. Geol. Soc. Am.* **100**, 72-95 (1988).

25. S. Mitra. Fault-propagation folds: geometry, kinematic evolution, and hydrocarbon traps, *Bull. Am. Ass. Petrol. Geol.* **74**, 921-945 (1990).

26. S. Mitra and J. Namson. Equal-area balancing, *Am. J. Sci.* **199**, 563-599 (1989).

27. R.P. Nickelsen. Fold patterns and continuous deformation mechanisms of the central Pennsylvania folded Appalachians: In: *Tectonics and Cambrian-Ordovician Stratigraphy in the central Appalachians of Pennsylvania, Guidebook*, Pittsburgh Geological Society with the Appalachian Geological Society, 13-29 (1963).

28. R.L. Rich. Mechanics of low-angle overthrust faulting as illustrated by Cumberland thrust block, Virginia, Kentucky, and Tennessee, *Bull. Am. Ass. Petrol. Geol.* **18**, 1584-1596 (1934).

29. D.J. Sanderson. Models of strain variation in nappes and thrust sheets: a review, *Tectonophysics*, **88**, 201-233 (1982).

30. L. Strayer and P.J. Hudleston. Numerical modeling of fold initiation at thrust ramps, *J. Struct. Geol.* (in press).

31. J. Suppe. Geometry and kinematics of fault-bend folding, *Am. J. Sci.* **283**, 684-721 (1983).

32. J. Suppe and J. Namson. Fold-bend origin of frontal folds of the western Taiwan fold-and-thrust belt, *Petroleum Geology Taiwan*, **16**, 1-18 (1979).

33. J. Suppe and D.A. Medwedeff. Fault-propagation folding, *Geol. Soc. Am. Abstr. with Prog.* **16**, 670 (1984).

34. R. Zoetemeijer and W. Sassi. 2-D reconstruction of thrust evolution using the fault-bend fold model. In: K. R. McClay (ed), *Thrust Tectonics*, Chapman and Hall, 133-140 (1992).

PART 3

EXTENSIONAL TECTONICS

Proc. 30th Int'l. Geol. Congr., Vol 14, pp. 121-148
Zheng et al. (Eds)
© VSP 1997

Mechanisms of Extension and their Influence on Basin Geometry: The St. George's Channel Basin, Offshore UK

MICHAEL J WELCH

School of Earth Sciences, University of Birmingham, Birmingham B15 2TT, UK

Abstract

Over the past 20 years much work has gone into understanding the mechanisms of lithospheric extension, and producing models to explain the geometry and evolution of rift basins. These models have tended to simplify the deformation occurring within the upper crust by assuming that extension occurs either by pure shear (such as the uniform extension model) or exclusively along a few large basin-bounding faults (such as the flexural cantilever model). However, the stratigraphic geometry of many basins cannot be explained by these simple models; for example, the common occurrence of thick sequences on the footwalls of basin-bounding faults. This requires the generation of accommodation space by subsidence. However, the flexural cantilever model predicts uplift, while the uniform stretching model cannot model the effects of faults at all. This paper looks at the St. George's Channel Basin, which has been mapped in detail using an extensive seismic dataset, and attempts to explain the structure and stratigraphic geometry of the basin by applying forward modelling techniques to four cross sections taken from different areas of the basin. This has allowed the determination of the detailed mechanisms by which extension has occurred within the upper crust, and of variations in these both spatially and through time. Results from the modelling indicate that a combination of extension on basin bounding faults and pure shear distributed across the basin is required to explain the geometry of the basin, although the pure shear component of extension probably represents extension on numerous sub-seismic scale faults across the basin. The relative influence of these different mechanisms of extension varies considerably across the basin, and throughout its history. During the Triassic, for example, up to 80 % of total extension in the brittle upper crust occurred by pure shear. During the Middle and Upper Jurassic, however, extension in the upper crust occurred mainly on a single fault (the St. George's Fault) in the basin centre, but on the edges of the basin, where the fault died out, pure shear extension was dominant.

Keywords: St. George's Channel, Mesozoic, rifting, extensional basin, forward modelling

The St. George's Channel Basin is a Mesozoic extensional basin lying offshore southwestern Britain (Fig. 1). Previous studies of the basin [13, 15, 17] have shown that there are several large normal faults running through the basin, which appear to exhibit a strong control on basin stratigraphy, however there has been no attempt to quantitatively assess their role in basin evolution. To do this, we must determine the mechanisms by which extension occurred in the upper crust in this region, in particular whether it occurred by movement on these faults, or was accommodated by other means. This study is an attempt to do so by mapping the basin structure and stratigraphy and forward modelling key sections across the basin.

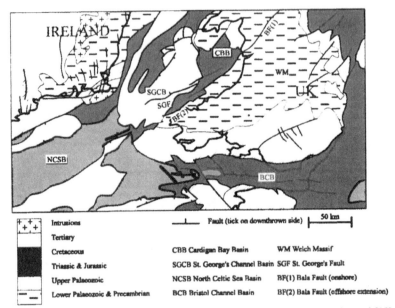

Figure 1. Geological map of the St. George's Channel Basin region. Simplified from Tappin *et. al.* [15]

FORWARD MODELLING

We can reconstruct the tectonic evolution of a basin from observations and measurements of the current structure and stratigraphy of the basin (the observed basin attributes), by looking for the effects of tectonic events and processes which made up these histories (the mechanisms of basin evolution). Forward modelling does this by postulating a specific mechanism of basin evolution (the modelled scenario), and using theoretical models to predict the basin geometry that this will produce. This is then compared with the observed basin geometry, and if a reasonable match is found, we can conclude that the modelled scenario is a plausible account of the true mechanism of basin evolution (although not necessarily the only one). Generally a range of possible mechanisms of evolution are modelled in order to identify those which are plausible, or that which is the most likely (the best-fit case).

This study uses theoretical models of the behaviour of the lithosphere during extension to predict the vertical movement (subsidence or uplift) of the surface, given a set of mechanical and structural parameters such as amount of extension, mechanism of extension, amount of movement on faults, crustal thickness, etc. The theoretical models must account for the influence of all the processes which occur during lithospheric extension (such as mantle and lower crustal extension, fault movement, isostatic compensation and flexural strength within the lithosphere) on the vertical movement at the surface. The calculated (sediment loaded) vertical displacement can then be compared to the thickness of sediment deposited over a particular time interval. If they are compatible, then the set of parameters used describes a plausible mechanism of extension during that period (although not necessarily the only one). In practice, a two-dimensional cross section of the basin is modelled, generating a (sediment loaded) subsidence profile across the basin; this can be compared to the two-dimensional stratigraphic geometry of the appropriate unit. However, we cannot simply equate the two as the thickness of sediment deposited over a particular time interval is not directly proportional to the subsidence at that point; other factors will influence it. Among these are initial topography (or palaeobathymetry) (particularly if this is variable across the basin), sedimentary factors (such as rate of sediment supply, type of depositional system active, or erosion), sea level changes, and the flexural response of the lithosphere to sediment load. The influence of these factors (some of which may themselves be dependent on tectonic activity) must be taken into account when determining what predicted stratigraphic geometries are consistent with observed stratigraphic geometries.

In order to carry out this procedure accurately we need two things. Firstly, we need to know the stratigraphic geometry of the unit which we are modelling, across the section. This can be derived from a depth converted seismic cross section by a decompaction calculation to remove the effects of compaction under the weight of overlying strata [19]. Secondly, we need a mathematical model which can calculate the surface vertical displacement generated by lithospheric extension, given a set of structural and mechanical parameters. The simplest such model is the McKenzie model [11], which was further refined in [12]. This model calculates sediment loaded surface subsidence during both syn-rift and post-rift phases, generated by extension of an amount β (calculated proportionally). This model has several disadvantages, however. The initial model [11] is a one-dimensional model; it can calculate subsidence at a single point, but not as a profile across a basin (although this is remedied in [12]). It is an instantaneous model (i.e. it assumes that extension occurs instantaneously rather than over a finite time) and it assumes extension occurs entirely by pure shear; it cannot model the effects of large faults on vertical displacement (which can be considerable locally). However, several models have been produced which can calculate subsidence profiles combining extension by pure shear within the mantle and lower crust with extension on faults in the upper crust; e.g. the Chevron model [5, 20], dealing with listric faults, or the flexural cantilever model [9] dealing with planar faults.

Specialist software is available for forward modelling of basins, a prominent example of which is the program STRETCH, developed by Prof. Nick Kusznir of the University of Liverpool [9]. This program can calculate subsidence along a two-dimensional profile across a basin, during both syn-rift and post-rift phases, using either a pure shear model, a Chevron model, or a flexural cantilever model. It can model a basin containing multiple faults, fully incorporating the effect of lithospheric flexure in supporting loads, and can take into account the effects of partial fill of the basin, erosion of uplifted blocks, changes in sea level, and post-rift subsidence following previous rift phases. Its main weakness, however, is that it cannot calculate subsidence profiles where extension occurs by a combination of mechanisms (e.g. partially by pure shear and partially on planar faults). For this reason I have developed my own model which can calculate subsidence profiles where this is the case. The derivation of this model and an explanation of its methods of calculation are given in the appendix. This model has several limitations, in particular that it can only partially compensate for the effect of lithospheric flexure in supporting loads, and can only accurately model planar faults where both the hangingwall and the footwall have undergone net subsidence (i.e. when sediment cover is present on both); full details of its limitations are given the appendix.

To fully model a basin cross-section it must be divided into chronostratigraphic units, which must then be decompacted to determine the original thickness of sediment deposited during that particular period. The constraints on changes in the position of the sediment surface (i.e. palaeobathymetric or topographic changes) must be noted for each case. Modelling of each section starts with the lowest unit in the basin which is assumed to be entirely generated by active extension. A series of calculated subsidence profiles is generated for varying configurations of structural and mechanical parameters, to find the scenario or range of scenarios which best match the decompacted stratigraphic geometry. This is then used to calculate a predicted profile of post-rift subsidence over the period of deposition of the overlying unit, and this is subtracted from the observed stratigraphic geometry of that unit. Any remaining subsidence can then assumed to be the result of further extension, which can be modelled as previously described. This procedure is repeated for each chronostratigraphic unit, allowing the amount and distribution of extension through time to be constrained and studied.

COMPARISON OF EXTENSIONAL MECHANISMS

Before attempting to model the St. George's Channel Basin, preliminary modelling was carried out using both packages in order to contrast subsidence profiles generated by different mechanisms of extension. Firstly, models were run to compare syn-rift subsidence profiles generated by equal amounts of extension occurring by different mechanisms (all other parameters were kept identical in each case). Profiles were generated for 2 km of pure shear extension (in a basin 50 km wide), and 2 km of extension on a planar fault (dipping at 60°). The results show a strong contrast (Fig. 2a). The total (net) amount of subsidence across the basin was similar in all cases, but pure shear extension produced a much wider, shallower basin than fault extension; the basin covered a width of c. 50 km, but the maximum depth was only 0.5 km. The fault-bounded basin had a maximum depth of over 2 km, but a width of only c. 20 km; it showed over 1 km of uplift on the footwall of the fault. The effects of footwall uplift can be removed by superimposing pure shear extension (centred on the fault). In the case of a planar fault with 1 km of extension (and > 0.5 km footwall uplift) an additional 3 km of pure shear extension is required to counteract footwall uplift (Fig. 2b).

Models were also run to compare the subsidence generated during syn-rift and post-rift phases. Two cases were run, modelling 2 km of extension by pure shear and on a planar fault, and post-rift subsidence calculated after 20 ma and 100 ma. It was found that the amount of post-rift subsidence generated was small; a maximum of 75 m after 100 ma and 200 m after 100 ma in both cases (compared to 500 m of syn-rift subsidence for pure shear extension and 2 km for extension on a fault) (Fig. 3). Post-rift subsidence did cover a wider area than syn-rift subsidence, so that on the extreme margins it formed the largest component of subsidence. Although post-rift subsidence did occur on the footwall of the planar fault, nowhere was it great enough to counteract the effects of syn-rift uplift, so the entire footwall stayed above base level.

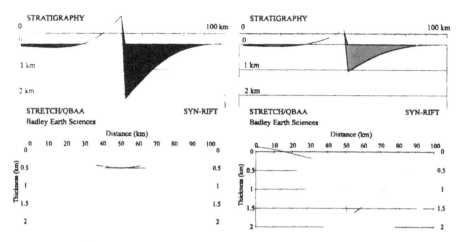

a) Subsidence profiles produced by 2 km of extension on a single fault (top) and by pure shear (bottom). The fault controlled basin is nearly 5 times deeper, but much narrower. Note also the uplift on the fault footwall.

b) About 0.5 km of footwall uplift is produced by 1 km of extension on a single fault (top). 3 km of additional pure shear extension centred on the fault is required to preventfootwall uplift (bottom).

Figure 2. Comparison of subsidence profiles produced by pure shear extension and extension on a fault.

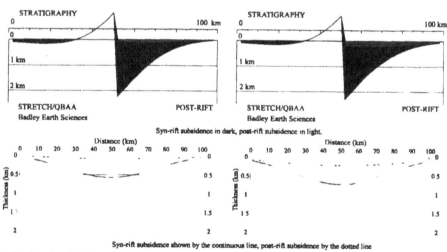

a) Stratigraphy after 20 ma post-rift subsidence, following 2 km of extension on a single fault (top) and by pure shear (bottom).

b) Stratigraphy after 100 ma post-rift subsidence, following 2 km of extension on a single fault (top) and by pure shear (bottom).

Figure 3. Comparison of subsidence profiles during the syn-rift and post-rift stages of basin evolution. Post-rift subsidence is small compared to syn-rift subsidence, particularly after a short time, or in a fault-controlled basin.

The general conclusions would appear to be as follows:

1. A small amount of extension on a single fault can produce the same amount of subsidence as a large amount of pure-shear extension. Basins produced by pure shear have undergone as much as 5 times more extension than fault-controlled basins of a similar depth.

2. Footwall uplift always accompanies planar faulting. If a basin has significant syn-rift strata on the footwalls of its basin-bounding faults, then it must have undergone a significant degree of pure shear extension.

3. The amount of post-rift subsidence generated is generally small compared to the amount of syn-rift subsidence; < 20 % in the short term (< 30 ma), and < 50 % in the long term (> 100 ma), particularly if extension occurs on large faults. Most thick sediment sequences must be at least partly generated by rifting.

THE ST. GEORGE'S CHANNEL BASIN

The stratigraphy of the St. George's Channel Basin has been well studied [1, 15, 16, 17], and ranges from Permo-Triassic to Tertiary. The basin strata reach a thickness of up to 12 km, and unconformably overlie Carboniferous rocks, which exhibit a deformation fabric imprinted by two orogenic episodes, the Caledonian and the

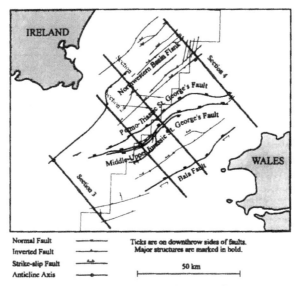

Figure 4. Structural map of the St. George's Channel Basin showing sections used for modelling. Note the major faults (bold) and the minor faults (feint).

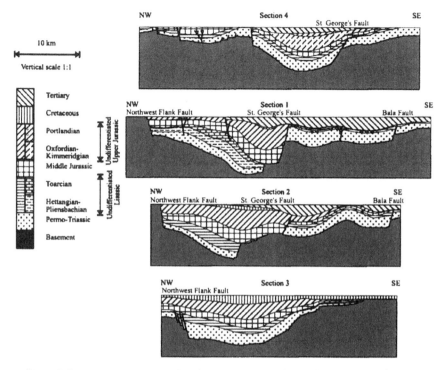

Figure 5. Cross sections at various points along the basin (see Fig. 4 for locations). Note the extreme variations in profile along the length of the basin.

Variscan [16]. The overall structure of the St. George's Channel Basin is that of a half-graben whose geometry varies along the basin axis (Fig. 4). The major basin-bounding faults are the St. George's Fault and the offshore extension of the Bala Fault. Almost all the structures in the basin, particularly these basin-controlling faults, are oriented NE-SW, along the axis of the Caledonian orogeny, and may have been formed by reactivation of Caledonian basement structures [10, 15]. In the central part of the basin the St. George's Fault defines the southeast margin of a large half-graben which forms the deepest part of the basin (Fig. 5); it migrated southwest during the Liassic, with the Middle Jurassic fault plane set back up to 3 km from the Permo-Triassic St. George's Fault. In the northwest, the St. George's Fault is offset dextrally by about 15 km, along an E-W axis (parallel to the axis of the Variscan orogeny). At this point, displacement on the Middle and Upper Jurassic fault plane dies out. There are numerous parallel minor folds and faults, particularly on the flank of the half-graben, few of which are laterally continuous and many of which show evidence of strike-slip motion, in the form of flower structures, or of inversion. Displacement on both the St. George's Fault and the Bala Fault decreases away from the centre of the St. George's Channel, with displacement taken up on other faults, often on the opposite side of the basin, changing the basin profile markedly (Fig. 5).

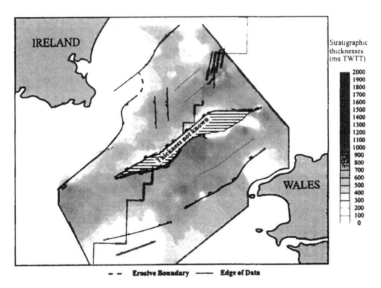

Figure 6. Triassic structure and stratigraphy. Note the thickening stratigraphy on the footwalls of major faults, and the influence of minor faults on the northwest flank.

Data from 2 seismic surveys, comprising 38 dip lines shot perpendicular to the basin axis and 14 strike lines shot parallel to the basin axis, along with 12 wells, were used in this study. Initially, chronostratigraphic horizons marked by discordant reflectors (e.g. onlap and downlap surfaces or erosive unconformities) were interpreted independently on individual dip lines. These were then mapped across the basin, by correlating horizons on separate dip lines along connecting strike lines, and were dated by tying in with well logs. It was found that horizons identified independently on single dip lines tied in well along strike lines, and generally corresponded with chronological boundaries identified in wells. It was then possible to produce isochron maps of the basin, showing variations in stratigraphic thickness (measured in two-way travel time) of units deposited over a known time period (i.e. bounded by dated chronostratigraphic horizons). It has also been possible to interpret the major structures which cut chronostratigraphic horizons on the seismic dataset, and thus produce maps of the structure of each unit. By overlaying these on isochron maps of the same unit, it is possible to assess the influence of these structures, particularly extensional faults, on sediment deposition at different times, which can help in understanding the tectonic and structural evolution of the basin. In addition, four dip lines from different parts of the basin have been depth-converted and decompacted, for use in modelling (Figs 4 & 5).

Triassic
The Triassic sequence (Fig. 6) consists almost entirely of continental sediments, although the uppermost Rhaetian shows a marine incursion from the South [16]. The major fault structures do appear to exert some control over sedimentation, with sediments thickening into the hangingwalls, but there is still a considerable thickness of

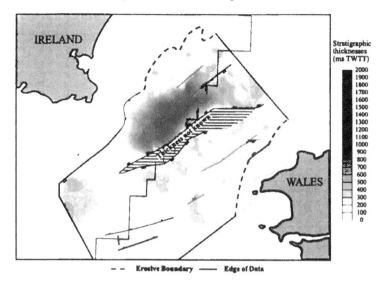

Figure 7. Liassic structure and stratigraphy. Note that the thickest sediment is seen in the centre of the basin, not adjacent to the major faults.

sediment on the footwalls of all faults, suggesting that subsidence was not entirely controlled by movement on these major faults.

Lower Jurassic

Lower Jurassic strata (Fig. 7) consist exclusively of marine clays, which may have been deposited in a considerable depth of water [16]. The sediment geometry does not seem to relate to the fault geometry; the thickest succession is found in the centre of the basin, and the unit thins into the footwall of the St. George's Fault; south of the St. George's Fault the sequence is condensed. Seismic stratigraphic analysis suggests that the Lower Jurassic succession may contain large sediment wedges, downlapping to the southeast and toplapping to the northwest; these may be interpreted as sediments prograding from the northwest flank into the main part of the basin. Some of these units have been mapped out and can be up to 20 km in size.

Middle & Upper Jurassic

Middle and Upper Jurassic strata (Fig. 8) show mainly shallow marine facies, consisting of clays (often containing traces of lignite) and oolitic limestones [16]. The stratigraphic geometry appears to be very much controlled by the St. George's Fault, with the thickest sediment found in the immediate hangingwall of this fault, thinning towards the northwest basin flank. On the footwall of the St. George's Fault, a very condensed sequence is seen (although in places this appears to have been eroded out by post-Jurassic inversion).

A brief examination of the four cross sections selected for modelling shows a few very striking features. The first is that large faults do appear on each section, most notably

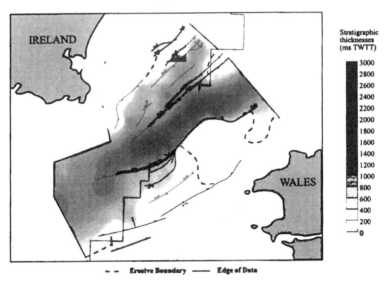

— —   Erosive Boundary   ——   Edge of Data

Figure 8. Middle-Upper Jurassic structure and stratigraphy. Note the influence of the St. George's Fault on the stratigraphy, and the thin succession seen south of it

the Bala Fault, the St. George's Fault, and the Northwest Flank Fault. Often these appear to be active throughout the basin history, with individual units from throughout the stratigraphic column thickening into their hangingwalls and thinning over their footwalls (although as mentioned earlier care must be taken when interpreting this as evidence for fault activity). These faults have obviously influence the subsidence profile of the basin, and have therefore been included in the modelled scenarios wherever appropriate. In many cases however these faults still show thick (over 1 km) sediment cover on their footwalls (particularly in the Triassic). This cannot be due to passive infilling of a pre-existing depression (i.e. a rising sediment surface level), as the sediment surface level remains constant (at sea level) throughout the Middle and Upper Jurassic, and actually falls during the Triassic as we move from a terrestrial to a marine environment [16]. Nor is it likely that changes in eustatic sea level (to create more accommodation space) are responsible, as these can only account for at most 500 m of sediment in the short term, and in the long term their cyclic nature will ensure that accommodation space generated during a period of rising sea level will be destroyed when sea level later falls. Extension on further large faults outside the area of study could cause additional subsidence within the basin, but no such faults have ever been mapped, and it is unlikely that faults large enough to counteract the effects of footwall uplift within the basin would remain unnoticed. It therefore seems likely that an additional mechanism of regional subsidence within the basin is required to explain the stratigraphic geometry. To account for this, a variable component of pure shear extension has been included in most modelled scenarios; the extent of this, along with the amount of extension on major faults, are the main parameters which have been altered in order to obtain a good fit for the observed stratigraphic geometries (the fact that net footwall subsidence has occurred on these faults   allows   my   model   to   calculate

Table 1. Values taken for constants used in modelling. The values quoted here were used throughout the modelling except where otherwise otherwise stated in the text.

a	= thickness of asthenosphere	= 125 km	[11]
t_c	= thickness of crust	= 29 km	[4]
Z_d	= thickness of brittle upper crust	= 10 km	[9]
ρ_m	= density of mantle at 0° C	= 3300 kg/m^3	[11]
ρ_c	= density of crust at 0° C	= 2800 kg/m^3	[11]
ρ_f	= density of sediment fill	= 2680 kg/m^3	[6]
T_1	= temperature of the asthenosphere	= 1333° C	[11]
α	= thermal expansion coefficient of the mantle	= 3.28×10^{-5} °C^{-1}	[11]
T_e	= elastic thickness of the lithosphere	= 5 km	[9]
ε	= Young's modulus for the lithosphere	= 70 Gpa	[5]
σ	= Poisson's ratio	= 0.25	[5]
g	= gravitational field strength	= 9.8 m/s^2	[5]
ψ	= angle of major planar faults to surface	= 60°	[9]

Numbers in brackets indicate reference.

subsidence profiles accurately; it cannot accurately model planar faults where net footwall uplift has occurred). The 'pure shear' component of extension inferred may in fact be the result of a mechanism other than true pure shear. The most likely such mechanism is extension on a large number of minor faults distributed across the basin. Several large minor faults, with maximum throws of up to 500 m, can in fact be seen on many seismic sections (particularly in the north of the basin) and some have been mapped (Fig. 4). It is likely that many more smaller faults also exist, below the level of seismic resolution (c. 200m) and therefore not able to be modelled individually. Faults of this size within Lower Jurassic, Triassic, and basement rocks are common in the Bristol Channel Basin (a related Mesozoic basin) and on the Pembrokeshire coast [3, 14]. Work in the Triassic Cheshire basin has revealed that fault sizes follow a fractal distribution [2], and hence the presence of large faults strongly suggests the existence of many more smaller faults; in some areas these small faults are estimated to accommodate up to 70 % of total extension.

Another obvious feature of the cross sections is that, while to the southeast of the St. George's Fault we see many relatively thin units, northwest of the fault all units tend to be of comparable thickness. Preliminary modelling suggested that sequences deposited during the post-rift phase will be much thinner than those deposited during the syn-rift phase, so it appears that we may need to invoke active extension throughout the Mesozoic to account for the stratigraphy we see northeast of the St. George's Fault. Finally, although it is now generally believed that most extensional faults are planar [7], the geometry of the St. George's Fault (in particular the 'step back' from the Triassic fault plane to the Middle and Upper Jurassic fault plane) suggests that this may be better modelled as a listric fault, at least later in its history.

With these considerations in mind, units from throughout the basin stratigraphy have been modelled on each section. In each case, subsidence profiles have been calculated for all sensible configurations of parameters, the main variables being throw on individual faults, and amount and distribution of pure shear extension. Mechanical properties of the lithosphere have generally been kept constant, with values taken from previous research (Table 1); however in some cases different configurations of these parameters have been tested to determine their influence on subsidence. For each case, the configuration of parameters which best fits the observed stratigraphic geometry has been selected and described, along with a brief description of why other configurations fit less well. In some cases, due to lack of stratigraphic constraints or for other reasons, several different configurations of parameters are equally plausible; where this is the case, all have been described.

RESULTS

Section 1

This section is taken across the centre of the basin, where displacement across the St. George's Fault appears to be at a maximum. The Triassic basin consists of two graben separated by a horst, but there is a considerable sediment cover on both the horst and the basin margin (Fig. 9a); up to 2 km covering the southeast margin, and a minimum of 1.5 km covering the horst (although the maximum thickness is not known). The best fit simulation incorporates four planar faults with throws of between 0.5 and 1.2 km, giving a total fault extension (heave) of 3 km. In order to account for the sediment covering the footwalls of these faults, 12 km of pure shear extension is required across a width of 50 km, with extension centred on the horst (Fig. 9a). If sediment thickness on the horst is greater than 1.5 km, extension on the horst-bounding faults may be considerably less. This is a minimum estimate of extension as it assumes that the sediment surface level remains constant; in fact, as the Triassic strata are terrestrial while the Lower Jurassic strata are marine, it would appear that the sediment surface level fell during the Triassic [16], in which case further (largely pure shear) extension would be required to account for this subsidence.

Triassic extension of this magnitude would produce a maximum of 0.8 km of post-rift subsidence in the basin centre during the Lower Jurassic, and about 0.5 km on the margins. This is sufficient to explain the Lower Jurassic strata seen to the southeast of the St. George's Fault, but cannot explain the 4 km of Lower Jurassic strata to the northwest of the St. George's Fault (Fig. 9b). If this were formed by passive infilling of a pre-existing basin, water depth would need to have reached 2-3 km by the end of the Triassic (falling to around 0 km by the end of the Lower Jurassic). It is more plausible that this is at least partially due to active rifting during the Lower Jurassic, with extension on faults northwest of the Triassic horst combined with pure shear extension centred on that horst to prevent footwall uplift southeast of the St. George's Fault. With no change in palaeobathymetry throughout the Lower Jurassic (the scenario with maximum extension), 0.5 and 2 km of extension respectively is required on the major faults, coupled with 3 km of pure shear extension over a width of 30 km (Fig. 9b).

The Middle and Upper Jurassic has been divided into three units for modelling: the Bathonian, the Oxfordian & Lower Kimmeridgian, and the Upper Kimmeridgian & Portlandian, each of which is bounded by distinctive chronostratigraphic horizons which can be easily mapped across the basin. Bathymetry remained constant, at about sea level,

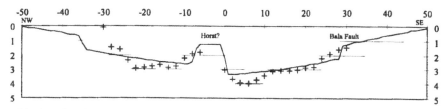

a) Triassic. 12 km of pure shear extension is required to account for the thick succession observed on the footwalls of faults, in particular that of the Bala Fault.

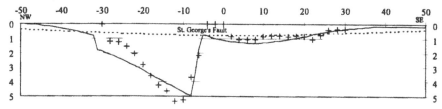

b) Lower Jurassic. Post-rift subsidence can explain the observed stratigraphy southeast of the St. George's Fault, but 5.5 km of active extension is required to explain that to the northwest.

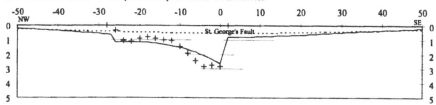

c) Bathonian. 2.5 km of fault extension is inferred, with no pure shear extension (as there are no strata on the footwalls of the faults). The faults are listric.

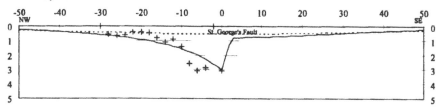

d) Oxfordian & Lower Kimmeridgian. 2.5 km of extension is inferred on the (listric) St. George's Fault. Extension by other mechanisms is hard to constrain due to erosion of strata.

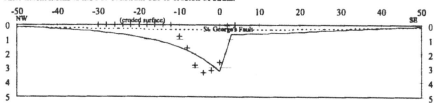

e) Upper Kimmeridgian & Portlandian. 3 km of extension is inferred on the (listric) St. George's Fault. Extension by other mechanisms is hard to constrain due to erosion of strata.

Figure 9. Observed stratigraphy and calculated subsidence profiles for Section 1. Observed profiles marked by crosses, calculated post-rift subsidence by dotted lines, calculated total subsidence by solid lines. Scales are marked in km.

throughout this period [16], so thick sediment deposits imply the generation of accommodation space by tectonic subsidence. During the Bathonian, post-rift subsidence following Triassic and Lower Jurassic rifting can account for a maximum of 0.5 km of subsidence; in order to account for the 2.5 km of Bathonian strata we see northwest of the St. George's Fault, 2 km of extension is required on the St. George's Fault, with a further 0.5 km on the Northwest Flank Fault (Fig. 9c). The rollover geometry seen on the cross section is best matched if these faults are listric (the fact that the Middle Jurassic St. George's Fault plane appears to lie 2-3 km southeast of the Triassic fault plane is further evidence of listric geometry). If the major faults are listric, the condensed Bathonian sequence southeast of the St. George's Fault can be explained entirely by post-rift subsidence; if they are planar footwall uplift will be greater, and to compensate for this extension on faults would need to be halved, and combined with up to 1 km of pure shear extension.

Post-rift subsidence from all previous rifting can account for a maximum of 0.5 km of subsidence in the Oxfordian & Lower Kimmeridgian and 0.3 km in the Upper Kimmeridgian and Portlandian. 2.5 km of Oxfordian & Lower Kimmeridgian extension and 3 km of Upper Kimmeridgian & Portlandian extension is required on a listric St. George's Fault to account for the observed thickness of these units (Figs. 9d & 9e). The amount of extension on the Northwest Flank Fault and by pure shear is hard to ascertain, as much of the basin margin stratigraphy which could be used to constrain this has been removed by subsequent inversion.

Section 2

This section is taken from further southwest, where displacement on the St. George's Fault is considerably less than in the centre of the basin (section 1), and the fault has splayed to become an extended fault zone (although for simplicity it has been modelled as a single fault in this study).

The Triassic basin again comprises two graben separated by a horst, with sedimentary cover on both the horst and the southeast basin flank (Fig. 10a). Extension on the major faults is less than on section 1, varying between 0.5 and 0.7 km on individual faults (with a total of 2.2 km), with 8 km of pure shear extension required to account for the sediment thickness on the footwalls of major faults (Fig. 10a) (as on section 1, if sediment thickness on the horst is greater than that assumed here, extension on the horst-bounding faults may be less). Again this is a minimum estimate, assuming no fall in sediment surface level during the Triassic. In contrast to section 1, pure shear extension in this part of the basin appears offset from the structural basin centre, and is centred around 10 km southeast of the horst.

Extension of this degree during the Triassic will generate approximately 0.5 km of Lower Jurassic post-rift subsidence southeast of the St. George's Fault, and 0.3 km northwest of it (Fig. 10b). This can explain the Lower Jurassic strata observed southeast of the St. George's Fault, but to account for the 3 km observed to the northwest requires up to 1 km of Lower Jurassic extension on the St. George's Fault and 0.5 km on the Northwest Flank Fault, combined with 5 km of pure shear extension centred on the Triassic horst (10 km northwest of the Triassic centre of extension) (Fig. 10b).

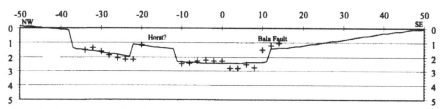

a) Triassic. As on section 1, 8 km of pure shear extension is required to explain the strata observed on the footwalls of the basin-bounding faults.

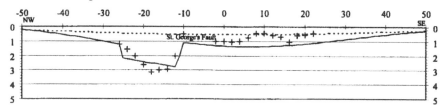

b) Lower Jurassic. As on section 1, post-rift subsidence can explain the strata southeast of the St. George's Fault, but 6.5 km of extension is required to explain that to the northwest.

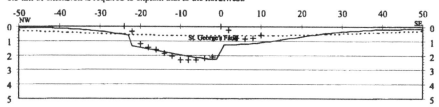

c) Bathonian. Unlike on section 1, thick strata are observed on the footwall of the St. George's Fault here. 1.5 km of pure shear is required to explain this.

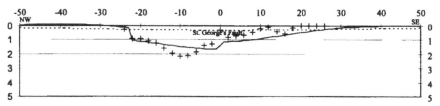

d) Oxfordian & Lower Kimmeridgian. As in th Bathonian, 1.5 km of pure shear is required to explain the strata observed on the footwall of the St. George's Fault.

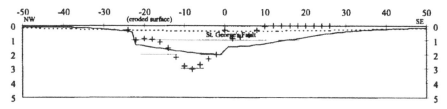

e) Upper Kimmeridgian & Portlandian. A similar scenario will fit the observed strata of this period, but it is hard to constrain due to erosion on the basin flanks.

Figure 10. Observed stratigraphy and calculated subsidence profiles for Section 2. Observed profiles marked by crosses, calculated post-rift subsidence by dotted lines, calculated total subsidence by solid lines. Scales are marked in km.

The Middle and Upper Jurassic sequence has been subdivided into the same three units as section 1. The best fit scenario for the Bathonian (after taking into account 0.6 km of post-rift subsidence) was achieved with two planar faults, the Northwest Flank Fault (with 0.4 km extension) and the St. George's Fault (with 0.6 km extension), but 1.5 km of pure shear extension over a width of 30 km centred on the St. George's Fault is necessary to account for the Bathonian strata on the footwall of the St. George's Fault (Fig. 10c). Even thicker Oxfordian & Lower Kimmeridgian strata on the footwall of the St. George's Fault can be explained by 1.5 km of pure shear extension (over a width of 30 km) combined with 0.5 km of extension on the Northwest Flank Fault, and 0.3 km of extension on the St. George's Fault (both planar), along with 0.3 km of post-rift subsidence. The Upper Kimmeridgian & Portlandian sequence could be explained by a similar scenario (and 0.25 km post-rift subsidence), but is hard to constrain due to erosion of strata on the basin flanks (Fig. 10d & 10e).

Section 3

This section is taken from the extreme southwest where the St. George's Channel Basin joins the North Celtic Sea Basin. The St. George's Fault has completely died out, with all major fault displacement occurring on the Northwest Flank Fault, although in general the basin appears to be much less fault controlled than further north.

The thickness of Triassic and Lower Jurassic strata on the footwall of the Northwest Flank Fault is not well constrained on this section (no wells on the footwall penetrate to this depth), so we are not able to constrain the proportion of extension occurring on this fault compared with that occurring by pure shear; estimates of extension range from 12 km of pure shear extension and 2.8 km of extension on the fault, to 3 km of pure shear extension and 4.5 km of extension on the fault (throughout the Triassic and Lower Jurassic combined) (Fig 11a). Pure shear extension was probably centred on the fault, as otherwise the basin would have been much wider.

The footwall sediment thickness is better constrained for the Middle & Upper Jurassic, although post-rift subsidence following Triassic and Lower Jurassic rifting is hard to constrain (estimates range from 0.3-0.6 km during the Bathonian, 0.2-0.4 km during the Oxfordian & Kimmeridgian, and 0.05-0.1 km during the Portlandian). The best fit scenario for the Bathonian combines 2.5-3.5 km of pure shear extension, occurring over a width of 30 km centred 15 km southeast of the Northwest Flank Fault, with 0.3 km of extension on the (planar) Northwest Flank Fault (Fig. 11b). During the Oxfordian, Kimmeridgian, and Portlandian, we see no evidence of any fault movement, with 4 km of pure shear extension during the Oxfordian & Kimmeridgian and 3.5 km during the Portlandian, over the same area as during the Bathonian (Fig. 11c & 11d).

This is the only part of the St. George's Channel Basin which has significant Cretaceous cover (of up to 1.5 km); however, this can be explained purely by post-rift subsidence

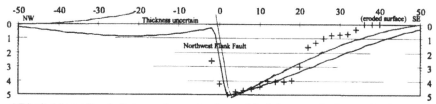

a) Triassic & Lower Jurassic. Strata are not mapped northwest of the fault, so extension is uncertain. The grey line shows 4.5 km of fault extension and 3 km of pure shear, the black line 2.8 km of fault extension and 12 km of pure shear.

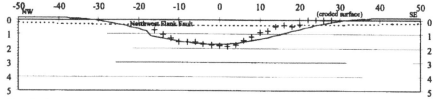

b) Bathonian. The basin has a sag profile with little fault control. Only 0.3 km of fault extension occurred, combined with 2.5-3.5 km of pure shear.

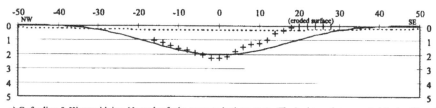

c) Oxfordian & Kimmeridgian. No major faults are seen in these strata. The basin can be accounted for by 4 km of pure shear.

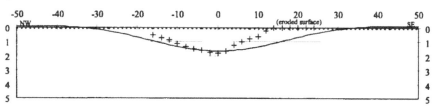

d) Portlandian. Again, there appears to be no major fault extension during this period, but 3.5 km of pure shear occurred.

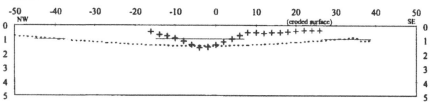

e) Cretaceous. This is the only section on which Cretaceous strata are observed. They can be explained purely by post-rift subsidence following previous extension.

Figure 11. Observed stratigraphy and calculated subsidence profiles for Section 3. Observed profiles marked by crosses, calculated post-rift subsidence by dotted lines, calculated total subsidence by solid lines. Scales are marked in km.

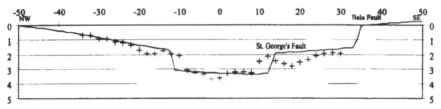

a) Triassic. The horst seen further southwest is not seen here. 2.4 km of fault extension combined with 10 km of pure shear can account for the strata observed.

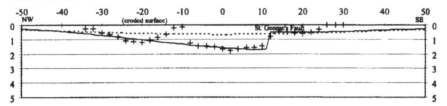

b) Lower Jurassic. A further 0.7 km of fault extension combined with 2 km of pure shear is required to explain the observed Lower Jurassic strata.

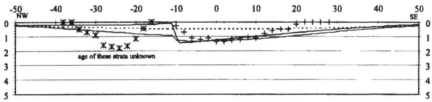

c) Bathonian. The St. George's Fault is not active, and fault extension switches to a southeast-dipping fault. Strata northeast of this are unknown, so extension could be largely pure shear (grey line) or largely fault-controlled (black line).

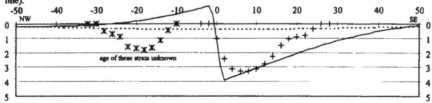

d) Oxfordian & Lower Kimmeridgian. The thickness of strata observed here requires 2 km of fault extension and 2 km of pure shear to explain.

Figure 12. Observed stratigraphy and calculated subsidence profiles for Section 4. Observed profiles marked by crosses, calculated post-rift subsidence by dotted lines, calculated total subsidence by solid lines. Scales are marked in km.

following Triassic and Jurassic extension, and there is no evidence for further extension (Fig. 11e).

Section 4

Section 4 is taken across the northeast of the St. George's Channel Basin, where it adjoins the Cardigan Bay Basin. It is a structurally complex area in which major faults from both basins appear to control subsidence and stratigraphy.

Triassic subsidence is controlled by three major faults: the St. George's Fault, the Bala Fault (although its role is hard to ascertain as the stratigraphy of its footwall is not well constrained), and a southeast dipping fault forming the southernmost extension of a major fault within the Cardigan Bay Basin (Fig. 12a). 0.8 km of extension on each of these faults combined with 10 km of extension by pure shear is required to account for the 3 km of Triassic strata in the basin centre and up to 2 km on the flanks (Fig. 12a). Some of this 'pure shear' extension occurs on faults which are large enough to be observed and mapped on the seismic dataset (although they cannot be resolved well enough to model individually). Of these, some follow a Caledonoid trend while others appear to form en-echelon structures, possibly accommodating rotation caused by extension being oblique to the trend of the major (Caledonoid) faults (see Fig. 4). In order to account for the Lower Jurassic strata, an additional 0.7 km of extension is required on the St. George's Fault combined with 2 km of pure shear extension (Fig. 12b).

During the Middle & Upper Jurassic the northwestern fault appears to have been the only active fault, with no displacement observed across the St. George's Fault. Movement on this fault is hard to determine as there is no Middle or Upper Jurassic stratigraphy remaining on the immediate footwall (although there may be some further northwest). Bathonian strata could be explained (after 0.45 km of post-rift subsidence) by 0.8 km of extension on this fault combined with 0.8 km of pure shear extension (with no footwall elevation remaining constant), or by 0.2 km of extension on the fault combined with 3 km of pure shear extension (giving 800 m of footwall subsidence), or a combination of the two (Fig. 12c). The Oxfordian & Lower Kimmeridgian profile could be explained (after 0.35 km of post-rift subsidence) by 2 km of extension on the fault and 2 km of pure shear extension (extension cannot have occurred exclusively by pure shear as this would result in too wide a basin) (Fig. 12d). Upper Kimmeridgian & Portlandian extension is hard to constrain due to heavy erosion, particularly on the basin flanks, but requires (in addition to 0.1 km of post-rift subsidence) at least 2.5 km of extension on the fault, or 10 km of pure shear extension (or a combination of both) to account for the observed sediment thickness.

CONCLUSIONS

In order to account for the stratigraphic thicknesses observed throughout the basin, active extension must have occurred throughout most of the Mesozoic; none of the stratigraphic units studied can be explained completely by post-rift subsidence following earlier extension episodes (except possibly the Aalenian & Bajocian, and the Cretaceous). This does not mean that extension was continuous; it may have been episodic but over shorter intervals than those modelled. Furthermore, extension did not always occur over the whole basin; some areas (particularly southeast of the St. George's Fault) may have experienced periods of post-rift subsidence while extension was active elsewhere.

Throughout the basin, syn-rift sedimentation is observed on the footwalls as well as on the hangingwalls of major extensional faults. In order to account for this, it is suggested that a significant proportion of extension was not accommodated by movement on these faults, and may instead have been accommodated by a large number of minor faults distributed evenly across the basin.

The distribution of extension between major and minor faults appears to vary spatially and through time. Extension in the Triassic appears to have been accommodated more on minor faults than was the case in the Lower Jurassic, while extension in the Middle & Upper Jurassic appears to have been very much concentrated on the St. George's Fault in the centre of the basin, but largely on minor faults towards the northeast and southwest. This spatial variation may be due simply to segmentation of the major faults, with the centres of fault segments being able to accommodate large amounts of extension while the tips of the segments could not; hence in the transitional areas between fault segments (and between the separate basins that they define structurally) extension was largely taken up by smaller faults. The temporal variation in the distribution of extension is harder to explain. It may be due to changes in the extensional field over time, so that when extension is oriented perpendicular to pre-existing fault structures (e.g. in the Middle and Upper Jurassic), these are reactivated, whereas when extension is oriented oblique to these structures it tends to occur on new faults formed perpendicular to extension, or by the opening of en-echelon structures associated with transtension and rotation of blocks.

Acknowledgements

I would like to thank Marathon UK (Ltd.) and partners for providing funding and data for this research, and for all the help and assistance I have received from them. I would also like to thank Dr. J. P. Turner at the University of Birmingham for invaluable advice and encouragement.

REFERENCES

1. K.W. Barr, V.S. Colter, R .Young. The Geology of the Cardigan Bay - St. George's Channel Basin. In: *Petroleum Geology of the Continental Shelf of Northwest Europe*, Institute of Petroleum, London, 432-443. (1981).
2. R.A. Chadwick. Fault analysis in the Cheshire Basin, NW England [in press]. (1996).
3. C.J. Dart, K.R. McClay, P.N. Hollines. 3D analysis of inverted extensional fault systems, southern Bristol Channel Basin, UK In: *Basin Inversion*. J.G. Buchanan, P.G. Buchanan (eds.), Geological Society Special Publication **88**, 393-413. (1995).
4. J. Dyment. SWAT and the Celtic Sea Basins - Relations with the Variscan crust, recent formation of the Moho. *Bulletin de la Societe Geologique de France* **5**, 477-487. (1989).
5. S.S. Egan. The flexural isostatic response of the lithosphere to extensional tectonics. *Tectonophysics* **202**, 291-308. (1992).
6. D.A. Falvey, M.F. Middleton. Passive continental margins: evidence for a prebreakup deep crustal subsidence mechanism. In: *Colloquium on Geology of Continental Margins (C3)*, Oceanologica Acta **4** (supplement), 103-114. (1981).

7. J.A. Jackson. Active normal faulting and crustal extension. In: *Continental Extensional Tectonics.* M.P. Coward, J.F. Dewey, P.L. Hancock (eds.), Geological Society Special Publication **28**, 3-17. (1987)

8. G.T. Jarvis, D.P. McKenzie. Sedimentary basin formation with finite extension rates. *Earth and Planetary Science Letters* **48**, 42-52. (1980).

9. N.J. Kusznir, P.A. Zeigler. The mechanics of continental extension and sedimentary basin formation: A simple-shear/pure-shear flexural cantilever model. *Tectonophysics* **215**, 117-131. (1992).

10. T. McCann, P.M. Shannon. Late Mesozoic reactivation of Variscan faults in the North Celtic Sea. *Marine and Petroleum Geology* **11**, 94-103. (1993).

11. D.P. McKenzie. Some remarks on the development of sedimentary basins. *Earth and Planetary Science Letters* **40**, 25-32. (1978).

12. D.P. McKenzie, N.J. White. Formation of the 'Steer's Head' geometry of sedimentary basins by differential stretching of the crust and mantle. *Geology* **16**, 250-253. (1988).

13. D. Naylor, P.M. Shannon. *The Geology of Offshore Ireland and Western Britain.* Graham and Trotman, London. (1982).

14. M. Nemcock, R. Gayer, R. Miliorizes. Structural analysis of the inverted Bristol Channel Basin; implications for the geometry and timing of fracture porosity In: *Basin Inversion* J.G. Buchanan, P.G. Buchanan (eds.), Geological Society Special Publication **88**, 355-392. (1995).

15. S.H. Petrie, J.R. Brown, P.J. Granger, J.P.B. Lovell. Mesozoic History of the Celtic Sea Basins In: *Extensional tectonics and stratigraphy of the North Atlantic margins.* Tankard, Balkwill (eds.), American Association of Petroleum Geologists Memoir **46**, 411-424. (1989).

16. D.R. Tappin, R.A. Chadwick, A.A. Jackson, R.T.R. Wingfield, N.J.P. Smith. *The Geology of Cardigan Bay and the Bristol Channel.* British Geological Survey Offshore Report. (1994).

17. R.M. Tucker, G. Arter. The tectonic evolution of the North Celtic Sea and Cardigan Bay Basins with special reference to basin inversion. *Tectonophysics* **137**, 291-307. (1987).

18. D.L. Turcotte, G. Schubert. *Geodynamics: Applications of continuum physics to geological problems.* Wiley, New York. (1982).

19. J.E. Van Hinte. Geohistory analysis - application of micropalaeontology in exploration geology. *Bulletin of the American Association of Petroleum Geologists* **62**, 201-222. (1978).

20. P. Verall. Structural interpretation with applications to North Sea problems *in* JAPEC course notes **3**. (1982)

APPENDIX: CALCULATION METHOD EMPLOYED IN FORWARD MODELLING

The model that was used to create theoretical subsidence profiles across the basin was a modification of the McKenzie pure shear model [11] to incorporate simple shear extension along major faults in the brittle upper crust.

The calculation used a finite element technique to calculate subsidence at discrete points along a section across the basin given the amount of extension on each major fault (which could be either listric or planar) and the amount of pure shear extension (distributed across the basin). The lithosphere was divided into three layers (mantle, ductile lower crust, and brittle upper crust). For at each point on the section, subsidence or uplift generated by extension in each layer was calculated independently, and the three values summed to derive the net vertical movement at the surface at that point (Fig. A1). An explanation of the symbols used is given below. Numbers in square brackets give references.

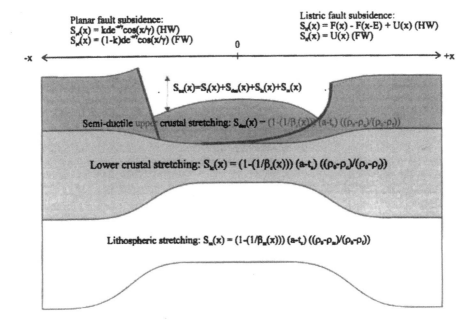

Figure A1. Mathematical model used for calculating subsidence across basin.

ρ_a' = density of asthenosphere (temperature dependent) $= \rho_m\left(1 - \alpha T_1\right)$ [11]

ρ_m' = density of lithospheric mantle (temperature dependent) $= \rho_m\left(1 - \dfrac{\alpha T_1}{2}\left(1 + \dfrac{t_c}{a}\right)\right)$ [11]

ρ_{lc}' = density of ductile lower crust (temperature dependent) $= \rho_c\left(1 - \alpha T_1\left(\dfrac{t_c + Z_d}{2a}\right)\right)$ [11]

ρ_{uc}' = density of brittle upper crust (temperature dependent) $= \rho_c\left(1 - \dfrac{\alpha T_1 Z_d}{2a}\right)$ [11]

a = thickness of asthenosphere	t_c = thickness of crust
Z_d = thickness of brittle upper crust	ρ_m = density of mantle at $0°$ C
ρ_c = density of crust at $0°$ C	ρ_f = density of basin fill
T_1 = temperature of the asthenosphere	α = mantle thermal expansion coefficient
σ_m = width of mantle extension / 2	σ_c = width of crustal extension / 2
T_e = elastic thickness of the lithosphere	ε = Young's modulus for the lithosphere
σ = Poisson's ratio	g = gravitational field strength
ψ = angle of major planar faults to surface	τ = age since rifting occurred (ma)

Mantle Layer:

In the lithospheric mantle, all extension was assumed to be accommodated by ductile pure shear with a Gaussian distribution of width $2\sigma_m$ around the basin centre. σ_m is variable. Uplift at the surface is generated by replacement of cool lithosphere with hot asthenospheric material, as given by the equation

$$S_m(x) = \left(1 - \frac{1}{\beta_m(x)}\right)(a - t_c)\frac{(\rho_a - \rho_m')}{(\rho_a - \rho_f)} \qquad [11]$$

$$\beta_m(x) = \text{distribution function of } \beta \text{ within mantle} = 1 + \left(\frac{E_T}{\sigma_m\sqrt{2\pi}}\right)e^{-\frac{x^2}{2\sigma_m^2}} \qquad [12]$$

Ductile Lower Crust:

Extension in the ductile lower crust is also assumed to be entirely ductile with a Gaussian distribution of width $2\sigma_c$ around the basin centre (σ_c need not be equal to σ_m [12]). σ_c is variable. Extension in the ductile lower crust causes subsidence at the surface due to replacement of crustal material with lighter sediment or water, as given by the equation

$$S_{lc}(x) = \left(1 - \frac{1}{\beta_{lc}(x)}\right)(t_c - Z_d)\frac{(\rho_a - \rho_{lc}')}{(\rho_a - \rho_f)} \qquad [11]$$

$$\beta_{lc}(x) = \text{distribution function of } \beta \text{ in lower crust} = 1 + \left(\frac{E_T}{\sigma_c\sqrt{2\pi}}\right)e^{-\frac{x^2}{2\sigma_c^2}} \quad [12]$$

Pure Shear in Brittle Upper Crust:

Extension in the brittle upper crust is partitioned between extension on specific major faults, and extension on minor faults, which is approximated as pure shear spread across the basin. This component is assumed to have a Gaussian distribution across a width $2\sigma_c$ around the basin centre, and subsidence is given by the equation

$$S_{duc}(x) = \left(1 - \frac{1}{\beta_{duc}(x)}\right)Z_d\frac{(\rho_a - \rho_{uc}')}{(\rho_a - \rho_f)} \quad [11]$$

$$\beta_{duc}(x) = 1 + \left(\frac{E_D}{\sigma_c\sqrt{2\pi}}\right)e^{-\frac{x^2}{2\sigma_c^2}} \quad [12]$$

Extension of the Basin:

So far we have calculated the amount of syn-rift subsidence due to crustal thinning, but we have not taken into account lateral movements due to crustal extension in the case of pure shear extension mechanisms (in both the upper and lower crust and the mantle). Since the lateral length parameters used (σ_m, σ_c, and fault positions) refer to the basin prior to extension, the equations given above will calculate subsidence due to pure shear at a point measured relative to the basin centre prior to extension. In order to calculate subsidence at a point x measured relative to the basin centre after extension (i.e. its current position) we must substitute x' for x in these equations (and also when calculating β factors), where x' is the location of point x prior to extension, and is given by the approximation

$$x' = \begin{cases} x - E_*\left(\Phi\left(\dfrac{x}{\sigma_*}\right) - \dfrac{1}{2}\right) & \text{for} \quad x \geq 0 \\ x + E_*\left(\Phi\left(\dfrac{x}{\sigma_*}\right) - \dfrac{1}{2}\right) & \text{for} \quad x < 0 \end{cases}$$

where $\sigma_* = \sigma_m$ for mantle calculations and σ_c for crustal calculations and $E_* = E_T$ for mantle and lower crustal calculations and E_D for upper crustal calculations, and

$$\Phi(x) = \frac{1}{\sqrt{2\pi}}\int_{-\infty}^{x}e^{-\frac{u^2}{2}}du \qquad\qquad \text{(taken from tables)}$$

The equations given below for subsidence due to extension on faults already take into account the effects of lateral extension (assuming the footwall remains stationary and extension occurs entirely on the hangingwall), so no similar correction need be made for them.

Listric Faults:

Subsidence caused by movement on listric faults is calculated using the Chevron Construction [5, 20] with a correction for flexural isostatic uplift due to the weight of the basin (which is assumed to be a line load in this model). For a fault breaking the surface at the basin centre (x = 0), dipping to the right, subsidence is given by the equation

$$S_{lf}(x) = F(x) - F(x - E_F) + U(x)$$

$$F(x) = \text{geometric fault function} = \begin{cases} Z_d \left(1 - e^{-\frac{x}{Z_d}} \right) & \text{for} \quad x \geq 0 \\ 0 & \text{for} \quad x < 0 \end{cases}$$

$U(x)$ = flexural isostatic response of lithosphere

$$= -\frac{E_F Z_d (\rho_{uc}' - \rho_w)}{8D} e^{-\frac{|x|}{\gamma}} \left(\cos\frac{|x|}{\gamma} + \sin\frac{|x|}{\gamma} \right)$$

where

$$D = \frac{\varepsilon T_e^{\,3}}{12(1 - \sigma^2)}$$

and

$$\gamma = \left(\frac{4D}{(\rho_a' - \rho_f)g} \right)^{\frac{1}{4}}$$

[18]

Planar Faults:

Subsidence due to movement on planar faults was calculated by assuming that each plate will flex as a cantilever (the lower plate uplifting and the upper plate subsiding), and that the integrated vertical displacement across the basin will be equal to the integrated vertical displacement which would be produced across the basin by an identical amount of pure shear extension, as the total isostatic restoring force will be the same in each case. Hence for a fault breaking the surface at $x = 0$, dipping to the right, subsidence is given by the equation

$$S_{pf}(x) = \begin{cases} kd\,e^{-\frac{|x|}{\gamma}}\cos\left(\frac{|x|}{\gamma}\right) & \text{for} \quad x \geq 0 \\ -(1-k)d\,e^{-\frac{|x|}{\gamma}}\cos\left(\frac{|x|}{\gamma}\right) & \text{for} \quad x < 0 \end{cases} \qquad [18]$$

d = total vertical displacement on fault = $E_F \tan\psi$

k = hangingwall subsidence / total vertical displacement of fault

k can be calculated by equating integrated vertical displacement across the basin for planar fault extension and pure shear extension:

$$\int_{-\infty}^{\infty} S_{planar\,fault}(x)\,dx = \int_{-\infty}^{\infty} S_{pure\,shear}(x)\,dx$$

where

$$\int_{-\infty}^{\infty} S_{planar\,fault}(x)\,dx = \int_{-\infty}^{0} kd\,e^{-\frac{|x|}{\gamma}}\cos\left(\frac{|x|}{\gamma}\right)dx$$
$$-\int_{0}^{\infty}(1-k)d\,e^{-\frac{|x|}{\gamma}}\cos\left(\frac{|x|}{\gamma}\right)dx + \frac{(2k-1)d\,E_F}{2}$$

(this must be calculated numerically), and

$$\int_{-\infty}^{\infty} S_{pure\,shear}(x)\,dx = E_F\,Z_d\,\frac{(\rho_{uc}'-\rho_f)}{(\rho_a'-\rho_f)}$$

$$\Rightarrow (2k-1)d\int_{0}^{\infty} e^{-\frac{|x|}{\gamma}}\cos\left(\frac{|x|}{\gamma}\right)dx + (2k-1)d\,\frac{E_F}{2} = E_F\,Z_d\,\frac{(\rho_{uc}'-\rho_f)}{(\rho_a'-\rho_f)}$$

$$\Rightarrow (2k-1)d\left(\int_{0}^{\infty} e^{-\frac{|x|}{\gamma}}\cos\left(\frac{|x|}{\gamma}\right)dx + \frac{E_F}{2}\right) = E_F\,Z_d\,\frac{(\rho_{uc}'-\rho_f)}{(\rho_a'-\rho_f)}$$

$$\Rightarrow (2k-1) = \frac{E_F Z_d}{d\left(\displaystyle\int_0^\infty e^{-\frac{|x|}{\gamma}} \cos\left(\frac{|x|}{\gamma}\right)dx + \frac{E_F}{2}\right)} \frac{\left(\rho_{uc}'-\rho_f\right)}{\left(\rho_a'-\rho_f\right)}$$

$$\Rightarrow k = \frac{E_F Z_d}{2d\left(\displaystyle\int_0^\infty e^{-\frac{|x|}{\gamma}} \cos\left(\frac{|x|}{\gamma}\right)dx + \frac{E_F}{2}\right)} \frac{\left(\rho_{uc}'-\rho_f\right)}{\left(\rho_a'-\rho_f\right)} + \frac{1}{2}$$

Post-Rift Subsidence:

Post-rift subsidence occurs after tectonic extension has ceased, due to cooling of the upwelled hot asthenosphere under the basin. It occurs across the basin, and since it has no tectonic origin it is not associated with any fault movement. It is given by the formula

$$S_{PR}(x,t) = S_0 r(x')\left(1 - e^{-\frac{t}{\tau}}\right)$$

where

$$r(x') = \frac{\beta_m(x')}{\pi} \sin\frac{\pi}{\beta_m(x')}$$

$$S_0 = \frac{4a\rho_m \alpha T_1}{\pi^2(\rho_m - \rho_f)}$$

[11]

This model makes several assumptions and approximations. Firstly, it assumes that extension occurs instantaneously, i.e. that post -rift subsidence does not start to occur until after tectonic extension has ceased. This assumption is valid only for short periods of moderate rifting when the duration of extension $< 60/\beta^2$ ma; in the case where extension occurs over a duration of 90 ma with a total β of 1.5, the true post-rift subsidence will be 50% less than that predicted by the instantaneous model [8]. The model also makes some approximations in the flexural response of the lithosphere (it assumes that listric fault basins can be treated as line loads, and that planar fault plates can be treated as cantilevers). Comparison of the results produced by this model with those produced by more rigorous models (e.g. STRETCH developed by Prof. Nick Kusznir of Liverpool University, UK [9]) suggest that these approximations do not produce unacceptable discrepancies.

The major assumption in the model is that the basins are filled to a constant level at all times. In practice, changes in sediment supply rate and sea level changes may invalidate this assumption, so care must be taken when comparing theoretical profiles with observed profiles to account for the possible effects of this. Furthermore, in its treatment of planar faults, by equating integrated vertical displacement across the basin with that produced by an equal amount of pure shear extension, the model assumes that uplifting footwall blocks are displacing basin fill rather than leading to subaerial exposure and possibly erosion; hence it is only valid when the footwall crest remains below the level of basin fill (e.g. when there is syn-rift sedimentary cover on the footwall).

Proc. 30ᵗʰ Int'l. Geol. Congr., *Vol.* 14, *pp.* 149—157
Zheng et al. (Eds)
© VSP 1997

Mesozoic Geothermal Anomaly in Western Hills of Beijing and Origin of Fangshan Metamorphic Core Complex

HONGLIN SONG
China University of Geosciences, Beijing ,China 100083
NING ZHU
*Anhui institute of Geophysical and Geochemical exploration ,Hefei, China ,*230022

Abstract

The configuration of the Fangshan metamorphic core complex consists of three layers, i. e. between the Archean basement and the upper part of the cover consisting of Jurassic and Cretaceous sedimentary sequence there is a ductilely deformed Middle Proterozoic to Lower Paleozoic low-grade metasedimentary sequence. The structures of this sequence, such as intrafolial to interformational recumbent folds, bedding subparallel shear zones and mylonites, bedding subparallel foliation, were formed by subhorizontal ductile shear under an extensional regime. The spatial distribution of these ductile structures is inhomogeneous, but generally coincident with an area of paleo-geothermal anomaly with tempratures above 300℃ . Stretching lineations in the ductilely deformed sequences plunge to the SEE except around the Cretaceous Fangshan pluton where the lineations dip outward from the pluton, a pattern that indicates reorientation by bollooning of the pluton . The Mesozoic geothermal anomaly and SEE-directed extension are the main factors causing the formation of these ductile flow structures and the thinning of the crust, while the thermal uplift and plutonism during the Yanshanian stage have played an important role in the uplift of the metamorphic core complex, and a thermal event of mantle origin may be the main cause of the paleo-geothermal anomaly.

Keywords: *geothermal anomaly, metamorphic core complex , extension structure , Fangshan*

INTRODUCTION

The cause of uplift of metamorphic core complexes is a major concern in studies of the formation of metamorphic core complexes. Uplift has been described in terms of isostatic rebound in areas of upper crust thinned by extension[5,11]. A clear spatial and temporal link between core complex formation and plutonism activity has been noticed [1]. Syntectonic intrusions have been considered as the underlying cause for differential uplift of the footwall duing tectonic denudation of metamorphic core complexes[4]. It is proposed by them that short-lived abnormally thermal pulse caused by igneous intrusions may trigger transient metamorphism and ductile deformation in shear zones duing episodes of continental extension. The area adjacent to the Fangshan metamorphic core complex located 40 km southwest of Beijing offers a good example showing relationships between a geothermal anomaly, plutonism, ductile strain localization and formation of the Fangshan metamorphic core complex.

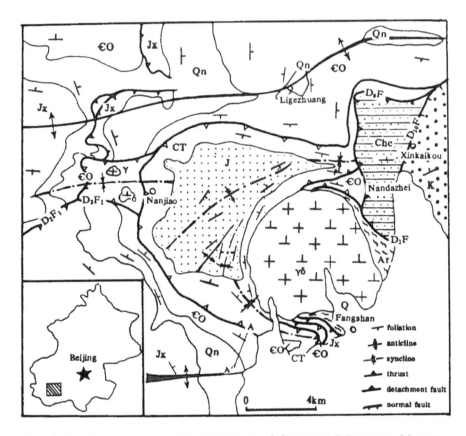

Figure 1. Simplified structural map of the Fangshan area. Q Quaternary; K Cretaceous; J Jurassic; CT Triassic-Carboniferous; ∈O Ordovician-Cambrian; Qn Upper Proterozoic Qinbakou system; Jx Middle proterozoic Jixian system; Chc Middle Proterozoic Changcheng system; γ Granite; γδ Granodiorite; δ Diorite; D_2F_1, Xiayuanling thrust; D_3F_1, Nanjiao detachment fault; A-A', location of Fig. 5

TECTONIC SETTING

The Fangshan metamorphic core complex lies within the North China (or Sino-Korean) craton. It is a dome-like basement uplift, surrounded on the north, west and south sides by a cover of Middle Proterozoic-Jurassic strata forming the crescent-shaped Beiling syncline (Fig. 1). To the east the core complex is cut by the Nandazhai thrust (D_8F) trending N-S and then turning to ENE in the north. Farther east, at its boundary with the Beijing plain, there is the Xinkaikou normal fault ($D_{10}F$) dipping steeply to the east[9,10].

Compared with Cordilleran complexes, the Fangshan metamorphic core complex was mainly formed during the Mesozoic and suffered multiphase deformation. The exposed part of the crust of the Fangshan metamorphic core complex consists of three tectonic layers, i. e. a metamorphic core, an upper unmetamorphic cover and, between them, a middle ductilely deformed slab. The metamorphic core consists of Archean rocks, mainly dynamo-metamorphic rocks of the top ductile shear zone, such as felsic and mafic blastomylonites. The upper part of the cover consists of unmetamorphic Jurassic coal-bearing and clastic sediments that have been deformed as Jura type folds trending NE and detached from the top of Triassic slates. High angle normal faults are present in Cretaceous conglomerates in the front of the hills. They reflect the Cenozoic extensional event that caused differential uplift of the Western Hills and the depression of the North China plain.

An interesting feature is that between the basement and unmetamorphic cover there is a ductilely deformed middle slab, the lower part of the cover, which consists of a Middle Proterozoic to Lower Paleozoic sedimentary sequence that has suffered inhomogeneous low-grade metamorphism and strong ductile deformation. The contact between the Archean core and younger cover is a basal detachment fault (D_1F). The blastomylonites of the footwall near the detachment fault have been retro-metamorphosed and shattered, becoming a chlorite-epidote cataclasite or gouge. The central part of the metamorphic core is occupied by the Fangshan pluton consisting of earlier quartz-diorite (150 Ma) and later granodiorite (132-120 Ma). It is a ballooning-type intrusion. The mylonitic foliation of basement rocks, the basal detachment surface and the schistosity of adjacent upper- plate rocks around the pluton all dip steeply outward to form a domelike feature. The lineations of the rocks also dip steeply outward, unlike the regional stretching lineation which trends SEE. This is the result of reforming of fabrics by the active emplacement of the ballooning-type pluton, and indicates that the final formation of the Fangshan metamorphic core complex is closely related to the Cretaceous plutonism in this area.

CHRACTERISTICS OF DUCTILE FLOW STRUCTURES

A series of ductile flow structures and interformational low angle normal faults have developed in the middle slab, the lower part of the cover. These ductile flow structures have been studied in detail by Shan et al. [7]and Shan and Fu [8]. They include: (1) bedding subparallel ductile shear zones (Fig. 2); (2) intrafolial to interformational recumbent folds ranging from microscopic to mapping scale (Fig. 3); all of the folds are asymmetric indicating that the upper plate moved to the SEE; (3) sheath folds in limestone or marble; (4) transposition of bedding by a slaty cleavage or schistosity (S_1) which represents both the axial surface foliation of recumbent folds and a bedding subparallel regional foliation (Fig. 4); (5) the boudinization of competent beds , such as dolomite layers in limestone sequence, and the local pinching out of some competent units. It should be pointed out that the entire gently dipping stratigraphic sequence is still in order, although the ductile deformation within

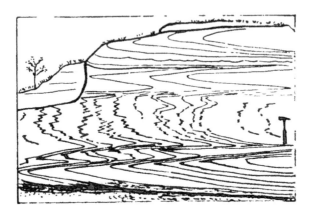

Figure 2. Bedding subparallel ductile shear zones and recumbent folds in the Upper Proterozoic marble, Ligezhuang

Figure 3. Recumbent folds in the Cambrian limestone, Nanjiao

each unit of this sequence is very strong and the local succession is disordered by strongly overturned multiple folding (Fig. 5). Another important feature is that although the local strata are folded several times, the whole sequence is thinned ,as compared with the adjacent area, by 0. 42, 0. 21 and 0. 27 for the thicknesses of Upper Proterozoic, Cambrian and Ordovician sequences respectively. This is because of the strong thinning of layers by ductile flow. Therefore, the thinning might be the result of a subhorizontal laminar shear flow of strata under ductile extension of the crust, if so, it constitutes a new kind of tectono-stratigraphic unit which may be the common pattern for low-grade metasedimentary terranes. It is very similar to the ice folds in the front of glaciers.

Figure 4. Transposition of bedding by schistosity in the Cambrian oolitic limestone and rootless folds of pelitic bands, 2 km south of Ligezhuang

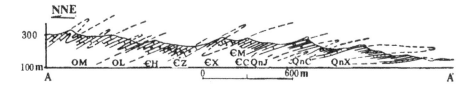

Figure 5. A cross-section showing intraformational to interformational recumbent folds, 6 km west of Fangshan County. OM — OL, Ordovician; ∈H, Upper Cambrian; ∈Z — ∈X, Middle Cambrian; ∈M — ∈C, Lower Cambrian; QnJ, QnC, QnX, Upper Proterozoic Jingeryu, Canglongshan and Xianmaling Fm.

The direction and sense of shear inferred from the stretching lineation, the vergence of recumbent folds, and other kinematic markers in the ductilely deformed sequence indicate that the upper layers have moved SEE relative to the lower layers, this is consistent with the direction and sense of movement of the basal detachment fault and interformational low angle normal faults inferred from their cross-cutting relationships to the strata of their hanging walls.

The spatial distribution of these ductile flow structures is inhomogeneous, not along the layers, but around the metamorphic core within a radius of about 25 km. From the center outward the ductile deformation becomes weaker and gradually transitions to brittle deformation (Fig. 6).

TEMPERATURE CONDITION DURING DUCTILE DEFORMATION

Temperature conditions during the ductile deformation in this area have been

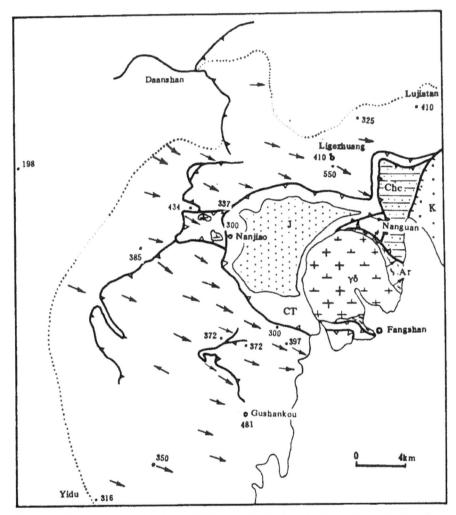

Figure 6. Sketch map showing the distribution of ductile flow structures around the Fangshan metamorphic core complex. The arrows show the plunge direction of stretching lineations, the numbers show the temperature during its deformation in Centigrade degrees, and the dotted line outlines the area of development of ductile flow structures

calculated by means of : (1) assemblages of syntectonic metamorphic minerals; (2) paleo-geothermometers, such as the pair of amphibole and plagioclase in gneisses, garnet and biotite in schist, calcite and dolomite in marble; (3) the ratio of calcium to magnesium within calcite in syntectonic veins; and (4) the homogenization temperature of inclusions in syntectonic quartz veins. The data are shown on Figure 6, and indicate that there is a strong paleo-geothermal anomaly around the metamorphic core. Considering the thickness of strata in this area during the Late Triassic when the ductile shear happened, the thickness of the crust column above the Ordovician was ca. 765 m, above the Upper Proterozoic was ca. 1180 m and above the Middle Proterozoic was ca. 2470 m. Then the paleo-geothermal gradient at the Gushankou station would be $481/2.47 = 195°C/km$ for the Middle Proterozoic marble, at the Ligezhuang village $410/1.18 = 348°C/km$ for the Upper Proterozoic marble and at the Nanjiao village $300/0.765 = 392°C/km$ for the Ordovician limestone. Even if considering the influence of tectonic thinning of the crust, for example, by using the original thinkness of the strata mentioned above, the paleogeothermal gradient for the Upper Proterozoic marble at the Ligezhuang would be $410/2.5 = 160°C/km$, this gradient is still anomalously high for metamorphic terranes. It is indicated from the paleo-geothermal map that the temperature was high in the metamorphic core area and lowered gradually toward south, west and north. In the east of the Nandazhai thrust the Middle Proterozoic sandstone and dolomite of its hanging wall were deformed brittlely. According to the balanced cross-section of the thrust this wall was transferred from 24km farther SEE. Figure 6 indicates that the centre of the thermal anomaly triggering the ductile deformation was in the area encircled by Nanguan, Nanjiao and Fangshan. The diorite sills, veins and small plutons (207Ma) pervasively distributed in this area reflect the close relationship between this thermal anormaly and mantle-origin magmatism.

A correlation analysis of the strain, temperature and confining pressure has been calculated. The axial ratio X/Z of the strain ellipsoid has been taken as the factor of strain. Confining pressure is calculated from the syntectonic metamorphic minerals by using the paleogeobarometer. The result shows that the strain is temperature sensitive. The correlation factor between the strain and temperature is 0.83 to 0.84, while between strain and confining pressure is only 0.50 to 0.52. Figure 6 illustrates that the 300°C isotherm is coincident with the outer boundary of the distribution area of the ductile flow structures.

MESOZOIC IGNEOUS EVENTS

After a tectonic and magmatic quiescence in the North China craton during the Paleozoic era strongly active magmatism began during the early Mesozoic era. Igneous events in the research area reflect partially this active magmatism, which has been studied by a number of Chinese geologists [2, 3, 12]. The earliest Mesozoic igneous event includes intrusions of the Shangweidian gabbro-diorite pluton (229 Ma K/Ar), the Nanjiao diorite pluton and sheets (207 Ma K/Ar), and syntectonic spessartite vein involved in the recumbent folds. The diorite sheets along detachment faults were deformed by ductile shear and became gneissic diorite before the Jurassic. The igneous event of early Yanshanian stage was a mantle-derived basalt eruption, a fissure eruption along E-W direction, during the Early Jurassic (191-198.5 Ma). The middle Yanshanian igneous event was a pervasive eruption of andesite and trachyandesite with a thickness up to 4971m during 160-152 Ma and the intrusion of intermediate pluton. The latest Yanahanian igneous event was an eruption of rhyolite and rhyodacite and multiple plutons commonly consisting of earlier intermediate-acid rocks in the rim and later acid rocks in the centre with an age range of 135. 9-100 Ma. The Fangshan granodiorite pluton is one of these intrusions. It has been suggested that the evolution of magmatism in this area during Mesozoic was from a mantle-derived mafic magma, to a mixed magma due to partial melting of the lower crust, and to a mainly crust-derived magma mixed with mantle-derived material in the earlier stage of every event [6,12]. This indicates that the heat sources had migrated upwards from the mantle during the Mesozoic era.

CONCLUSIONS

The Mesozoic geothermal anomaly and SEE directed subhorizontal shear under an extensional regime are the main factors causing the formation of ductile flow structures in the lower part of the sedimentary cover and the thinning of the crust in this area. Bouyancy of thermal rock material caused by the Mesozoic geothermal anomaly and plutonism during the Yanshanian stage have played an important role in the uplift of the metamorphic core, in addition to the isostatic rebound of the thinned crust. Considering the Mesozoic igneous events in this area, the thermal event of mantle origin may be the main cause of this paleo-thermal anomaly and the main dynamic factor of the uplift of the Fangshan metamorphic core comlex.

Acknowledgments

Supported by the important basic reseach grant of the Ministry of geology and Mineral Resources of P. R. China (8502207). Shan Wenlang, Fu Zhaoren and Zhang Changhou have coworked in the reseach and provided invaluable discussion. Thanks to Gregory A. Davis for a helpful review of the original manuscript.

REFERENCES

1. Crittenden, M. D. , Coney, P. J. , and Davis, G. H. , eds. Cordilleran metamorphic core complexes, *Geol. Sci. Am. Memor.* 153, 490 (1980).
2. Bai Zhimin, Xu Shuzhen and Ge Shiwei, Badaling Granitoid complex, Geol. Pub. House, Beijing, (in Chinese), 142—157 (1991).
3. Bao Yigang, Bai Zhimin, Ge Shiwei and Liu Cheng, Yanshanian Volcanic Geology and Volcanic Rocks in Beijing, Geol. Pub. House, Beijing (in Chinese), 152—162 (1995).
4. Lister, G. S. , and Baldwin, S. L. , Plutonism and the origin of metamorphic core complexes, *Geology* 21, 607—610 (1993).
5. Lister, G. S. , and Davis, G. A. , The origin of metamorphic core complexes and detachment faults formed during Tertiary continental extension in the northern Colorado River region, U. S. A. , *J. Struct. Geol.* , 11, 65—94 (1989).
6. Ma Changqian, Wang Renjing and Yang Kunguang, Magmatic Thermodynamic Structures of The Zhoukoudian Granodioritic Intrusion in The Western Hills of Beijing, 30th IGC Field Trip Guide T207, Geol. Pub. House, Beijing (1996).

7. Shan Wenlang, Fu Zhaoren and Ge Mengchun, The folding layer and its bedding rheid tectonic community in West-Hill, Beijing, *Earth Science* (in Chinese),2, 33—42 (1984).

8. Shan Wenlang and Fu Zhaoren, A preliminary analysis of the horizontal laminar shear flow structure in the West-Hill, Beijing, *Earth Science* (in Chinese), 2, 113—120 (1987).

9. Song Honglin,Fu Zhaoren and Yan Danping, Extensional Tectonics and Metamorphic Core Complex of The Western Hills, Beijing, 30th IGC Field Trip Guide T210, Geol. Pub. House, Beijing (1996).

10. Song Honglin and Wei Bize, Metamorphic core complexes and their significance in the continental crustal evolution, *J. China Univ. Geosciences*, 1, 111—121 (1990).

11. Wernicke, B. ,and Axen,G. J. , On the role of isostasy in the evolution of normal fault systems, *Geology*, 16, 848—851 (1988).

12. Yu Jianhua, Fu Huiqin, Zhan Fenglan and Guan Meisheng, The Plutonism of Beijing Area, Geol. Pub. House, Beijing (1993).

Proc. 30th Int'l. Geol. Congr., Vol. 14, pp.158-172
Zheng *et al.* (Eds)
© VSP 1997

The Xiaoqinling Detachment Fault and Metamorphic Core Complex of China: Structure, Kinematics, Strain and Evolution.

JINJIANG ZHANG   YADONG ZHENG
(Department of Geology, Peking University, Beijing 100871, China)

QUANZENG SHI   XIANGDONG YU   LIANGWEI XUE
(Geological Institute of Henan Province, Zhengzhou 450053, China)

Abstract

The Xiaoqinling metamorphic core complex (XMCC) is located at the southern margin of the North China Craton. Along the boundary of the XMCC is a detachment fault system: mylonitic rocks, chloritized breccias, microbreccias and fault surface. Many retrograded shear zones, normal faults and contraction faults developed within the XMCC. There are two groups of linear features in the XMCC according to their occurrences: one represents the earlier ESE-WNW extension and the other shows the normal faulting toward the outsides of the XMCC. Fabrics, strain and kinematic vorticity in the related mylonitic zone present the characteristics of a detachment fault. The strain increases from ESE to WNW and geometry of quartz c-axis fabrics changes progressively from crossed girdles on the ESE boundary to single girdle on WNW. In the same direction, the kinematic vorticity number increases, *i.e.*, the ratio of pure shear rate to that of simple shear decreases. The ESE-WNW extension and the detachment began at ~75.9 Ma, as a result of crustal thickening caused by the crustal shortening during Yanshanian orogeny. The following uplift due to denudation resulted in the emergence of the XMCC. The retrograded shear zones and normal faults within the XMCC were formed by the collapse of the XMCC during late- and post-orogenic event.

Keywords: Xiaoqinling metamorphic core complex, Detachment fault, Kinematics, Strain, Kinematic vorticity

INTRODUCTION

Geologic studies in the Basin and Range Province [2, 4, 5] gave rise to the term "metamorphic core complex" (MCC) and a new definition for detachment faults. A metamorphic core complex refers to an isolated domed terrain that is composed of moderate to high-grade rocks, and is structurally over-lain by an upper plate of lower grade rocks from an initially higher crustal level. The large-scale low angle normal fault juxtaposing the upper and lower plates is called a detachment fault [3, 4, 14]. The detachment faults root into the middle crust and their upper front or sprays reach the surface. Mylonitic rocks, chloritized breccias and microbreccias develop in the footwall of the detachment fault from deep to shallow structural levels [9]. Progressive extension results in uplifting of the fault and its lower plate, causing the overprinting of different tectonites in order and forming the sharp fault boundary of the metamorphic core complex.

The Xiaoqinling metamorphic core complex (XMCC), covering an area of ~2,500 km² and taking the shape of a hockey stick, is located in an extensional corridor along the southern margin of the North China Craton (Fig.1). There are several MCCs in this corridor and the XMCC is the largest. It is chiefly composed of high-grade rocks and younger granitic plutons. The high-grade rocks (upper amphibolite to granulite facies) include late Archean plagioclase-amphibole gneisses and gneissic diorites, and paragneisses, quartzites and marbles of earlier Proterozoic age. These units experienced very strong ductile deformation. The large intrusions in XMCC are mainly late Yanshanian granitic plutons. A mylonitic zone up to hundreds of meters thick occurred on the top and along the boundary of the core complex and developed across both metamorphic and plutonic units (Fig.2). Nonmetamorphic and very low-grade rocks make up the upper plate. They are Cretaceous-Paleogene sandstones and conglomerates , Cambrian carbonates, middle Proterozoic quartzosandstones and volcanic rocks. The upper-plate rocks are faulted brittely and take the shape of tilted blocks with small basins. While Shi et al (1993) [15] thought the XMCC was formed by a single detachment fault, Hu et al (1994) [7] held the idea that it was a symmetric MCC produced by extension along two boundary normal faults like a horst system. Further study (this paper), however, suggests that the XMCC formed by progressive deformation: earlier ESE -to - WNW extension formed a detachment fault and the

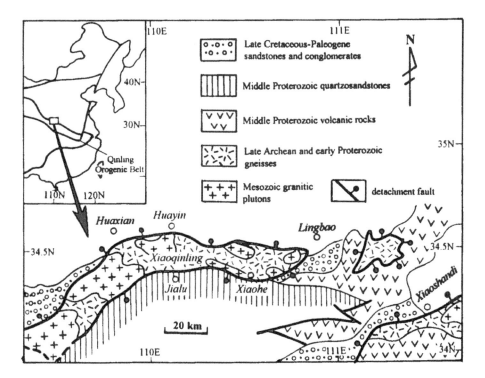

Fig.1. Sketch map of the extensional corridor in the southern marginal area of the North China Craton.

J. Zhang et al.

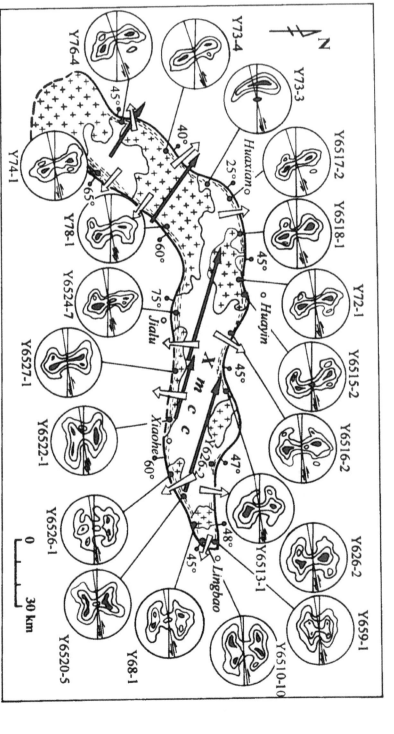

Fig. 2. Structural map of the XMCC. The black arrows represent the trend of the penetrative ESE – WNW lineation and their heads indicate the moving direction of the upper plate. The hollow arrows stand for the down – dip linear features within the XMCC. The dashed lines represent the footwall mylonitic zone. Quartz c – axis fabric diagrams with sample numbers were measured on XZ plane. The diameter with shear sense represents the shear plane and the other diameter is the S – foliation.

uplift of the lower plate due to tectonic denudation produced the XMCC. Subsequent uplift and extension resulted in collapse of the MCC and induced outward-dipping normal faults.

DETACHMENT FAULT AND TECTONITES

The detachment fault and its related footwall tectonites are excellently exposed along the boundary of the XMCC. The components of the fault system include, from bottom to top, a mylonitic zone, a chloritized breccia zone, a microbreccia zone and the fault surface.

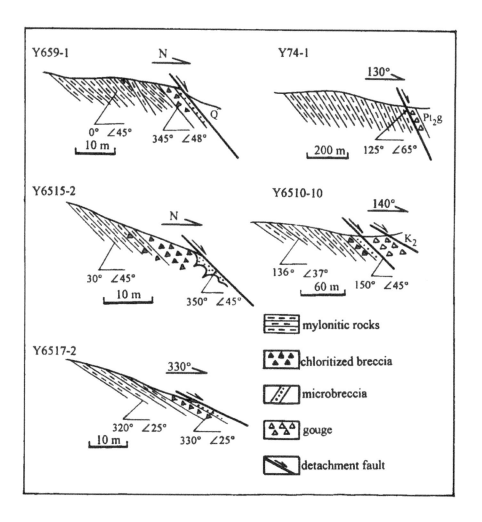

Fig.3. Cross-sections of the detachment fault and tectonites along the boundary of XMCC. The number for each gives the locality where the sample was collected (illustrated in Fig.2).

Mylonitic zone

Mylonitic gneisses and granites, felsic mylonites and some carbonate mylontites form a mylonitized zone along the boundary of XMCC. This zone is called the boundary mylonitic zone (Figs. 2, 3). Its upper part was shattered and chloritized. The thickness of the boundary mylonitic zone is variable. On the south flank, it is more than 100 m thick and locally exceeds 500 m. It is less than 100 m on the north flank and absent locally. However, where footwall mylonitic rocks are absent clasts exist in the preserved breccias.

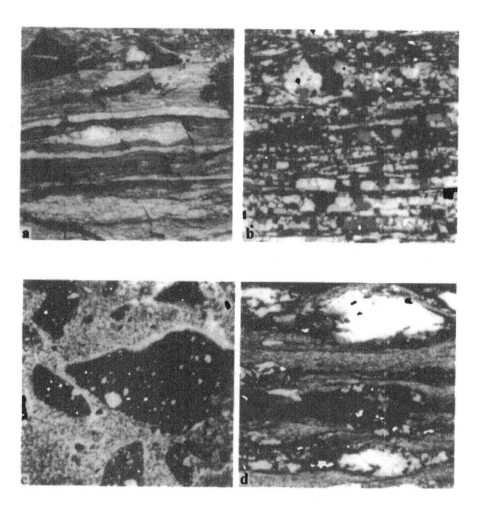

Fig.4. a. Felsic mylonite bands interbedding in mylonitic gneisses on the southeast flank of the XMCC. Some of the bands were boudinaged and rotated during the progressive shear. The geometry of the boudins shows the upper part moved to WNW. b. Microphotograph of the mylonite at the same locality as Fig.4a. c. Microphotograph of microbreccias with pseudotachylytes (the semitransparent clasts). d. Microphotograph of phyllonite from the retrograded shear zone with quartz as porphyroclasts and sericite as matrix.

The rocks within the zone show features of mylonitization. Penetrative mylonitic foliation and ESE-WNW mineral lineation developed in them. Felsic mylonites lie within the mylonitic gneisses and granites as fine-grained bands and progressive deformation is expressed by the boudinage and rotation of the bands (Fig.4a). Grains of quartz and mica in the mylonitic rocks underwent ductile deformation and became quartz-ribbons and mica fishes, while feldspars exhibited brittle behavior and became clasts (Fig.4b). Carbonate mylonites, exposed locally on the north flank, show banded structure with serpentinized interbeds. Prismatic minerals such as tremolite have a preferred orientation parallel to the extensional lineation.

The mylonitic foliation dips outward at angles a little less than the fault surface (Fig.5a). It is steeper (60~85°) on the south flank and gentler (25~50°) on the north flank. The lineation in the mylonitic zone has an ESE-WNW trend, plunging to ESE on the south flank and to the WNW on the north flank (Fig.7).

Chloritized cataclastic zone

A 10-50 m thick chloritized cataclastic zone lies above the mylonitic zone and was developed across it (Fig.3). The rocks in it were strong chloritized and have a gray-green color. The degree of chloritization depends on the lithology of the parent rocks. Those from amphibole-gneisses were chloritized stronger than those from granites. Coarse cataclasts developed in the lower part of the zone and the rocks evolved into finer grained breccias upwards. The clasts in breccias are fragments of mylonitic rocks and ductilely deformed quartz. The matrix contains very fine grained materials produced by cataclasis. Quartz and calcite veins penetrate the matrix and fill cracks in the clasts.

Microbreccia-pseudotachylyte zone

Microbreccia occurs under the surface of the detachment fault as a thin resistant layer with a thickness of 0.08-1.5 m (Figs.3, 5a). It has a sharp border with the underlying breccia zone; in some places the microbreccias have been injected into the breccias (Fig.3d). Usually, the microbreccias have a gray-green color, but some are brown or gray white. They are of several generations and exhibit widely different clast grain size. Both the clasts and matrix are composed of ultracataclastic materials (Fig 4c). Pseudotachylyte veins are common in the microbreccias as are pseudotachylyte clasts. These irregular veins have a maximum width of 2cm. The pseudotachylytes are brown and semitransparent in plane light (Fig.4c) and show complete extinction in crossed nicols. Abundant lithoclasts and fragments of minerals with resorption borders are present in the pseudotachylytes.

Microbreccias are produced by ultracataclasis [13] and pseudotachylytes are formed by friction-induced melting or extreme cataclasis on fault surface during seismic slips [20]. Pseudotachylytes are reported in other detachment faults, *e.g.* , US Cordillera [3] and Inner Mongolia of China [18]. The occurrence of pseudotachylytes is probably evidence for seismic events along the detachment fault during extension. The characteristics of multi-phase deformation such as dissections of veins and clasts suggest that the fault behaved in a stick-slip mode.

Detachment fault surface

The surface of the detachment fault is well exposed along the boundary of the XMCC, except in part of the south flank (Xiao-Jialu in Fig.2). The fault surface dips outward and gives rise to the topography of the XMCC. The dip-angle on the north flank has a range of 25~50° and is generally around 45° . On the south flank, the surface is steeper, usually over 60° (Fig.3). There are two groups of striae on the fault surface. The earlier group is sub-parallel to the ESE-WNW extensional lineation in the mylonitic zone. The younger group is sub-parallel to the dip of the fault surface on which it lies.

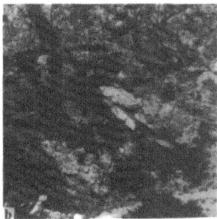

Fig.5 a. Detachment fault with a dip-angle of 45° on the north flank of XMCC. Under the fault surface (the polished surface on the hill side) are the mylonitic rocks whose foliation is shallower than the fault surface. b. A listric normal fault developed within the XMCC.

Gouge/breccia zone

A gouge/breccia zone with a thickness of 5~30 m lies directly above the detachment fault (Fig.3). It constitutes the basal part of the upper plate and composed of breccias and layers of clay gauge of 0.3~0.4 m thick. Striae are developed on the layer surfaces. These striae are down-dip and parallel to the younger ones on the surface of the detachment fault.

It is obvious that the XMCC is similar in essence to the metamorphic core complexes of the Southwestern U.S. Cordillera [2,3,4]. The structures and the assemblage of tectonites in the detachment fault suggest progressive uplift of the lower plate.

DUCTILE SHEAR ZONES AND FAULTS WITHIN THE XMCC

There are many retrograded shear zones, ductile-brittle and brittle faults within the XMCC. It is worth noting that contraction faults developed locally in the extensional system and they formed during the extension.

Retrograded shear zones

Retrograded shear zones, with a thickness of 0.5-10 m, are distributed within the XMCC. They dip outward like the detachment fault, but a little steeper than the latter. Phyllonites with foliation and down-dip lineation developed in the shear zones. Quartz and sericite are the chief components of the phyllonites. The quartz forms porphyroclasts and sericite makes up the matrix (Fig.4d). The structures and the mineral behavior demonstrate that the retrograded shear zones were formed at shallower levels than those of the boundary mylonitic zone. On the south and northwest flanks of the XMCC, phyllonites are superimposed on and redeform the boundary mylonitic zone.

Ductile-brittle and brittle faults

Near the boundary of the XMCC, listric and domino normal faults (Fig.5b) tilt the rock blocks and thin the boundary units locally. These faults,with 0.2-5 m thick fractured zones, have similar attitudes to the retrograded shear zones. "Hot" striae and steps, foliated and cleavaged rocks developed in ductile-brittle faults. "Cold" striae and steps, gouge and breccias were formed in the brittle faults.

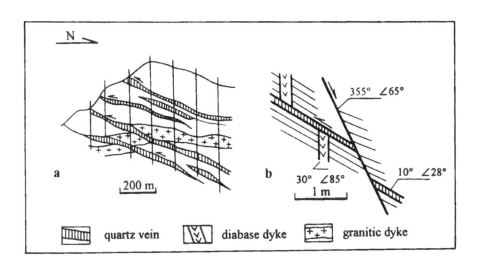

Fig.6. Dyke offset by the quartz-vein filled thrust faults. a. Section with drill holes. b. Small section drawn from a tunnel in the Tonggou gold mine. Both localities are in north part of the XMCC in east area.

Contraction faults

Thrusts filled with gold-bearing quartz veins resemble the retrograded shear zones in distribution ,but their dips are shallower than the latter. In the Tonggou gold mine, the retrograded shear zones have a dip angle of ~60° , whereas the veins dip at about 30° . They cut the the retrograded shear zones and are dissected by the brittle normal faults (Fig. 6b). Kinematic indicators in the north part of the XMCC,such as thrusted dykes (Fig.6) and the geometry of the veins, show southward thrusting. The veins filling the thrusts take a step-like shape with characteristics of alternate shallow-thick and steep-

thin segments. The thick segments occur in the non-phyllonitized units and thin ones
go along the retrograded shear zones.

KINEMATICS OF DETACHMENT FAULT

There are two groups of linear features in the XMCC in term of their attitudes: regional
ESE-WNW lineation and down-dip plunging linear features.

The regional ESE-WNW extensional lineation developed within the XMCC and the
boundary myolnitic zone. It increases outwards from the center of the XMCC and is
most strongly developed in the boundary mylonitic zone. The attitude of the lineation in
the mylonitic zone is similar to that in the gneisses below the mylonitic zoon and the
penetration of it changes gradually from gneisses to mylonitic zone. Other ESE-WNW-
trending features include the weak orientation of the clasts in microbreccias and the
early striae on the surface of the detachment fault. The attitudes of these features are
consistent in the studied area, plunging to WNW on the north flank and to ESE on the
south flank (Fig.7). The nailhead striae on the surface of detachment fault, macro- and
micro-structures in mylonitic rocks (Fig.4b) provide kinematic indicators that show the
upper plate moved WNW-ward.

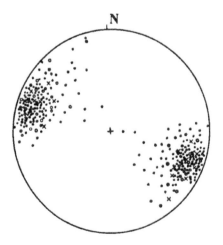
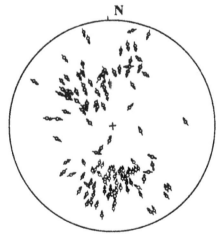

Fig.7. Stereographic projections of linear features of the XMCC (equal area). Dots and small circles stand for the lineation penetratively developed in the XMCC and the boundary mylonitic zone respectively, and crosses for the early striae on the detachment fault.

Fig.8. Stereographic projections of the late faulting within the XMCC (equal area). Small circles are the normals to the retrograded shear zones and normal faults, and dots represents the detachment fault. Arrows show the moving direction of the hanging wall.

The down-dip plunging linear features (Fig. 8) include stretching lineation and "hot"
striae in the retrograded shear zones, "cold" striae on the surfaces of normal faults and

the younger striae on the surface of the detachment fault. These features have a wide range of attitudes, but most of them plunge down dip on the surfaces they lie. Normal faulting is evidenced by kinematic indicators such as the microstructures in phyllonites (Fig.4d) , the relationship of the cleavages to the fault surfaces and the steps on the fault surfaces. Fig.8 shows that the hanging walls of the shear zones and faults within the XMCC slipped downward and away from the XMCC; the detachment fault also had such a movement during late time. Because these structures within the XMCC formed at a higher level and a late time, they probably resulted from the gravity collapse of the core complex during progressive uplift due to continuing extension. The small thrusts cut the retrograded shear zones and they were in turn dissected by the normal faults (Fig.6b). This means short periods of contraction possibly occurred during the collapse.

SHEAR TYPE OF THE DETACHMENT FAULT

Most shear zones in nature are probably generated by general shear which combines pure shear and orthogonal simple shear parallel with the boundary of the shear zone [6, 11, 17]. Different tectonic settings produce different shear, the characteristics of which are reflected by the fabrics, stain and vorticity in the shear zones. The changes of strain and vorticity in the shear direction, therefore, indicate the shear type and the

Table 1. Strains (Rs) , kinematic vorticity number (Wk) , S-C angle (α) , angle between eigenvectors (υ) and the ratio of pure shear rate to that of simple shear (ε / γ) from measurement on the samples illustrated in Fig.2.

No. sample	Rs=X/Z	$\alpha(°)$	$\upsilon(°)$	Wk	$\varepsilon \ \gamma$
Y76-4	4.4/1	17	19	0.95	0.17
Y6517-2	3.4/1	18.5	24	0.91	0.22
Y73-4	4.2/1	16	24.5	0.91	0.23
Y73-3	3.9/1	16	31	0.86	0.30
Y6518-1	3.0/1	17	32	0.85	0.31
Y74-1	3.8/1	12	32.5	0.84	0.32
Y78-1	4.0/1	11	40	0.76	0.42
Y6524-7	3.9/1	10	46	0.69	0.52
Y72-1	3.3/1	10	49.5	0.65	0.59
Y6515-2	3.1/1	10	51	0.63	0.62
Y6513-1	2.6/1	11.5	51	0.63	0.62
Y6510-2	3.2/1	9	52.5	0.62	0.65
Y626-2	2.5/1	10.5	56.5	0.55	0.75
Y68-1	2.5/1	10	57	0.54	0.76
Y6520-5	2.7/1	9	58.5	0.52	0.82
Y6527-1	2.4/1	10	60	0.50	0.87
Y6510-10	2.6/1	8.5	61.5	0.48	0.92
Y6526-1	2.2/1	9	62	0.47	0.94
Y659-1	2.4/1	8	63	0.45	0.98
Y6522-1	1.95/1	7.5	67.5	0.38	1.21

tectonic setting. For example, strain increases and quartz c-axis fabrics change from crossed girdle to single girdle in the moving direction of a detachment fault [8, 10]. The strain analysis with polar Mohr construction can identify the extensional or compressive regime [19, 21, 22].

Fabric analysis

Quartz c-axis fabric diagrams measured on the samples collected from the boundary mylonitic zone are illustrated in Fig.2. All the diagrams have asymmetric patterns with respect to the S-foliation. The main girdles of the diagrams incline to ESE and refer to an ESE-to-WNW movement of the upper plate. The fabric patterns in ESE part show a crossed-girdle. To WNW, the patterns change gradually into single main girdle with or without a minus girdle. The asymmetric and crossed girdle patterns demonstrate that the boundary mylonitic zone was formed by combination of pure shear and simple shear, *i. e.*, general shear. The change of the diagrams in the shear direction show that the component of pure shear was getting larger in an ESE direction.

Strain analysis

Strains were measured on the same samples used for fabric analysis, using De Paor's R_f / ϕ_f method. The strain ratios (Rs=X/Z) are listed in Table 1. From the data, we can see that the shear strain increases in a WNW direction, from a low strain of 1.95/1 in ESE areas to a high one of 4.4/1 in WNW areas. Increase of strain in the shear direction is one of characteristics of a detachment fault [8,10]. Usually, a detachment fault has many splays that develop up dip and form an imbricate structure. Down dip, slip accommodated by the splrays accumulate in the detachment fault and higher strain is obtained by the rocks in deeper parts of the shear zone.

Kinematic vorticity analysis

Kinematic vorticity number (Wk) is a measurement of non-coaxility during progressive deformation. It indicates the shear type experienced by the rocks:Wk=0 for pure shear, Wk=1 for simple shear, and 0<Wk<1 for general shear [12]. According to the formulas of Wk=cos[tg^{-1}(2 ε γ)] and v =tg^{-1}(2 ε γ) [1, 19, 21, 22], the Wk and the ratio of pure shear rate ε to that of simple shear γ can be calculated from v which is the angle between the two eigenvectors in the XZ plane. Our procedure is as following:

a. Obtain the angle α of C-foliation to S-foliation on the fabric diagrams.
b. Construct polar Mohr diagram for the shear zone using Rs and α, measure the angle v on the Mohr diagram [19, 21, 22] (Fig.9).
c. Calculate the Wk and ε γ.

The results of the measurements and calculations are listed in Table 1. The magnitudes of Wk fall on a range of 0.38~0.95, which means a general shear. The ε γ has a range of 0.17~1.2. The Wk increases whereas ε γ decreases from ESE to WNW. These changes demonstrate that the pure shear had a high component in ESE areas and become smaller to WNW areas. From the Mohr diagrams (*e.g.*,Fig.9), we know the boundary mylonitic zone is a thinned shear zone[19, 21, 22]. This kind of shear zone was formed by combination of simple shear and pure shear with maximum stretch direction parallel to the simple shear direction, *i.e.*, it was formed under an extensional setting.

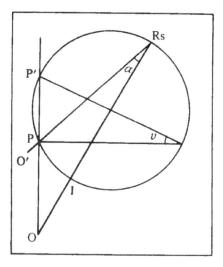

Fig.9. Polar Mohr diagram of sample Y6517-2. It presents a thinned shear zone formed under an extensional setting in which the maximum stretch direction of the pure shear component is parallel with that of the simple shear. If line Rs-O' intersects the circle and reference axis at point P', the diagram will represent a thickenned shear zone under contraction setting (thrust)

AGE OF THE EXTENSIONS

Synkinematic lamprophric sills developed in the boundary mylonitic zone. They are parallel to the mylonitic foliation and their borders were mylonitized. Their age is ~75.9 Ma (biotite, K-Ar). At the boundary of the XMCC, the Huashan granitic pluton (south of Huayin in Figs.1,2) was mylonitized. However, apophyses from it are intruded into boundary mylonitic gneisses away from the detachment fault and were deformed by the retrograded shear zones. The pluton has an age range of 100~60 Ma (biotite Ar-Ar, whole-rocks K-Ar). There are clasts of microbreccias in the Paleogene basal conglomerates on the southeast side of the XMCC. From these data, we suggest that extension began after 100 Ma and, more definitely, developed during 75.9~60 Ma to form the detachment fault system; the lower plate rocks of the XMCC emerged on the surface during Paleogene time. The collapse of the XMCC happened after 60 Ma, producing the structural zones within the XMCC.

TECTONIC EVOLUTION

During Yanshanian Orogeny, southward thrusting on a large scale took place along the southern margin of the North China Craton [16]. The crustal shortening made up a thick orogenic wedge and late orogenic extension due to unstability of the wedge formed an extensional corridor on its back edge. From 75.9 Ma to 60 Ma, when the N-S compressive regime was still working, the unstable wedge had to escape laterally and ESE-WNW extension resulted. The detachment fault reached the earth surface on the north flank of the Xionger Shan (Xiaoshandi in Fig.1) and formed a half-horst MCC (a MCC which has a detachment fault boundary only on one flank. See Fig.10). The upper plate were faulted by the sprays of the detachment fault and took the shape of tilted blocks with basins [16]. Extensional denudation led to the doming of the lower plate and its detachment fault which resulted in the emergence of the XMCC and other MCCs in the extensional corridor during the Paleogene (Fig.10a). The overprinting of mylonitic rocks-chloritic breccia-microbreccia happened in the course of footwall uplift. After 60 Ma, continuing uplift and extension resulted in the collapse of the XMCC. Its upper plate moved outward and formed the down-dip striae on the surface of the detachment fault. Continuing uplift led to erosion of the XMCC and exposed the structural zones within it (Fig.10b). To the north of the XMCC, the combination of gravity and extension generated the largest displacement, accommodated the biggest subsidence and led to the upper plate being buried by loess. The asymmetry of the XMCC with a steep south flank and shallower north flank may be the result of a renewal of the south-directed thrusting.

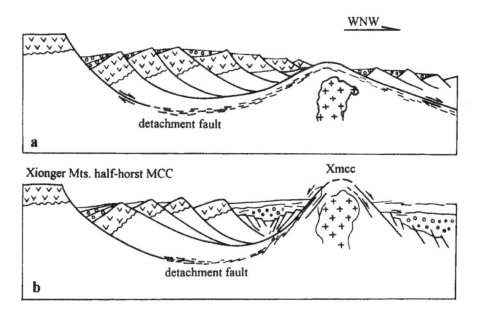

Fig.10. The model of the XMCC's evolution. a. Before 60 Ma. b. After 60 Ma.

CONCLUSION

The XMCC, located in south margin of the North China Craton, is similar to the MCCs of the Southwestern U.S. Cordillera. Its domed lower plate is composed of mylonitic gneisses and plutons. The upper plate is made up of nonmetamorphic or very low-grade rocks. The detachment fault overlies a zone of progressively younger mylonites, chloritized breccias and microbreccias. Strain, fabrics and vorticity in the mylonites are indicative of extensional tectonics. Many retrograded shear zones and normal faults developed within the XMCC. There are two groups of linear elements in the XMCC. One represents early ESE-WNW crustal extension and the other stands for the collapse of the XMCC. The XMCC was formed by the late and post orogenic extension related to crustal shortening. The extension, beginning at ~75.9 Ma, produced the detachment fault and resulted in the top to the WNW movement. Extensional denudation exhumed the XMCC in the Paleogene. Progressive uplift led to the collapse of the XMCC and induced the outward and down-dip movement of upper plate rocks after 60 Ma.

Acknowledgment

We express our deep appreciation to Prof. Gregory A. Davis. He reviewed this paper carefully and gave many important suggestions for improving the paper both in content and English usage. Special thanks are given to Mrs. Xu who drew all the figures for this paper. This research is financially supported by the National Science Foundation of China and the Doctoral Program Foundation of Institution of Higher Education of China.

REFERENCES

1. A. R. Bobyarchick. The eigenvalues of steady flow in Mohr space. *Tectonophysics*, 122, 35-51(1986).
2. M. D. Jr. Crittenden, P. T. Coney and G. H. Davis (Eds). Cordilleran metamorphic core complex. *Mem. geol. Soc. Am.* 153 (1980).
3. G. A. Davis and G. S. Lister. Detachment faulting in continental extension: perspectives from the southwestern U.S. Cordillera. *John Rodgers Symposium vol. Spec. pap. geol. Soc. Am.*, 218,133-159 (1988).
4. G. H. Davis. Characteristics of metamorphic core complexes, southern Arizona. *Abstr. with programs. geol. Soc. Am.* 9, 944 (1977) .
5. G. H. Davis and P. J. Coney. Geological development of Cordilleran metamorphic core complexes. *Geology*, 7, 120-124 (1978).
6. D. G. De Paor. Orthographic analysis of geological structures-I. Deformation theory. *Jour. Struct. Geol.* 5, 55-278 (1983).
7. Z. Hu and Z. Qian. A new idea of geological tectonics in the Xiaoqinling Region. *Geological review*, 40 (4),289-295 (1993) (in Chinese with English abstract).
8. J. Lee, E. L. Miller and J. F. Sutter. Ductile strain and metamorphism in an extensional tectonic setting: a case study from the northern Snake Range, Nevada, USA. In: M. P. Coward, J. F. Dewey & P. L. Hancock (Eds). *Continental Extensional Tectonics*. Blackwell Scientific Publications, 267-298 (1987).
9. G. S. Lister and G. A. Davis. The origin of metamorphic core complexes and detachment faults formed during Tertiary continental extension in the northern Colorado River region, U.S.A. *Jour. Struct. Geol.* 11, 65-94 (1989).
10. J. Malavieille. Late orogenic extension in mountain belt: insights from the Basin and Range Province and the late Paleozoic Variscan Belt. *Tectonics*, 12, 115-130 (1993).
11. P. E. Matthews, R. A. B. Bond and J. J. Ven den Berg. An algebraic method of strain analysis using elliptical markers. *Tectonophysics*, 24, 31-67 (1974).

12. W. D. Means, B. E. Hobbs, G. S. Lister and P. F. Williams. Vorticity and non-coaxility in progressive deformation. *Jour. Struct. Geol.* **2**, 371-378 (1980).

13. J. C. Phillips. Character and origin of cataclastic rocks developed along the low-angle Whipple detachment fault, Whipple Mountains, California. In: E. G. Frost & D. L. Martin (Eds). *Mesozoic-Cenozoic tectonic evolution of Colorado River region, California, Arizona and Nevada.* San Diego California Cordillera publishers, 109-116 (1982).

14. K. Sefert. Cordilleran metamorphic core complex. In: Sefert (Ed). *Encyclopedia of Structural Geology and Plate Tectonics.* New York. Van Nostrand Reinhold Co. , 113-130 (1987).

15. Q. Shi, G. Qin, M. Li, X. Zhou and Y. Feng. The late orogenic detached extension and mineralization in West Henan Province. *Henan geology,* **11** (1), 26-36 (1993) (in Chinese with English abstract).

16. Q. Shi. The study of thrusts and nappes in southern margin of North China Craton, Henan Province, China. *Report of issue on development of geological science and technology* (1996) (in Chinese).

17. C. Simpson and D. G. De Paor. Strain and kinematic analysis in general shear zones. *Jour. Struct. Geol.* **15**, 1-20 (1993).

18. Y. Wang, Y. Zheng, Q. Zang and Q. Zhang. Classy matrix of pseudotachylite in Yagan extensional detachment fault in Sino-Mongolian Boundary and kinematic implication. *Chinese Science Bulletin,* **39**, 1895-1897 (1994).

19. J. Zhang and Y. Zheng. Kinematic vorticity, polar Mohr circle and their application in quantitative analysis of general shear zones. *Jour. Geomechanics,* **1** (3), 55-64 (1995) (in Chinese with English abstract).

20. J. Zhang, J. & Y. Zheng. A general survey of pseudotachylyte and its formation process and mechanism. *Geological Science and Technology Information.* **14** (3), 22-28 (1995) (in Chinese with English abstract)

21. J. Zhang and Y. Zheng. 1996. Polar Mohr constructions for strain analysis in general shear zones. *Jour. Struct. Geol.* (in press).

22. J. Zhang and Y. Zheng. Basic principles and applications of kinematic vorticity and polar Mohr circle. *Geological Science and Technology Information* (in press).

Proc. 30th Int'l. Geol. Congr., Vol.14, pp.173-184
Zheng *et al.* (Eds)
© VSP 1996

Ductile Extension and Uplift of Granulites in the Datong-Huaian Area, North China

J.S. ZHANG[1], P.H.G.M. DIRKS[2] AND C.W. PASSCHIER[3]

1) Institute of Geology, State Seismological Bureau, P.O.Box 634, 100029 Beijing, China

2) Dept. of Geology, University of Zimbabwe, P.O.Box: MP 167, Mount Pleasant, Harare, Zimbabwe

3) Dept. of Geology, University of Mainz, 55099 Mainz, Germany

Abstract

Archaean granulites in Datong-Huaian area were strongly ductile deformed in response to an extensional detachment reacted in lower crust in about 2500-2400 Ma following the crustal thickening. Three litho-tectonic domains can be recognised by their distinct deformation history and kinematic structural patterns formed during the detachment process. The lower structural domain comprise more mafic TTG gneiss, dominated on mapping scale by domal structures. The intermediate structural domain consist predominately of biotite-rich gneiss and is characterised by a strongly foliated tectonic melange which acted as a major macro-scale decollement. The upper domain comprises Khondalites characterised by lower peak-metamorphic pressures and later granulite grade structures. In this rocks, a syn-tectonic S-type granites are widespread. The granulites are characterised by a gneissic fabrics. Foliation and lineations within the lower domain were strongly recrystallized and are now random in orientation, whilst mineral elongation and fold axis are co-lineated and consistently reoriented in the intermediate and upper domains, parallel to the extension direction all with ~ 25°plunges. Sense of shear indicators across the entire terrain indicate that the hanging walls of the detachment were displaced downward to SW in the lower and intermediate domains, and to WSW in the upper domain. Microlithons of garnet-bearing high-pressure mafic granulites are structurally widespread in the lower and intermediate domains. Development of symplectite textures in these rocks, produced by a near isothermal decompression process, is varying in relationship with their parent mineral assemblages and deformation stages through the uplifting history. Some of the symplictite textures were elongated parallel to the extensional direction.

Keywords: TTG gneiss, high-pressure granulite, khondalite, detachment, lineation, co-linear

INTRODUCTION

Tectonic processes involving crustal thickening followed by extensional collapse and tectonic denudation are thought to be the main geodynamic controls allowing the rapid exhumation of deeper crustal material including granulites [6, 23, 14, 18, 5]. Textures and geometries formed at low crustal levels, as well as related kinematics have been described from high-grade metamorphic core complexes exposed in orogenic domains [1, 7, 16, 3]. The central part of the north China granulite belt (Fig. 1A, B) is a suitable area for structural-metamorphic correlations, because of the juxtaposition of a granodioritic basement overlain by allochtonous poly-metamorphic metasediments with different, but related structures and P-T histories [28]. The Datong-Huaian granulites was recognised as a high-P

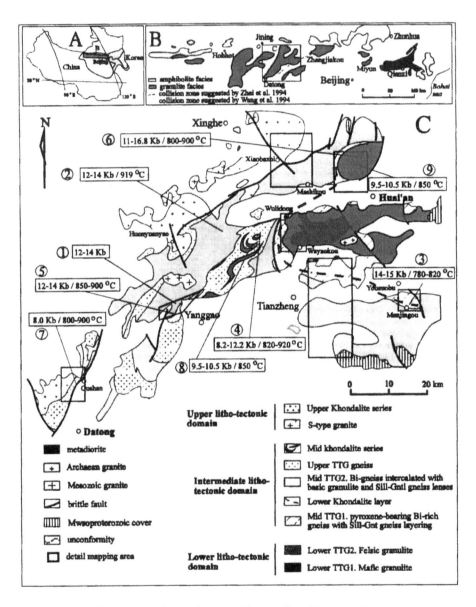

Figure 1. A & B-index maps showing location of the granulite belt in northChina and study area. C-geological map showing three litho-tectonic domains in the Datong-Huaian granulites. The intermediate domain lies on top of the Lower TTG gneiss and is mostly distributed around a domal uplift at the eastern of Wayaokou-Mashikou area. A sillimanite-garnet gneiss layer in the lower unit of this domain can be traced continuously around the dome. The presented P-T conditions are estimated by 1-W.Y. Cui, 1982; 2-X.L. Qian et al., 1985; 3-M.G. Zhai et al., 1992; 4-J.Z. Liu et al., 1992; 5-J.S. Zhang et al., 1994; 6-R.M. Wang et. al., 1994; 7-Y.H. Yan, 1994; 8-H.L. Mei et al., 1994; 9-Y.S. Geng et al., 1994

granulite terrain over the last ten years when it has received much attention. Studies mostly from metamorphic petrology have tried to document the nature and distribution pattern of the high - P assemblages by compressional tectonic activity in different collisional setting [27, 26]. To explain the different P-T characteristics between both units, Zhang et al. [28] proposed an extensional model in which a low-angle detachment allowed the unroofing and exhumation of the high-P TTG gneiss, which was consequently brought in contact with the supracrustal khondalite series. The above tectonic models not necessarily mutually exclusive, but reflect a lack of exact knowledge concerning the structural geometries and the actual mechanical processes. In this paper, we present the precise kinematics and deformation geometries recorded in the rocks to constrain the extension and uplifting model deduced mainly by P-T histories from a previous petrologic and thermodynamic studies.

Geologic setting

The granulitic rocks in Datong-Huaian area was systematically investigated since the mid-60's and have been collectively classed as the Jining Group of Archaean age, and sub-divided in a lower basement and an upper cover unit of metasediments respectively referred to as TTG gneiss and khondalite series [15, 19]. The TTG gneiss is dominated by pyroxene-bearing mafic to intermediate rocks, which gradually becomes more felsic and biotite-amphibole-rich (charnockitic) towards its top. The upper Jining Group is dominated by sillimanite-garnet-graphite-bearing metapelites interbedded with garnetiferous felsic gneiss, graphitic gneiss and minor mafic (2-pyroxene) gneiss.

Age constraints

Although exact ages for the Archaean and Palaeoproterozoic thermo-tectonic events in the Datong-Huaian granulites are less certain, a minimum age in the area can be obtained from unmetamorphosed Mesoproterozoic cover sediments that were deposited between 1840-900 Ma [22]. Subdivision of published age data for the various lithological units is illustrated in figure 2.

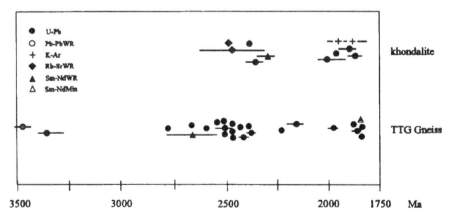

Figure 2. Diagram show age distribution in the Datong-Huaian granulites

Combining the above dates, the following conclusions may be drawn regarding the thermal-tectonic history of the area. An early crust forming episode occurred during the mid-Archaean (3300-3500 Ma) and was followed by a further episode of crustal formation between 2700 - 2600 Ma. A first high - grade metamorphic event, associated with the formation of the high-P granulites followed by extensional collapse and most of the structural geometries in the area occurred between 2500-2400 Ma. This event affected both basement and khondalite units. A further thermal event and associated granite intrusion may have occurred around 2200 Ma although the exact nature of this event is unclear. Around 1900 Ma the area experienced a pervasive high-grade metamorphic overprint, visible as a second generation of granulite grade assemblages [28]. By 1800 Ma the area had stabilised and was exhumed to surface.

P-T estimates

In the Datong-Huaian granulites, manifestations of multiple high- to low-pressure granulite facies events have been recognised [2, 12, 20, 25, 27, 28]. In the TTG gneiss, The peak assemblage yielded P-T conditions of 12-15 kbar and 800-900 °C (Fig. 1C). The garnet in the matrix assemblage is generally replaced by a reaction corona including opx-plag-hbl, which yield estimates of 7-9 kbar and 750 - 820 °C [10, 27, 28]. Late, post-kinematic reaction rims of px-hbl-plag formed at 4-6 kbar and 650-700 °C [28]. In the khondalite, a prograde inclusion assemblage record conditions of *at least* 6-7 kbar at 600-700 °C [13]. Peak-assemblages formed during the main granulite event and record P-T conditions that range from 7-10 kbar at 700-900 °C [10, 11, 13, 28]. Retrograde corona assemblages of plag-hbl around garnet in a mafic garnet-bearing lens in the khondalite were recorded at 7-9 kbar and 750-800 °C [28] suggest that decompression occurred in these rocks. This is also confirmed by the retrograde replacement of peak-kyanite by sillimanite [13]. Later reaction rims formed during a second metamorphic overprint, probably around 1900 Ma [28], and yield P-T estimates of 4-6 kbar at 650-750 °C [11, 13, 28].

The reconstructed P-T paths for the dominant (2400-2500 Ma) events in both the TTG gneiss and khondalites are characterised by clockwise loops dominated by cooling in the khondalites, and by 6-8 kbars of near-isothermal decompression in the TTG gneiss [28]. The differences between these paths may be real, or could be an artifact of the difference in bulk composition and closing temperatures and pressures for the mineral systems used in the calculations. The second generation, lower-P assemblages recorded in both the TTG gneiss and the khondalites reflect a reheating event, associated with granite intrusions and minor decompression [11, 28].

LITHO-TECTONIC DOMAINS

Three litho-tectonic domains (Fig.1) can be recognised in the Datong-Huaian granulites on the basis of their distinct deformation history and kinematic structural patterns, formed during the detachment process.

The lower structural domain is restricted to the stratigraphically lower part of the TTG gneiss including a more mafic granulite unit (L-TTG1) at the base and became more felsic rock association (L-TTG2) upward. This domain is dominated by domal or synform structures with complex lineation patterns, multiple foliations. Uplift related fabrics were found to be an important feature in this domain. The lower domain are characterised by the follow-ing structural trends: 1) Highly variable foliation and lineation orientations; 2) The occurrence of more than one gneissic layering, and hence truncations and transpositions of different gneiss foliations; 3) The occurrence of numerous interfering fold episodes, many of which are collinear in outcrop; 4) The strongly recrystallised nature of the foliation and lineation, generally obliterating mineral lineations defined by individual grains.

Rocks in *The intermediate structural domain* are complex and variable from place to place. A predominate rock association in the centre and eastern part of the terrain consists of biotite-rich gneiss, including a lower pyroxene-bearing unit (M-TTG1) and an intermediate pyroxene-free gneiss unit (M-TTG2). A ~50 m thick sillimanite-garnet gneiss layering (L-khond.) occurred intermittently between the L-TTG1 and L-TTG2 (Fig. 1C); but become more felsic to the west. The intermediate domain has the following structural characters: 1) A strongly foliated tectonic melange which acted as a major macro-scale decollement. 2) A lineation with constant orientation over the outcropping area; 3) The occurrence of only one gneissic layering with some isoclinal intrafolial folds; 4) Mineral lineations defined by indi-vidual grains are generally preserved because recrystallisation of the dominant fabric ele-ments is not pervasive. High-pressure mafic granulites (about 12-14 kbar) with Gnt-Cpx-Plag-Q mineral assemblage were found in the structurally lower and intermediate domain as tectonic fragment or lenses. These rocks must have passed the level of the lowermost crust during orogeny. Decompression textures are developed in these rocks to varying degree, depending on the parent bulk composition during exhumation.

The upper domain comprises khondalite series (Up-khond.) including meta-sedimentary sillimanite-garnet gneiss and quartz-garnet gneiss with minor quartzite, calc-silicate and graphite gneiss, interbedded with quartz-feldspar-biotite gneiss of metavolcanic origin. An S3-foliation completely replaced older one and syn-tectonic S-type granites are widespread in this domain.

GEOMETRY AND STRUCTURAL PATTERN

Rocks in the Datong-Huaian granulites are strongly affected by deformation related to extensional decollement and uplift of a lower crustal segment. The regional scale structural pattern is characterised by TTG gneiss domes and surrounding high angle and/or shallow dipping high-strain zones (Fig. 3A).

A major domal structure situated east of the Wayaokou-Wulidong road. Other mapping scale domes are found at north Mashikou, north Huaian and west Yanggao. Some NW trending older gneissic layering (S2) is preserved in the centre of these domes, where the rocks are not completely reworked by decollement (Fig. 3A,B). This layering is generally steeply dipping and was deformed in open folds with steeply plunging fold axes as a result of a lateral compression perpendicular to the extension direction, and refolded by a later generation fold with gently plunging folds axis parallel to the regional extension direction. The later generation of folds has been suggested to have formed as result of extension-related uplifting. The lower structural domain is dominated by progressively

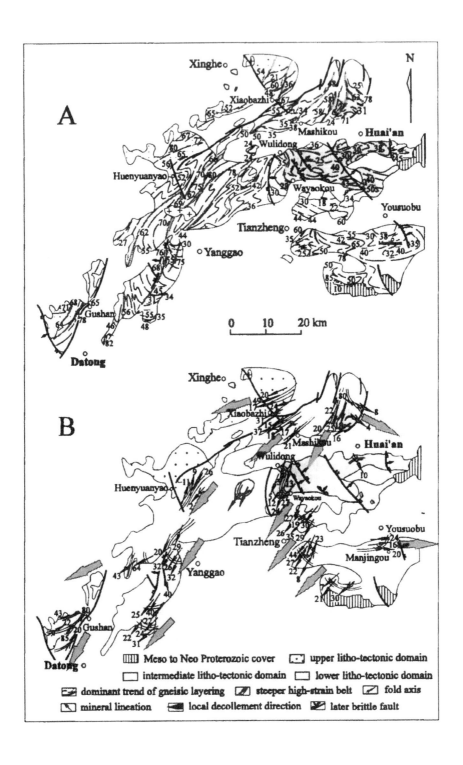

Figure 3. A) Simplified structural map of the Datong-Huaian granulites. It shows the principle foliation orientation across the entire area and regional structural patters. A major TTG gneiss domal uplift
situated east of Wayaokou-Mashikou is surrounded by a low-angle high-strain zone of biotite-rich
gneiss. Steeper NE trending high-strain belts are mainly developed at the western part of the terrain
where smaller TTG gneiss domes are uplifted between these belts. B) Structural map of the
Datong-Huaian granulites with simplified local foliation trends and projection of representative
lineations. D3 mineral elongation lineations and fold axes are sub-parallel in all three litho-tectonic
domains. The direction of extensional decollement as indicated by sense of shear indicators in outcrops is shown.

developing and flattening domes that penetrate the intermediate and upper domains where
structures are mainly related to SW or WSW directed lateral flow.

The shallow dipping high-strain zone comprises mainly biotite-rich gneiss as the described
for the intermediate structural domain. The main lithological units occur as cm- to m- scale
interbadded felsic and mafic granulite sheets that remained on the top of the TTG gneiss
domes. Investigations at N of Mashikou, SE of Tian Zhen and W of Yanggao show that they
are characterised by one straight foliation with intrafolial folds and a generally well developed mineral lineation. Fold axes are parallel to this lineation. It is likely that these features,
as well as the colinear fold episodes, resulted from non-coaxial constrictional flow. in a
sub-horizontal plain.

Steeper high-strain belts are mainly concentrated in the western part of the area, and are
several tens to a few hundreds of metres wide. These belts lie between the TTG gneiss
domes and acted as a strike-slip shear zones to adjust the differential displacement between
the domes during the decollement. Rocks in the steeper high-strain belts are strongly foliated with a cm- to dm- scale straight felsic layering intercalated with basic granulite lenses
or boudins. The foliation is near vertical in the central part of the belt but gradually
shallowing toward to both sides and is inclined towards the domes.

KINEMATICS OF THE EXTENSIONAL DECOLLEMENT

Although the orientation of foliations in the granulites is variable in response to the regional
structural patterns mentioned above (Fig. 3A), a shallowly plunging extensional lineations
show a consist orientation over the entire area (Fig. 3B).

As a consequence mineral shape lineations occur only in rare coarse-grained hbl-rich
dioritic lenses or opx-rich felsic gneisses. All other lineations are mineral aggregate
lineations, most of which consist of mm to cm-wide streaks or "sigars" of felsic material in
mafic gneiss or vise versa, and some deformed plagioclase-opx pseudomorphs after garnet.
In the intermediate and upper domains and the steeper high-strain belts, a shallowly plunging extensional lineation (~25 °) is constantly orientated, and mineral aggregate lineations
are parallel to crenulations and corrugations, and most of the fold axes are co-linear in outcrops. In the lower domain, three generations of lineation are recognised: an older southwards steeply plunging lineation related to NW-SE trending earlier gneissic layering (S2) is
visible, whilst orientations of second shallowly plunging (less 25 °) lineations that related to

domal structures are variable. Locally two generations of the lineation are preserved. In this case the steeply plunging lineations are commonly folded by open recumbent folds with fold axes that parallel the shallowly plunging lineation. It is likely that the S plunging steeper lineations are remnants of early transposed lineation produced by a simultaneously lateral compression during an earlier stage of the extension. Both lineations are strongly recrystallised in the lower litho-tectonic domain but clearly preserved in the intermediate and upper domains. The linear fabrics are plunging to SW in the intermediate-lower domain, to WSW in the upper domain in the central and western part of the area, but to SE or E in the eastern part (Fig. 3B).

Several types of sense of shear indicators are observed in the granulites, including C-S foliations, shear bands filled with opx-bearing anorthositic to granitic veins, large asymmetric delta-shaped clasts of garnet and orthopyroxene, asymmetric boudinaged mafic layers in a more felsic matrix and conjugate structures developing in a layering subjected to non-coaxial flow. In general, the SW to WSW plunging lineation is associated with a top to the SW or WSW, mostly normal sense of shear. Many recumbent folds that are colinear with the extension lineation are asymmetric when viewed normal to the fold axes. This could be interpreted as a secondary flow effect away from structural (gravitational) highs. It is likely that the rocks on the top of the detachment zones were moved away from a major domal structure which culminated in the area between Wayaokou and Mashikou.

OUTCROP STRUCTURES RELATED TO LATERAL COMPRESSION, EXTENSION AND UPLIFTING

Structures in outcrop scale showing an extensional related uplift are mainly preserved in the TTG gneiss domes where the decollement movement is relatively week. Outcrop from centre of the TTG domal uplift between Mashikou and Xiaobazhi (Fig. 4A, B) shows that an older steeper south-westwards dipping gneiss layering (S2) was folded, with steeper plunging fold axis and mineral elongation, by a simultaneously lateral compression during earlier stage of the extension. The envelope plane of 230 / 75 is suggested parallel to a previous orogenic belt. These folds were overprinted by a secondary fold with shallow plunging fold axis of 200 / 20, which is well co-lineated with the regional L3 mineral elongation lineation and parallel to the regional extensional direction. The secondary folds, therefore, were formed by an extensional related uplifting; These structures are also found on an outcrop (Fig. 4B) in the centre of the TTG gneiss domal uplift in north Huaian. It shows that an intrafolial fold (F2) was refolded during the extensional related uplift with a gentle plunging fold axis of 225 / 9 parallel to the regional extension direction.

Structures formed by extensional decollement and simultaneous lateral compression are well developed in the shallow and steeper dipping high-strain belts. Fig. 4C shows a 2-D view on an outcrop from a steeper high-strain belt in north Dongyanghe of north Huaian. Different kinds of linear structures, including asymmetric fold axes, are well co-lineated parallel to the extensional direction during the decollement movement. Whilst a 3-D view on an outcrop (Fig. 4D) from the shallow dipping high-strain zone of NE Tianzheng shows that crenulation lineations and asymmetric fold axes are reoriented parallel to mineral elongation direction formed by a contemporaneously compression and extension.

Evidences from the outcrop and mapping scale structures provide a geodynamic model (Fig. 4E) which shows that functions of both decollement, lateral compression and uplifting are contemporaneously acted during the extension in the lower crust.

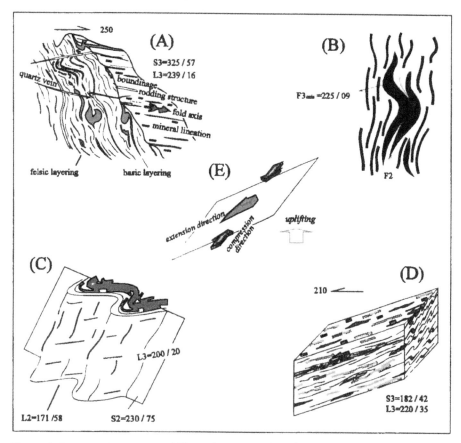

Figure 4. Structures in outcrop and kinematic interpretation of the extensional decollement in the lower crust. (A) Outcrop from the central part of the TTG domal uplift between Mashikou and Xiaobazhi; (B) Outcrop in the centre of the TTG gneiss domal uplift in north Huaian; (C) Two dimensional outcrop in a steeper high-strain belt in north Dongyanghe, north Huaian. Different kinds of linear structures are co-linear, parallel to the extension direction during the decollement movement; (D) 3-D view on an outcrop from the felsic gneiss NE of Tianzheng. It shows that the crenulation lineations are parallel to the mineral elongation direction, formed by a contemporaneous compression and extension; E) a geodynamic and kinematic model show that decollement, lateral compression and uplifting are all contemporaneously active during the extension in lower crust.

DISCUSSION

In the Datong-Huaian area, evidence exists for three major structural events. Early, NW-SE trending lineations associated with inclusion patterns may reflect prograde events and could be associated with the interleaving of TTG gneiss and laterally continuous slivers of khondalite during an episode of burial and/or crustal thickening. The second, and by far most dominant, event is related to the pervasive, shallowly SW plunging lineation and exhumation of the high-P rocks. This exhumation process may be genetically linked to the first event via a process of crustal thickening followed by extensional collapse [28]. The third episode of deformation occurred later and was associated with discrete deformation zones in the khondalites and a static thermal overprint of the TTG gneiss. We assume that all structures in the area associated with the regionally consistent SW-plunging mineral lineation formed during the second tectonic episode between 2500-2400 Ma ago. Although no direct ages on metamorphic minerals have been reported, charnockites emplaced during this event date at around 2450 Ma [9], suggesting that the above assumption is correct.

It is clear that foliation patterns in a granulite terrain like the Datong-Huaian area are not the most workable geometric feature to use when describing the deformation history, because of multiple transpositions and numerous overprinting folding episodes. In contrast, lineation patterns prove more useful since many of the overprinting foliations have similar mineral lineation directions and different fold episodes are colinear. This suggests that tectonically related events (or geotectonic settings) can be grouped on the basis of a common lineation direction, indicative of the principle direction of flow and a large range of fold styles and geometries that formed progressively during granulite metamorphism [21, 4]. Using this approach, the upper and intermediate structural domains are similar, while the lower domain shows a gradual transition into the intermediate domain as its more widely distributed foliation / lineation directions are gradually transposed. Textural evidence indicates that all three domains developed during exhumation of at least 6-7 kbars. The differences in deformational styles can therefore be interpreted as representing different geomechenical settings within the lower crust allowing the rise of deep crustal material.

It appears that during the exhumation of the high-P granulites, several processes occurred simultaneously. These include vertical doming, transposition in a horizontal foliation and lateral flow within a constrictional environment. All three processes operated simultaneously, but the dominance of each process varied at different crustal levels: processes in the lower structural domain were dominated by vertical flow; i.e. doming and simultaneous transposition; processes in the intermediate and upper domains were dominated by transposition and lateral flow. Geometries indicate that doming was a complicated process or, maybe, a simple process leading to complicated geometries. The presence of recumbent folds along the limb regions of the domes, the local progression of these recumbent folds in new transposition fabrics and the renewed doming of such transposition fabrics indicate that doming and transposition were dynamic and interacting processes during exhumation. The structures are similar to those described for mantled gneiss domes [24], and resemble solid state diapirs simultaneously subjected to gravity-driven collapse, i.e. transposition in horizontal fabrics. The random distribution of the domes confirms their diapiric origin rather than a fold interference pattern. There seem to be two driving mechanisms for doming or diapiric rise: (1) the existence of gravitational or thermal instabilities of hot and very ductile lower crust; (2) the removal of an overburden, i.e. gravitational collapse of the middle and upper crust. Both processes may have acted simultaneously. Concerning the second mechanism, it is important to consider the intermediate and upper structural domains.

Going up-stratigraphy, the domal structures are gradually transposed within the gneissic layering that characterises the higher structural domains. Although this gneissic layering is folded by open upright folds, the regional envelope dips shallowly and the lineation is shallow and plunges constantly to the SW. This fabric represents a crustal level where lateral flow with sheath folding and transposition of domal structures is more important. Movement indicators suggest that this flow was mainly SW-directed. Overall, the intermediate and upper structural domains may represent the deep crustal expression of a low angle detachment zone. Movement on this zone allowed unloading and the rise of deeper crustal levels, a process aided by (thermally-driven ?) solid state diapiric rise. The occurrence of high-P granulites within the intermediate structural domain indicates that rocks passed from the lower crustal domain into the intermediate crustal domain. Decompression melting during this exhumation process may account for the larger volumes of felsic material in the structurally higher domains. The difference in peak-metamorphic pressures between the TTG gneiss of the intermediate domain and the khondalites of the upper domain [28] suggests that as a direct result of extensional flow on the decollement zone, deep crustal rocks were exhumed and juxtaposed to shallower crustal rocks.

The colinearity of fold axes and lineation directions in the domains, characterised by supposedly extensional SW-directed flow, can only partly be explained by progressive rotation into sheath folds. Many open folds and crenulations are also colinear suggesting lateral compression or flow constriction in a NW- direction. We can only speculate about the reasons for this. One possible explanation could be that prograde crustal thickening, as reflected in the earlier lineation directions, occurred along a NW-SE axis, and that craton margins involved in this process trended parallel to the SW-plunging lineation.

Evidence from outcrop- scale structures provides a geodynamic exhumation model for the high-P granulites. This process involved unloading of the upper crust across a decollement within a constrictional environment, and simultaneous rise of deeper levels underlying the decollement. These deeper crustal rocks are progressively transposed as they rise and move into the decollement zone.

Acknowledgements

This work was supported by grants from the NSF of China (No. 49070133), the Joint Earthquake Science Foundation of China (No. 95098) and the Dutch Academy of Sciences.

REFERENCES

1. T.B. Andersen and B. Jamtnveit. Uplift of deep crust during orogenic extensional collapse: a model based on field studies in the Song-Sunnfjord region of western Norway, *Tectonics* **9**, 1097-1111 (1990).

2. W.Y. Cui. The P-T condition of mineralogy and crystallography for granulites in the Zhuozi-Yanggao area. Collected Papers on Geological research at Peking University, Publishing House of Peking University, 110-121 (1982).

3. N. G. Culshaw, J. W. F. Ketchum, N. Wodick a nd P. Wallace. Deep crustal ductile extension following thrusting in the southwestern Grenville Province, Ontario. *Can. j. Earth Sci.* **31**, 160-175 (1994)

4. P.H.G.M. Dirks, and C.J.L Wilson. Crustal evolution of the East Antarctic mobile belt in Pyda Bay: continental collision at 500 Ma ? *Precambrian Res.* **75**, 189-207 (1995).

5. P. Gautier and J.P. Brun. Crustal - scale geometry and kinematics of late-orogenic extension in the central

Aegean (Cyclades and Evvia Island). *Tectonophysics* 238: 1-4, 399-424 (1994).

6. S.L. Harley. The origins of granulites: a metamorphic perspective. *Geol. Mag.* 126, 215-247 (1989).

7. T.M. Kusky. Collapse of Archean orogens and the generation of late-to postkinematic granitoids, *Geology* 21:10, 925-928 (1993).

8. P. Li and M. Dai. The K-Ar absolute age determination of some pegmatites and granites in Inner Mongolia and Nan Ling mountain areas. *Geosciences China*, 1, 1-9 (1963).

9. D.Y. Liu, Y.S. Geng and B. Song. Archaean crustal evolution in Northwest Hebei Province. *Abstracts of 30th International Geological Congress.* 2, pp. 526 (1996).

10. J.Z. Liu. Study of cratonization of the Archaean granulite faces region and a lower crustal cross section, Inner Mongolia. *PhD. thesis Peking Univ* (1989).

11. J.Z. Liu, Y.P. Chen and X.L. Qian. Study on original tectonic environment of the Datong-Xinhe khondalite suite, *Lithospheric Geoscience, Beijing Univ. Press*, 61-68 (1992).

12. L.Z. Lu. Metamorphic P-T-t path of the Archaean granulite-facies terrane in Jining District, Nei Mongol, and its tectonic significance. *Acta Petrol.a Sinica* 4, 1-12 (1991).

13. L.Z. Lu, S.Q. Jin, X.C. Xu and F.L. Liu. The genesis of Early Precambrian khondalite series in southeastern Nei Mongol and its potential mineral resources. *Jilin Science and Technology Press, Changchun, China* (1993).

14. J. Malavieille. Late orogenic extension in mountain belts: insights from the Basin and Range and the late Paleozoic Variscan belt. *Tectonics* 12: 5, 1115-1130 (1993).

15. X.L. Qian, W. Cui and S. Wang. Evolution of the Inner Mongolia-Eastern Hebei Archaean granulite belt in the north China craton. *Geol. Res., Beijing Univ. Press.* pp.20-29 (1985).

16. J. Reinhardt and U. Kleemann. Extensional unroofing of granulitic lower crust and related low-pressure, high-temperature metamorphism in the Saxonian Granulite Massif, Germany. *Tectonophysics* 238, 71-9 (1994).

17. M. Sandiford & R. Powell. Isostatic and thermal constraints on the evolution of high temperature-low pressure metamorphic terrains in convergent orogens. *J. of Metamorphic Geology* 9, 333-340 (1991).

18. M. Seranne and J. Malavieille. Preface. *Tectonophysics* 238, vii-xi (1994).

19. Q.S. Shen, Y. Zhang, J. Gao and P. Wang. The Archaean metamorphic rocks in mid-southern Inner Mongolia. *Publ. Inst. Geol., Chinese Acad. Geol. Sci.* 21, 3-21 (1989).

20. J.D. Sills, K.Y. Wang, Y. Yan and B.F. Windley. The Archaean high-grade gneiss terrain in E Hebei Province, NE China; geological framework and conditions of metamorphism. *Geol. Soc. Spec. Publ.* 27, 297-305 (1987).

21. J.R. Sims, P.H.G.M. Dirks, C.J. Carson and C.J.L. Wilson. The structual evolution of the Rauer Group, east Antarctica; mafic dykes as passive markers in a composite Proterozoic terrain. *Antarctic Sci.* 6, 379-394 (1994).

22. D.Z. Sun (Editor). The early Precambrian Geology of the eastern Hebei. *Tianjin Science and Technology Press,Tianjin China* (1984).

23. Van den Driessche and J.P. Brun. Tectonic evolution of the Montagne Noire (french Massif Central): a model of extensional gneiss dome. *Geodinamica Acta* 5, 85-101 (1991).

24. B. van den Eeckhout, J. Grocott and R.L.M. Vissers. On the role of diapirism in the segregation, ascent and final emplacement of granitoid magma - Discussion. *Tectonophysics* 127, 161-169 (1986).

25. R.M. Wang, Z. Chen and F. Chen. Grey tonalitic gneiss and high - Pressure granulite inclusions in Hengshan, Shanxi Province, and their geological significance. *Acta Petrol. Sinic* 4, 36-45 (1991).

26. R.M. Wang, X.Y. Lai, W.D. Dong, J. Ma, B. Tang. Some evidence of the late Archaean collision zone in northwest Hebei Province. In Geological Evolution of the Granulite Terrain in North Part of the North China Craton. Edited by Qian Xianglin and Wang Renmin. *Seismological Press* (1994).

27. M.G. Zhai, J.H. Guo, Y.H. Yan, X.L. Han and Y.G. Li. High pressure basic granulite and disclosing of a lowermost crust of the early Archaean continental type in N. China. *Inst. Geol. Acad. Sinica, Advances in Geosc.* 2, 126-139 (1992).

28. J.S. Zhang, P.H.G.M. Dirks and C.W. Passchier. Extensional collapse and uplift of a polymetamorphic granulite terrain in the Archaean of north China, *Precambrian Res.* 67, 37-57 (1994).

PART 4

STRIKE-SLIP AND TRANSPRESSION TECTONICS

Proc. 30ᵗʰ Int'l. Geol. Congr., Vol.14, pp. 187-195
Zheng *et al.* (Eds)
© VSP 1997

Diachronous Uplift along the Ailao Shan-Red River Left-lateral Strike-slip Shear Zone in the Miocene and Its Tectonic Implication, Yunnan, China

WENJI CHEN, QI LI, YIPENG WANG, JINGLIN WAN

Institute of Geology, State Seismological Bureau, Beijing 100029, CHINA

MIN SUN

Department of Earth Sciences , University of Hong Kong, Hong Kong

Abstract

The Ailao Shan-Red River metamorphic belt in southern Yunnan province, China, was one of the largest left-lateral,ductil strike-slip shear zone developed during the mid-Tertiary in Asia. To better understand the nature and timing of this feature, $^{40}Ar/^{39}Ar$ age spectrum and fission track analyses were undertaken on 18 K-feldspars and 4 apatites from seven sampling sites (including three sections of cross metamorphic belt) from Yuanyang to Zhelong for about 225 km along this belt. Using the MDD model (Multiple-Diffusion Domain model) these results yield 18 curves of cooling history in the temperature range from 400 - 150 °C. The common characteristics of these curves were to have one (or two) rapid cooling (>100 °C/Ma), rapid uplift event, in the age range from 25.7 - 17.8 Ma. The fission track ages range from 10.6 to 5.61 Ma. Both results indicate dicreasing from SE to NW along the shear zone.

When ploted as a function of distance along the range, the starting time of rapid uplift and the fission track ages are getting younger linearly from Yuanyang (SE) to Zhelong (NW) at the slope of 34mm/yr and 30 mm/yr respectively. The surprising linear correlation can be interpreted by two probable tectonic models in different time ranges: (1) the slope of 34 mm/yr was probably just the speed of left-lateral strike-slip movement with uplift (slope angle of about 11°) and the starting time of the rapid uplift decrease linearly during 25.7 to 17.8 Ma from SE to NW; (2) The slope of 30 mm/yr was considered a homogeneous extended speed of tectonic uplift from SE to NW during 10 Ma (or 15.5 Ma) to 5.5 Ma, probably due to a resistance of unfree interface at the SE edge of the shear zone. These results also support that a 15.5 Ma unconformity separates a younger left-lateral transpressional regime from an older left-lateral transtensional regime in the Red River shear zone in the Tonkin Gulf, Vietnam.

Keywords: Ailao Shan-Red River Shear Zone, Multiple-Diffusion Domain Model, Fission Track Age, Tectonic Uplift.

INTRODUCTION

The northward convergence of India into Asia over the past 50 Ma has led to the development of thickened crust including the Himalaya and Tibetan plateau. Although recent works have confirmed that Ailao Shan-Red River metamorphic belt in southern Yunnan,China, was a left-lateral ductile strike-slip shear zone during the mid-Tertiary. In the continental-extrusion hypothesis, the South China Sea is interpreted to be a pull-apart basin

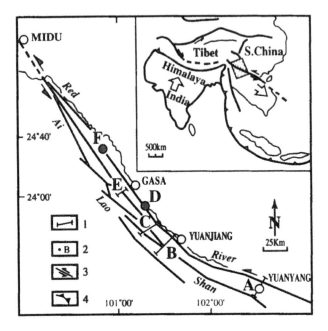

Figure 1. Sketch map of sampling localities along the RRSZ 1. Geological section; 2. Sampling site; 3. Strike - slip fault of Tertiary; 4. Reversed fault of Tertiary

at the southeastern termination of the Ailao Shan Red-River Shear Zone (RRSZ) that opened as a result of strike-slip motion [12,15,16]. Poles of rotation and rates of seafloor spreading computed from magnetic anomalies in the South China Sea suggest a total left-lateral offset along the RRSZ of about 550 km with slip rate of 5.5 cm/yr between 32 and 28 Ma and 3.5 cm/yr between 28 and 17 Ma [1,2,17,18]. The U-Pb mineral ages of 23.0 ± 0.2 Ma and 24.4 ± 0.2 Ma [14,9] from the Ailao Shan and the Dian Cang Shan provided an upper limit for the cessation of ductile behavior within this portion of the shear zone.

The MDD model (Multiple Diffusion Domain model) proposed by Lovera et al [10] is an extension of the Dodson closure model. It assumes a single sample of K-feldspar to have several different diffusion domains and hence several different closure temperatures. Therefore, a continuous cooling curve between 400-150 °C can be obtained by the MDD model calculation based on analyses of $^{40}Ar/^{39}Ar$ of a K-feldspar sample. In cooperation with T.M. Harrison (Department of Earth and Space Sciences, University of California, Los Angeles, USA), we have made $^{40}Ar/^{39}Ar$ dating and the MDD model treatment for 38 K-feldspar samples from granite, gneiss and mylonite in 7 areas of China. The results obtained confirmed the universality of nonstationary movements and rapid-cooling events in the tectonic-thermal history [4]. Using the MDD model, an rapid cooling event has been observed firstly from the cooling curve of K-feldspar sample FA-2-1 (see Fig.1 , site D). Therefore, Harrison et al [6] suggested the discontinuities in cooling rates reflect probably a transition in deformation style in shear zone during the early Miocene. The closure temperature (Tc) of fission track in apatite is taken to be 110±10°C. In combination with fission track results, the thermal history can be discussed between 150-100 °C. In this temperature range K-feldspar analysis does not provide a clear picture.

RESULTS OF ^{40}Ar/^{39}Ar AND FISSION TRACK THERMOCHRONOMETRY

^{40}Ar/^{39}Ar Thermochronometry

To better understand the nature and timing of the RRSZ, ^{40}Ar/^{39}Ar spectrum analyses were undertaken on 18 K-feldspars of seven sample sites, including four sections across the metamorphic belt (see Fig. 1 - A, B, C, E) from Yuanyang to Zhelong about 225 km along the RRSZ. All the ^{40}Ar/^{39}Ar spectrum analyses and the MDD modeling have been done in T.M. Harrison's lab. at UCLA. The detailed analytical methods have been given by Harrison et al. [6]. Interpreted using the MDD model these results yield 18 curves of cooling history. Two examples of this approach have shown in Fig. 2. The K-feldspar samples, YU-94 and YU-97, collected from the north of our research region (Fig.1, F), yielded very similar cooling histories, compatible with the ages of the coexisting hornblandes and indicating two rapid cooling events.

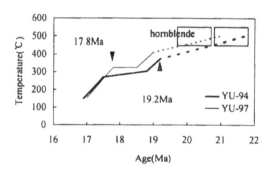

Figure 2. The cooling history of YU-94 and YU-97 K-feldspars

The starting time of first rapid cooling were 19.2 Ma and 19.0 Ma at same rate of 175°C/Ma for YU-94 and YU-97, respectively. The starting times of the second rapid cooling were 17.5 Ma and 17.8 Ma at the same rate of 200 °C/Ma for YU-94 and YU-97. The rapid cooling event may reflect ductile deformation ceased by a compoment of normal faulting along the shear zone [6]. The common characteristics of 18 cooling history curves are one (or two) rapid cooling event in the age range of 25.7-17.8 Ma. The oldest ages of the different starting time of the rapid cooling events in each area of seven sample sites and the relative starting temperature are considered as the starting time and temperature of the rapid cooling event in each area and shown in Table 1. We have considered that these times just represent the transitional time of the movement styles along RRSZ. For example, the 19.2Ma and 17.8 Ma were selected from sample YU-94 and YU-97 to be the starting time for the first and second rapid cooling events (in sample site F), respectively. The average starting temperatures were 374.3±26 °C and 324±27 °C for the first and second rapid cooling events, respectively. An indisputable fact is that the starting time of rapid cooling decreases from SE to NW along the RRSZ.

Table 1. Results of simulating the cooling histories by the MDD model for K-feldspar samples from the RRSZ

Sample location	Relative distance (km)	The first cooling stage			The second cooling stage		
		Starting time (Ma)	Starting temp. (°C)	Cooling rate (°C/Ma)	Starting time (Ma)	Starting temp. (°C)	Cooling rate (°C/Ma)
A	225	25.7	400	200	24.3	323	200
B	125	--	--	--	21.5	365	160
C	100	--	--	--	20.8	320	150
D	75	--	--	--	20.0	364	175
E	50	20.2	348	85	19.0	288	150
F	0	19.2	375	175	7.8	324	200

Results of Fission Track Thermochronometry

Four apatites separated from samples of 4 locations have been studied by the fission-track method of external detector following by A.J. Hurford [8]. The samples were irradiated in thermal neutron reactor (Atomic Institute, Academia Sinica, China). Track density measurements were performed by an OPTON microsrope, at a magnification of X1000, oil-immersion objective. Mean values, of spontaneous to induced track densities (ρ_s /ρ_i) were used to calculate ages . The standard sample used in our laboratory is apatite in the Fish Canyon tuff in the USA. The error of standard age of calibration method for the external detector method is ± 0.2% [5]. The closure temperature (Tc) of fission track in apatite is taken to be 110 ± 10 °C [11]. The fission-track results are listed in Table 2. Uncertainties in the chronologic date are quoted at 1σ.

Table 2. Apatite Fission Track Dating Results

Sample	Number of Grains	Spontaneous		Induced		ρ_s /ρ_i	Age±1σ, Ma
		N_s	ρ_s (10^7 /cm²)	N_i	ρ_i (10^7 /cm²)		
FA-2-1(D)*	58	208	3.5862	7309	252.034	0.0142	5.61±0.48
FA-5 (C)*	48	177	3.6875	5293	220.286	0.0167	6.59±0.60
FA-7 (B)*	49	114	2.3265	3045	124.286	0.0187	7.36±0.79
FA-9(A)*	58	2289	39.466	41705	1438.10	0.0274	10.6±0.6

*: sample localities (see in Fig. 1); ρ_s and ρ_i are spontaneous and induced track densities.

Apatites from four samples give fission track ages range from 5.61 Ma to 10.6Ma. When ploted with cooling curves obtained by the MDD modeling (Fig. 3), the cooling rates are 3.4°C/Ma and 8.1°C/Ma for sample FA-2-1 and FA-5 respectively. Four ages also decrease from SE (10.6 Ma) to NW (5.61 Ma).

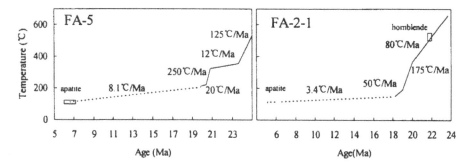

Figure 3. The cooling history of FA-2-1 and FA-5

The Surprising Linear Correlation

When ploted as a function of relative distance (see Table 1) along the RRSZ from Yuanyang to Zhelong , the starting time of rapid cooling, rapid uplifting, decrease linearly among these data of seven sites from Yuanyang (SE) to Zhelong (NW), at a slope between 33.8 mm/yr and 33.9 mm/yr for the first and seconed rapid cooling events, respectively (Fig.4). This function is obviously different from that proposed by Harrison et al.[7], in which "a plot of the times which various isotherms (400°C, 300°C and 200°C) were achieved as a function of distance along the RRSZ" based on a same date set. It is emphasized in this work that the starting times of rapid cooling just reflected the transitional time of the movement styles. The similarity between figure 4 and the RRSZ slip rate of about 35 mm/yr (28-17 Ma) estimated by Briais et al [2] from South China Sea magnetic anomalies suggests that the trend may reflect processes of plate motion. With new

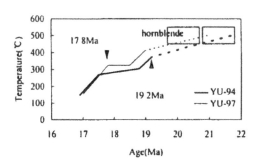

Figure 4. Plot of starting time of rapid cooling event and apatite fission track age versus relative distance. 1. First rapid cooling event; 2. Second rapid cooling event; 3. Apatite fission track age.

results from the apatite fission track ages in the range, a clear linear correlation was also obtained at a slope of 30.4 mm/yr. These new results may let us to discuss wider temporal and space range from 400 -100 °C in the temperature and from 27 - 5 Ma in the time.

TWO PREFERRED TECTONIC MODELS

Obviously, the most signifiacant feature of the data set is seen in a plot of the starting time of rapid cooling and the apatite fission track ages were achieved as a fuction of distance along the RRSZ from Yuanyang to Zhelong for about 225 km. The most remarkable feature of the data set is seen to have a slope between 34 mm/yr (25.7 - 17.8 Ma) and 30 mm/yr (10.6 - 5.6 Ma) which both are similar with the RRSZ slip rate of 35 mm/yr (28-17 Ma). To explain this feature, we suggest two probably tectonic models.

The Normal Fault-Strik Slip Motion between 28 and 17 Ma
The west side of fault have been moved from NW to SE along the RRSZ based on the continental-extrusion hypothesis. The magnitude of fault displacement can be calculated by the speed and time. If we assume the slope of 34 mm/yr as a slip rate of the west side, the samples present at site A were at site F at 25.7 Ma and moved to site A at 17.8 Ma (see Fig. 5). Because the rapid uplift was began at 25.7 Ma , the left-lateral strike -slip motion was associated with uplift by a slope angle of about 11° calculated by uplift and slip rate . According to the more apparent width of the shear zone and the more subdued topography in the southern portion of the range, the southern portion of the RRSZ have experienced a greater amount of extension relative to the northern portion. Our model also support that between 30 and 15.5 Ma there is a significant component of NE-SW distension within the strike-slip regime studied using seismic profiles calibrated by deep wells along the RRSZ in the Tonkin Gulf, Vietnam[13]. The normal strike-slip faulting between 28 and 17 Ma strongly supports the hypothesis of South China Sea opened as a result of strike-slip motion between both the time (28 - 17 Ma) and slip rate (35 mm/yr). If this indeed had been the case, the MDD model may prove a new powerful tool in studying horizontal movement in addition to vertical movement along strike-slip shear zones.

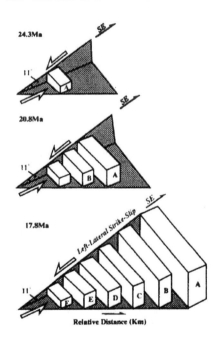

Figure 5. Sketch map of strike-slip movement with uplift

The Homogeneous Extended Speed of Tectonic Uplift between 15.5 and 5 Ma
In this model, the slope of 30 mm/yr between 10.6 and 5.6 Ma was considered a homogeneous extended speed of tectonic uplift from SE to NW (see Fig. 6). The west side of the fault

has started uplift at about 10.6 Ma , probably due to a resistance of unfree interface at the SE edge of the RRSZ which should result in a northwestward younging of the uplift processes, the extended speed of which should be similar to the plate velocities that control this movement. The uplift rates also decrease from SE (about 9 °C / Ma) to NW (about 3 °C / Ma) which also support the assumption of a resistance of unfree interface at the SE edge of the RRSZ. The 15.5 Ma unconformity is recorded in the Tonkin Gulf. Between 15.5 and 5.5 Ma, the narrow fault zone is affected by NE-SW compression [13]. This unconformity separates a younger left-lateral transpressional regime from an older left-lateral transtensional regime, which supports both models presented above for different time periods.

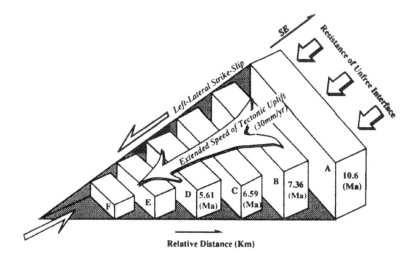

Figure 6. Sketch map of homogeneous extended speed of tectonic uplift

CONCLUDING REMARKS

A total of 18 cooling history curves were obtained using K-feldspars $^{40}Ar/^{39}Ar$ analysis with a MDD model treatment in the temperature range from 400 - 150 °C. These curves show one (or two) rapid cooling (>100°C / Ma), rapid uplift event, in the age range from 25.7 to 17.8 Ma.

When ploted as a function of distance along the RRSZ, the starting time of rapid uplift and the fission track ages dearease linearly indicated by these data from Yuanyang (SE) to Zhelong (NW) at a slope between 34 mm/yr and 30 mm/yr, respectively.

Two preferred tectonic models were suggested: (1) the normal strike-slip motion between 28 and 17 Ma; (2) the homogeneously extended speed of tectonic uplift from SE to NW between 10 (or 15.5 Ma) and 5.5 Ma. Both models are supported by 15.5 Ma unconformity separating a younger left-lateral transpressional regime from an older left-lateral transtensionsl regime in RRSZ.

The MDD model may provide a new tool in studying horizontal movement in addition to vertical movement along strike-slip shear zones.

REFERENCES

1. A. Briais. Cinematique d'ouverture de la mer de China du Sud (NanHai). Implications pour la tectonique Tertiaire de l'Asie. These de doctorat Universite Paris. 6, 239(1989).
2. A. Briais, P. Patriat and P. Tapponnier.Updated interpretation of magnetic anaomalies and seafloor spreading stages in the South China Sea: implication for the Tertiary tectonics of SE Asia. *J. Geophys. Res.* **98**, 6299-6328 (1993).
3. W.J. Chen, T.M. Harrison and O.M. Lovera. Thermochronology of the Ailo shan-Red River shear zone - a case study of multiple diffusion domain model. *Seismology and Geology*, **2**, 121-128 (1992),(in Chinese).
4. W.J. Chen, Q. Li and Z.J. Ma. Constrains of the MDD model to quantitative study of nonstationary tectonic movements, *Seismology and Geology*, **18**, 56-60 (1996).
5. L. Ding, D.L. Zhong, Y.S. Pan, X. Huang and Q.L. Wang. Fission track evidence of rapid uplift from Pliocene on tectonic joint of east Himalaya. *Chinese Science Bulletin*, **16**, 1497-1500 (1995),(in Chinese).
6. T.M. Harrison, W.J. Chen, P.H. Leloup, F.J. Ryerson, P. Tapponnier and U. Scharer. An early Miocene transition in deformation regine within the Red River fault zone, Yunnan and its significance for Indo-Asian tectonics, *J. Geophys. Res.* **97(B5)**, 7159-7182 (1992).
7. T.M. Harrison, P.H. Leloup, F.J. Ryerson, P. Tapponnier, R. Lacassin and W.J. Chen. Diachronous initiation of transtension along the Ailao Shan-Red River shear zone, Yunnan and Vietnam. *In: The Tectonic Evolution of Asia.* An Yin and T.M. Harrison (Eds). pp. 208-226. Cambridge Univ. Press, Cambridge (1996).
8. A.J. Hurford and P.F. Green. A users'guide to fission track dating calibration. *Earth Planet. Sci. Letts.* **59**, 343-354 (1982).
9. X. Liu, U. Schaerer and P. Tapponnier. Timing of Largescale strike-slip movements in the Diancang shan and Ailao shan sgear belts, Yunnan, China, *EOS*, **73**, 310 (1992).
10. O.M. Lovera, F.M. Richter and T.M. Harrison. The $^{40}Ar/^{39}Ar$ thermochronometry for slowly cooled samples having a distribution of diffusion domain sizes, *J. Geophys. Res.* **94**, 917-935 (1989).
11. C.W. Naeser and F.C.W. Dodge. Fission-track ages of accessary minerals from granitic rocks of the central Sierra Nevada Batholoth, California., *Geol. Soc. Amer. Bull.* **80**, 2201-2211 (1969).
12. G. Peltzer and P. Tapponnier. Formation and evelution of strike-slip faults, rift and basins during the India-Asia collision: An experimental approach, *J. Geophys. Res.* **93**, 15085-15117 (1988).
13. C. Rangin, M. Klein , D. Roques, X. Le Pichin and Le Van Trong. The Red River fault system in the Tonkin Gulf, Vietnam. *Tectonophysics.* **243**, 209-222 (1995).

14. U. Schearer, P. Tapponnier, R. Lacassin, P.H. Leloup, D. Zhong and S. Ji. Intraplate tectonics in Asia: a precise age for tertiary Large scale movement along the Ailao Shan-Red River shear belt, China. *Earth Planet. Sci. Lett.* **97**, 65-77 (1990).

15. P. Tapponnier, G. Peltzer, A.Y.L. Dain and R. Armijo. Propogating extrusion tectonic in Asia: new insights from simple experiments with plasticine. *Geology*, **10**, 611-616 (1982).

16. P. Tapponnier, G. Peltzer and R. Armijo. On the mechanics of the collision between India and Asia. In collision Tectonics, ed. M. P. Coward and A. C. Ries. Geol. Soc. London Spec. Publ. **19**, 115-157 (1986).

17. B. Taylor and D.E. Hayes. The tectonic evolution of the South China Basin, in the tectonic and geologic evolution of southeast Asian seas and islands (ed D.E. Hayes), *Geophys. Monog. Ser.* AGU. Washington, D.C. **23**, 23-56 (1980).

18. B. Taylor and D.E. Hayes. Origin and history of the South China Basin, in the tectonic and geologic evolution of southeast Asian seas and islands, Part 2 (ed. D.E.Hayes), *Geophys. Monog. Ser.* AGU. Washington, D.C. **27**, 9-104 (1983).

Proc. 30ᵗʰ Int'l. Geol. Congr. Vol. 14, pp. 196-202
Zheng et al. (Eds)

A New Model Deduced from Kalpin Transpression Tectonics and Its Implications to the Tarim Basin

HUAFU LU[1], DONG JIA[1], CHUMING CHEN[1], DONGSHENG CAI[1] SHIMIN WU[1], GUOQIANG WANG[1], LINZHI GUO[1], YANGSHEN SHI[1], DAVID GEORGE-HOWELL[2]

1)*Department of Earth Sciences, Nanjing University, Nanjing 210093, China*
2)*U.S.Geological Survey, Menlo Park, California 94051, U.S.A.*

Abstract

The redeformed thrust sheets in Kalpin Ranges, northwest Tarim, suggest the major scale sinistral strike slip of the Aheqi fault zone which is the northwest boundary of the Tarim Basin. The total amount of the Cenozoic sinistral strike slip displacement of Aheqi fault zone is estimated as 311km. It is of the same magnitude of the Altyn fault as the southeast boundary of Tarim basin. The sinistral strike slip of the Aheqi fault causes the formation of Aheqi-West Kunlun-Southwest Tarim depression-Altyn transpression tectonic system.

Keywords: Kalpin, strike slip fault, fault related contortion, crustal shortening rate, tranpression tectonics

INTRODUCTION

Kalpin Ranges are located on the northwest margin of the Tarim Basin (Fig. 1). The ranges trend in east-west direction and turn into east-northeast in its eastern part. The Pre-Sinian metamorphosed rocks[13], sedimentary strata of Sinian and Paleozoic exposed on the surface in Kalpin and Aksu (200km east to the town of Kalpin) areas. The whole Paleozoic sedimentary column of Kalpin Ranges is essentially the same as those from the drilling holes in north and central Tarim Basin[6]. Therefore the Kalpin area is part of Tarim continent geologically, and the difference is that the exposed Paleozoic strata have been uplifted 400-1000m higher than the Tarim Basin surface, while the Paleozoic strata are buried beneath more than 1,000m thick Cenozoic deposits.

The Kalpin Ranges are totally composed of thrust sheets, which comprise from Cambrian to Paleozoic strata generally dipping north. The Cambrian strata at the basal part of a sheet thrust over the Permian strata at the top part of another southerly situated thrust sheet. Therefore, the verging direction of the thrusting is obviously southward. These thrust sheets arrange parallelly with the length of 30 to 150km for a single sheet. The valleys between two sheets are usually filled with the present alluvial fans. The Miocene and Pliocene sometimes appear above the Permian strata being the top of the thrust sheets. These Tertiary sediments lie on the Permian parallelly or with only a few degrees difference, suggesting that the thrusts emplaced after or during the time of Miocene deposition. The Late Mioence-Early Pliocene strata, Kanchun formation exhibits nonmarine sandstones and mudstones, and in the Late Pliocene strata, Kuqa formation exhibits massif sandstones and conglomerates[6]. The sedimentary features of these two formations as an uniform Molasse assemblage indicate the uplifting and the thrusting of the rejuvenated Tianshan mountains during the Neogene time. The Mesozoic strata are lacking in Kalpin area except some minor

Cretaceous strata in the westmost of Kalpin Ranges, suggesting that the Kalpin area uplifted during Mesozoic, but the strata were not deformed and exhibited as a plateau with horizontal Paleozoic strata since that time until Tertiary. Then the Tianshan mountains rejuvenated since Miocene and the molasse deposition together with the thrusting started in Kalpin area.

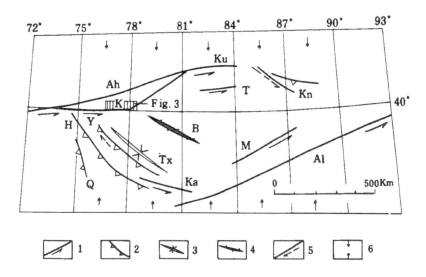

Figure 1. Aheqi-West Kunlun-Southwest Tarim depression-Altyn transpression tectonic system. Symbols: 1, strike slip fault; 2, thrusts; 3, rejuvenation foreland basin; 4, lithosphere bulge; 5, older strike slip fault; 6, regional compression direction; Ah, Aheqi fault; Al, Altyn fault; K, Kalpin faults; T, Tarim river fault; M, Mingfeng fault; Ka, Karakax fault; Ku, Kuqa faults; Y, Yecheng fault; H, Hongqilapu fault; Q, Qiaogeli fault; Kn, Kunan fault; Tx, Southwest Tarim depression; B, Eastern Bachu uplift.

In the Kalpin Ranges there are six rows but some times only two rows of thrust sheets. They were arranged straight or curved forming special micro-orocline pattern in map view.

It has been failed to explain the strange tectonic pattern by lateral ramps or tear faults. The north dipping thrust sheets result in the strata distributed regularly in each sheet, revealing an obvious polarity. For a single sheet, the Cambrian strata always locate at the bottom, the Ordovician, Silurian, Devonian, Carboniferous appear northward in sequence, and the Permian strata sometimes with a minor Neogene strata locate on the top, at the northmost of the sheet. Using the strata polarity of the sheet, considering the longitudinal strike slip faults along the thrust sheets, the mystic structure pattern of Kalpin Ranges is clearly solved, because the polarity records the complete information about the contortion processes of the thrust sheets.

MODELS OF STRIKE SLIP FAULT RELATED CONTORTION

Before explaining the tectonic pattern in Kalpin Ranges, it is necessary to define the terminology and models used below to analyze the strike slip faulting and the strike slip faults related contortions of the Kalpin thrust sheets. The structures like the folds in map view resulted from the displacement of the strike slip fault are termec the strike slip fault related contortions. The strike slip faults related contortions are divided into two groups, viz the symmetric type (Fig. 2A) and the asymmetric type (Fig. 2B, C).The symmetric type strike slip related contortion occurs in the case that the longitudinal strike slip fault is very steep or vertical, and the mechanical features of rocks on both sides of the fault are similar. In the case of the strike slip fault with a moderate or low dip angle, the emplacement of fault results in the contortion only in the hangingwall but not in the footwall, because gravitation makes it more difficult to be contorted in the footwall, which is the asymmetric type. The structures of Kalpin Ranges belong to the asymmetric type. For the symmetric type structure, there is a vertical digyre and the strike slip contortions occur on both sides of the fault.

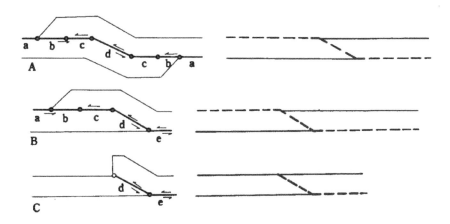

Figure 2. Model of strike slip related contortion, symmetric type (A) and symmetric type (B, C). Right part of the figure showing the situation before the emplacement of the faults. A, symmetric strike slip fault related contortion. symbols: a, front fault straight; b, front fault bend; c, middle fault straight; d, central fault bend. B, strike slip fault bend contortion. symbols: a, front fault straight; b, front fault bend; c, middle fault straight; d, rear fault bend; e, rear fault straight. C, strike slip fault propagation contortion. symbols: d, rear fault bend; e, rear fault straight.

Following Woodcock et al.[11,12], the segment of a longitudinal strike slip fault parallel to the main motion direction of the strike slip faulting is named fault straight (a, c and e in Fig. 2), while the segment of fault oblique to the motion direction of the fault is termed fault bend (d in Fig. 2). The segment b in Figure 2A, B originally is bend, but is directed straight after the displacement of faulting, it is also named fault bend. The strike slip fault straights are divided into three kinds, namely the front fault straight (a in Fig. 2A, B), the middle fault straight (c in Fig. 2A, B), and the rear fault straight (e in Fig. 2B, C). The geological bodies on both sides of all fault straights are parallel, but for the front and rear fault straights, the geological bodies have not been repeated in map view, while for the middle fault straight, the geological bodies have been repeated. Because there is no rear fault straight in the symmetric type structure, there only exist front and middle fault straights in both direction of the strike of the fault. For the asymmetric type structures, the front, middle, and rear fault straights arrange in order showing the motion direction of the contorted wall of the fault. In the asymmetric type structures, the strike slip fault bends are divided into two kinds, namely the strike slip front fault bend (b in Fig. 2B) and the strike slip rear fault bend (d in Fig. 2B,C). The rear fault bend keeps its original attitude while the front fault bend changes its attitude become parallel to the fault straights. The geological bodies on both walls of the fault bend are not parallel. In the case of the strike slip fault parallel to the thrust sheet, the strata of the hangingwall of the front fault bend are truncated by the fault and the strata of the footwall of the front fault bend is parallel to the fault plane. For the rear fault bend, the strata of the footwall are truncated by the fault and those of the hangingwall are parallel to the fault. For the asymmetric type structures, the strike slip fault bend contortion is composed of the rear fault straight, rear fault bend, middle fault straight, front fault bend and front fault straight forming micro-orocline structure of the hangingwall of the longitudinal strike slip fault (Fig. 2B). In the case that the fault displacement stops at the tip of the rear fault bend, the asymmetric type strike slip fault related contortion is named the strike slip fault propagation contortion, which is composed of the rear fault straight and the rear fault bend, and appears as an intensively asymmetric strike slip fault related contortion (Fig. 2C). For the symmetric strike slip fault related contortion, there are also the strike slip fault bend contortion (Fig. 2A) and the strike slip fault propagation contortion (not shown in Fig. 2).

THE LONGITUDINAL STRIKE SLIP FAULTS IN THE KALPIN RANGES

In the Kalpin ranges, two longitudinal strike slip faults (F_1 and F_2), two strike slip fault bend contortions (I and III) and a strike slip fault propagation contortion II of the asymmetric type structures are identified (Fig. 3).

The fault bend contortion   I is located to the north of the Kalpin valley. It is composed of the thrust sheet A, which is cut by longitudinal strike slip fault F_1 at point X in Fig. 3. To the east of point X, the fault F_1 takes the bottom surface of the thrust sheet A as the rear fault straight. At point X, the rear fault bend is formed. To the west of point X, the fault F_1 transfers to the top of the thrust sheet A, forming the middle fault straight. The segments A_1, A_2, A_3 of the sheet A are overlapped and repeated with segments A_4, A_5, A_6 (Fig. 3) in map view. The distance between the point X and X', the antipodal points, is the minimum sinistral strike slip displacement of the fault F_1, that is 80 km. The strike slip fault propagation contortion II is composed of the thrust sheets A and B. The thrust sheet B is cut by the fault F_2 at point Y (Fig. 3), and the rear fault bend is formed there. To the east of point Y, the fault F_2 occurs along the bottom of the sheet B, and to the west of point Y, the fault F_2 occurs on the top of sheet B. The sinistral strike slip movement of the fault F_2 stops at the fault tip (the open circle in Fig. 3), resulting in the recontortion of the fault bend contortion I , which then becomes part of the fault propagation contortion II . The segment B_2 of sheet B is overlapped and repeated with the segments B_3 and B_4 of the sheet B, and the distance between point Y and Y' is the minimum sinistral slip of the fault F_2 (55km) (Fig. 3). The fault propagation contortion B produces the complex bending of the sheet A and B exhibiting the distinctive micro-orocline tectonics of the Kalpin Ranges. Finally, there are several breakthrough strike slip faults which cut the contorted thrust sheets and cause some small displacement. It is clear that the fault F_1 emplaced earlier than fault F_2. The reconstruction of the fault F_1 and F_2 are shown in Figure 3B and C as restored maps.

The fault bend contortion III is located at the eastern segment of thrust sheet A outside of Figure 3, caused by the emplacement of a longitudinal fault F_3. The minimum sinistral strike slip of fault F_3. is 80km. The total amount of fault slip from F_1, F_2 and F_3 gives a composite sinistral strike slip as 215 km. Beside the thrust sheets A and B, there is at least one other thrust located to the north of sheet A outside of Figure 3, which is also cut and repeated by other longitudinal strike slip fault, with a minimum sinistral displacement of 15km. Therefore, the total amount of the sinistral strike slip in the Kalpin Ranges is 230km.

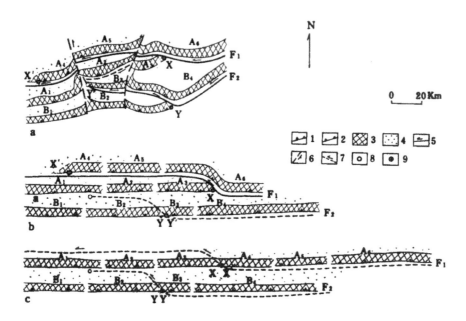

Figure 3. Kalpin strike slip tectonics of the redeformed thrust sheets (a) and restortion of the strike slip faults (b and c). Symbols: 1, thrust forming the thrust sheet A; 2, thrust forming the thrust sheet B; 3, Paleozoic; 4, Cenozoic; 5, strike slip fault F_1 and F_2; 6, breakthrough strike slip faults and antithetic strike slip faults; 7, potential fault; 8, fault tip.

The ENE trending Aheqi fault zone, being the boundary fault zone, separates the Paleozoic strata and structures of the South Tianshan and the Paleozoic, Mesozoic, Cenozoic strata and structures of the Tarim basin (Fig. 1), and is the northwest boundary fault zone of the Tarim basin. The Aheqi fault zone is a broad fault zone, the structures of the Kalpin Ranges are involved into the Aheqi fault system. The Aheqi fault zone's strike is N60°E. According to the slip line field caused by the India-Tibetan collision[8], the Aheqi fault zone falls into the position of sinistral strike slip line β. Closely located to the south of the Aheqi fault zone, the Kalpin Ranges faults get their E-W trending in their western part. Therefore, the Kalpin Ranges faults fall into the transpression[5] or restraining fault bend[12] position of the major sinistral strike slip system of Aheqi fault zone. In light of the relationship between the Aheqi fault zone and the Kalpin Ranges strike slip faults, the amount of sinistral displacement of the Aheqi fault zone is estimated as 311km[7]. According to Avouac[1,2], the wide spread Cenozoic tectonics occurred in Mid- Miocene (15-18Ma). As estimated above, the thrust emplaced during Miocene, coincides with the Avouac's idea. Taking 18Ma as the initial emplacement time of the thrusting in the Kalpin Ranges, the sinistral strike slip velocity of the Aheqi fault zone is 17mm/a ($311×10^6$mm/($18×10^6$a)).The amount of the sinistral strike slip (311km) and the slip velocity (17mm/a) of the Aheqi fault zone are close to the slip

(300km)[12] and slip velocity (30mm/a)[4,9] of the Altyn fault.

IMPLICATIONS TO THE TARIM BASIN

The Tarim basin is a rhomboid block but it is asymmetric obviously. Its northwest boundary fault (Aheqi fault) and the southeast boundary fault (Altyn fault) are much longer than its northeast and southwest boundary faults. Therefore, under the N-S compression by the India-Tibetan collision, the Aheqi and Altyn sinistral strike slip faults play more important role in the deformation mechanism. The Aheqi and Altyn faults are arranged parallel and in right step. As the sinistral strike slips occur along the two major faults, the West Kunlun mountains and the Tarim basin between them fall into the typical transpression tectonics[9].The West Kunlun becomes rejuvenated orogenic belt resulted from the transpression processes. The southwest Tarim depression is subsided by the crustal flexure resulting from the tectonic loading of the thrusting emplacement at the southwest Kunlun, being the deepest depression in the Tarim basin with more than 10km thick Neogene and Quaternary deposits. These sediments thin out toward the East Bachu uplift, the west segment of the Central Tarim uplift where the thickness of Neogene and Quaternary is only 1km. Structurally, there are extensional faults on the Central Tarim uplift. Located 300km to the northeast of the Kunlun mountains, the East Bachu is the lithosphere bulge resulted from the Kunlun thrusting and the southwest Tarim depression. Avouac[1,2] suggested that NWW trending Karakax(Fig. 1) Quaternary fault near the city of Hetian exhibited sinistral strike slip, indicating that some NW trending fault of the transpression belt also gets the nature of sinistral slip similar to the shearing sense of the Aheqi and Altyn faults which control the creation of transpression tectonics of Kunlun and Tarim. Therefore, it is clear that there exists the great Aheqi-Kunlun-Southwest Tarim-Altyn sinistral transpression tectonic system.

Avouac[1,2], Chen et al.[3] estimated that the Tianshan shortened in N-S direction for 420km since 15Ma. The 311km sinistral displacement of the Aheqi fault could give rise to the effect of 155km N-S shortening, which is equal to 37% of the 420km. This is much more than the amount suggested by Sengor[9]who estimated that 20% convergence would become strike slip in Himalayan. That implies that the strike slip is more important for crustal shortening than people thought. Probably, due to the lacking of comparable coordinate system and obvious comparing indicators, this kind of structures is still not being attended very much. But because the gravitation and the crustal buoyancy limitation for the thrusting in the case of continental collision, the extrusion of the crustal blocks seems to have much freedom, then the magnitude of the strike slip is as important as the thrusting, in some case, even more important than the latter.

Acknowledgments

This contribution is a part of results of the Eighth Five-Year Plan Research Program, Tarim Petroleum Province Tectonics, entitled "Study on the Structural Features of West Segment of North Margin, Kalpin and Yingmaili Areas of Tarim Plate". We thank China National Petroleum Corporation, Tarim Petroleum Exploration and Development Bureau(TPEDB) and Dr. C. Jia for their help in our study. We also thank Professor R. Wang and Professor Y. Zheng for their careful review and valuable suggestions to our manuscript.

REFERENCES

1. J.P. Avouac and P. T. Tapponnier. Kinematic model of active deformation in Central Asia. Jour. Geophs. Res. Lett. **70**, 895-989(1993).
2. J.P. Avouac. Applications des methodes de morphologie quantitative al a neotectonique: Modele cinematuqie des deformations actives en Asia Centrale. In: Ph.D. Thesis, Univ. de Paris(1991).
3. Y. V. Chen, J.P. Coourtillot, J.P. Cogne et al.. The configuration of Asia, prior to the collision of India: Cretaceous Paleomagnetic Constraints. Jour. Geophs. Res. **98:B12**, 2127-2194(1993).
4. M. Coward. Continental collision. In: *Continental Deformation*. P. L. Hancock(Ed). Pergamen Press(1994).
5. W.B. Harland. Tectonic transpression in Caledonian Spitzbergen. Geo. Mag. **108**, 27-42(1971).
6. C. Jia, H. Yao, J. Gao et al.. Stratigraphical system of the Tarim basin. In: *Symposium on the Oil and Gas Exploration in Tarim basin*. X. Dong and D. Liang (Eds). pp. 34-83. Xingjiang Science-Technology and Hygiene Press(1992)(in Chinese with English abstract).
7. H. Lu, D.G. Howell, D. Jia et al.. Kalpin transpression tectonics, Northwestern Tarim basin, Western China. Inter. Geol. Rev. **36**, 975-981(1994).
8. P. Molnar and P.G. Tapponnier. The relation of the tectonics of eastern Asia to the India-Eurasia collision: an application of slip line field theory to large scale continental tectonics. Geology. **5**, 212- 216(1977).
9. A.M.C. Sengor. Plate tectonics and orogenic research after 25 years: a Tethyan perspective. Earth Sci. Rev. **27**, 1-201(1990).
10. P.G. Tapponnier, G. Petzer and R. Armijo. On the mechanics of the collision between India and Asia. Geol. Soc. London Spec. Publ. **19**, 115-157(1986).
11. N.H. Woodcock and C. Schubert. Continental strike slip tectonics. In: *Continental Deformation*. P.L. Hancock(Ed). pp. 251-263. Pergamen Press(1994).
12. N.H. Woodcock and M. Fischer. Strike-slip duplexes. Jour. Struc. Geol. **8:7**, 725-735(1986).
13. X. Xao, Y. Tang, Y. Feng. *Geotectonics of North Xingjiang and Adjacent Area*. Geological Publishing house, Beijing(1992)(in Chinese with English abstract).

Proc. 30th Intl. Geol. Congr., Vol. 14, pp.203-216
Zheng et al. (Eds)
© VSP 1997.

Three-Dimensional Models of Transpressional Structures in China Coal Mine

GUILIANG WANG, HAIQIAO TAN, BO JIANG, HAI ZHOU

College of Mineral Resource and Environment Sciences, China University of Mining and Technology, Xuzhou, Jiangsu 221008, China

D. C. P. PEACOCK

Department of Geological Sciences, University of Plymouth, Drake Circus, Plymouth, PL4 8AA, U. K.

FENGYING XU

China National Petroleum Corporation, Beijing, 100724, China

Abstract

A range of transpressional structures in coal mines is described, for which three-dimensional data are available. Each structure is in a region which has undergone transpression. Differences between the structures are related to differences in the ratio of pure shear to simple shear. The fault network in Longquan coal mine, Shandong Province is dominated by contraction. The tight and narrow en echelon folds in Nantong coal mine, Sichuan Province and the oblique-slip duplex in Jiulonggang coal mine, Anhui Province show approximately equal amounts of simple shear and pure shear. The positive flower structure in Longwangdong coal mine, Sichuan Province and the faults and folds in Jingxing coal mine, Hebei Province are dominated by shear. Sigmoidal structure in Baishan coal mine, Anhui Province was caused by transpression and a simultaneous torsion, with a much higher proportion of shear than contraction. Examples are given of similar transpressional structures elsewhere in the world.

Keywords: transpressional structures, three-dimensional models, ratio of pure shear to simple shear, China coal mine

INTRODUCTION

Deformation by a combination of transcurrent shear and horizontal shortening across a shear plane is termed transpression by Harland [14] and by Sanderson and Marchini [24]. Transpression can occur locally, such as at restraining bends on strike-slip faults, and can also occur at a regional scale. S. Li describes transpressinal structures and recognizes a series of structural types which were formed by transpression and which form hydrocarbon traps [12, 20, 28,13].

The aim of this paper is to describe the three-dimensional geometries and evolution of six transpressional structures in Chinese coal mines. Coal mines are described because of the three-dimensional data available, but the role of coal in the deformation is not discussed. The regional context of each structure is given, each structure having been formed during regional transpression. A range of structures are described that show different ratios of shear to contraction. Each structure has good three-dimensional control. The structures described are the fault network at Longquan, the en echelon folds and faults at Nantong, the oblique-slip duplex at Jiulonggang, the positive flower structure at Longwangdong, the faults and folds at Jingxing, and the sigmoidal structure at Baishan.

Similar structures are common in coal mines in China and in coal mines elsewhere in the world. Understanding the geometries of such transpressional structures has, therefore, great economic importance.

REGIONAL GEOLOGY OF THE AREAS DESCRIBED

An overall account of the geology of China is given by Yang et al.[42], while Wang et al.[37] give a table of the main deformation events in eastern China. Figure 1 shows the locations of the areas described in this paper, with Figs. 2 to 7 giving more detailed maps.

Late Triassic (Indosinian Orogeny) to early Jurassic transpression along the southern boundary of the North China Block caused localized shear during overall contraction. This produced the bounding fault of the oblique-slip duplex in Jiulonggang Coal Mine.

The Tancheng-Lujiang fault system (Fig. 1) consists of a series of NNE-striking sinistral strike-slip faults in eastern China, along the margin of the Pacific Ocean [40]. Chen shows that the Tancheng-Lujiang fault system developed during Yanshan deformation of the late Cretaceous [6], during which the Pacific Plate was subduced NNW-wards beneath mainland China [19]. Yin and Nie [43 Figs. 2 and 6] present a model for the indentation of the South China Block into the North China Block between early Permian and late Jurassic time. This indentation caused sinistral transpression on the Tancheng-Lujiang Fault Zone, which has a maximum displacement of about 540 km. Transpression related to the Tancheng-Lujing fault system produced structures at the coal mines at Longquan, Jingxing and Baishan.

Oblique subductuon of the Pacific Plate caused the Xishan deformation during the Eocene to Neogene [3]. This movement produced local sinistral transpression when it interacted with preexisting NE- and N-striking structures, such as the Huayingshan Changshou faults respectively (Fig. 1). Such local transpression occurs at the Nantong and Longwangdong coal mines.

FAULT NETWORK IN LONGQUAN COAL MINE, SHANDONG PROVINCE

The Longquan coal field lies west of the Tancheng-Lujiang fault on the east limb of the 10° - trendings Zibo syncline, at Zibo in Shandong Province. Sinistral shear is related to the Zihe fault, which was initiated as a strike-slip fault in the Cretaceous and which is part of the Tancheng-Lujiang fault system. Wang et al. show that a network of extensional and strike-slip faults developed during the Jurassic to Cretaceous in the western part of Shandong Province [37].

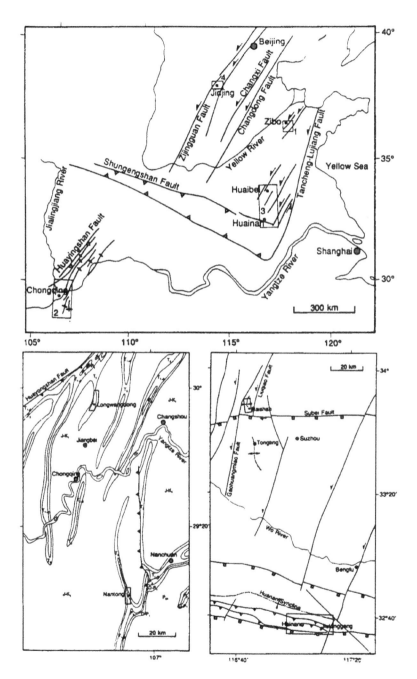

Figure 1 Regional geological map. 1a: Map of eastern China showing the location of the transpressional structures in coal mines which are describe in this paper. 1=Fault network in

Longquan coal mine, Shandong Province (Fig.2) : 2=En echelon structure in Nantong and the positive flower structure in Longwangdong coal field,Sichuan Province (Fig. 1b): 3= Oblique-slip duplex in Jiulonggang and sigmoidal structure in Baishan coal mine, Anhui Province (Fig. 1c): and 4= Faults and folds in Jingxing coal mine,Hebei Province (Fig. 6) . 1b : Map of the eastern part of Sichuan Province showing the location of the Nantong and Longwangdong area. 1c: Map of the northern part of Anhui Province, showing the Jiulonggang and Baishan areas.

The eastern border of the Longquan coal field is formed by the Zihe and Shangwujing faults [25, 4, 40]. In the coal field, a monocline of coal-bearing Permo-Carboniferous rocks that strikes between 30° and 40° , and that dips towards the NW at between 10° and 12° ,has been cut by various types of faults (Fig. 2). 704 faults have been located within the mine, which has an area of 6. 67km². Four groups of fault occur. A reverse fault dips gently towards 300° . The second group of faults strike approximately E-W and are transpressional. The third set have a NNW-strike, are transtenssional and have steep dips. The fourth group of faults are extensional and are

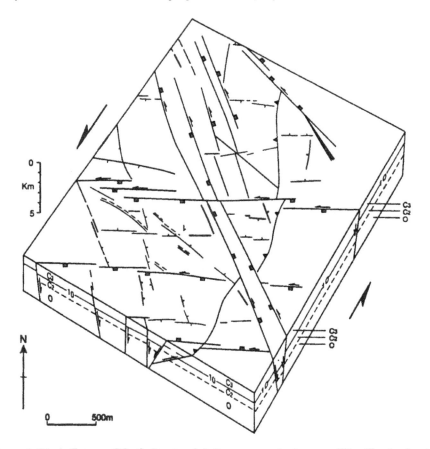

Figure 2. Block diagram of the fault network in Longquan coal mine, near Zibo. The dominant stress system was approximate NW-SE contraction with some N-S sinistral shear. C_3=Upper Carboniferous, C_2=Middle Carboniferous, O=Ordovician.

the most common, having average spacings of 380 m and 520 m respectively. They are sinistral, and dip towards 315° . The E-W- and NNW-striking faults (second and third groups) sinistral and are typically en echelon in map view, and intersect each other to cut the rock into rhombs. The acute angle of the rhombs·is towards 315° , which is approximately parallel to the bed dip direction [33].

The faults at Longquan represent a pure shear strain without rotation, although they formed by NW-SE contraction and N-S sinistral shear. This is indicated by the sinistral displacement on most of the faults, and by the contraction on the E-W faults and extension on the NNW-striking faults.

A similar pattern of deformation is shown by Shiells for faults at Dunstanburgh, Northumbria, U. K. , where a network of strike-slip faults, thrusts and folds occur [26].

EN ECHELON FOLDS IN NANTONG COAL FIELD

The Nantong coal field lies in the southern part of Sichuan Province (Fig. 1b), within the 10km long Nantong anticline. Its deformation occurred during the Xishan deformation of the Eocene to Neogene [3, 44, 30]. E-W contraction and sinistral shear were caused by regional NW-SE compression on N-S striking boundaries of the deformation zone.

The Nantong anticline is convex-eastwards, with the N-S trend in the central part and becoming NNE and SSE in the north and south respectively (Fig. 1b). It plunges northwards, where it decreases in width and amplitude. The anticline is composed of many smaller folds. In the middle and southern parts of the Nantong anticline are the Wuguishan and Miaoding anticlines, separated by the Pingtu syncline. Six smaller en echelon anticlines occur in the Wuguishan anticline. These third-order folds have an approximate NNW-trend and are right-stepping en echelon. Further south, however, the bending of trend towards the NNE causes left-stepping of the Shizhucao, Yuanyangshan and Tianchi anticlines. The oblique-slip contractional faults in the west limbs and axial regions of these anticlines are also en echelon. NW-striking strike-slip faults and approximately E-W striking extensional faults also occur (Fig. 3).

The N-S trend of the Nantong anticline implies approximate E-W contraction. In the region between Longwangdong and Nantong, however, a series of right-stepping en echelon anticlines (Fig. 1b) indicates N-S sinistral shear. This shear and contraction caused the convex-eastwards bend in the Nantong anticline and the right-stepping geometry of many of the larger (Fig. 1b) and smaller(Fig. 3) folds. Left-stepping en echelon folds and faults formed in the northern section of the anticline because of sinistral transpression, while left-stepping folds and faults formed in the southern section because of the local dextral transpression caused by the convex shape of the anticline. This mechanism is demonstrated by Wang, who uses small structures to analyses palaeostress orientations [34]. The low angle (5° to 15°) between the main structure(Nantong anticline) and the minor en echelon folds and faults indicates that contraction was more important than shear.

There are several detailed descriptions of structures similar to those at Nantong. Shiells [26, Fig. 15] gives a detailed map of en echelon folds with related thrusts and strike-slip faults at Spittal, Northumbria. Harding describes similar geometries in the Newport-lnglewood fault trend, California, where a series of en echelon folds, strike-slip faults and thrusts are related to dextral transpression [10, 11]. Nickelsen gives a detailed description of a similar pattern of deformation

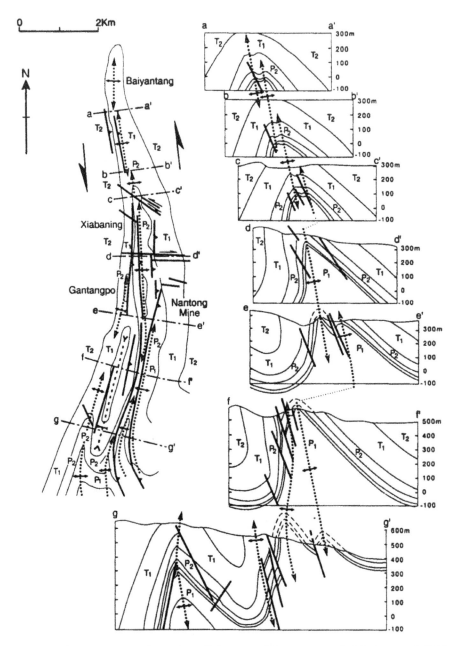

Figure 3. Map and cross-sections through the en echelon structure in Nantong coal field, Chongqing, Sichuan Province. The structure formed in approximate E-W contraction, with localized sinistral or dextral shear. T_2=Middle Triassic, T_1=Lower Triassic, P_{1f}=Upper Permian, P_2=Lower Permian, S=Silurian.

in coal-bearing deposits in Pennsylvania, where thrusts, strike-slip and normal faults occur within
en echelon folds [23]. Gayer and Nemcok describe Variscan transpression in SW Britain, which is
dominated by contraction, and where the South Wales coal field shows a similar deformation
pattern to the Nantong coal field [7].

OBLIQUE-SLIP DUPLEX IN JIULONGGANG COAL MINE

Jiulonggang coal mine is at Huainan, Anhui Province (Fig. 1c), within the thrust zone which is
north of the collision belt between the North China and South China blocks. The thrust zone
shows sinistral transpression and is dominated by faulting (Fig. 4). The Sungengshan thrust
passes through Jiulonggang mine, and a part of this thrust is here termed F_1. It dips at between
33° and 75° towards 200° , gently at depth and in the west, and steeply at shallow levels and
in the east. F_1 has a reverse and sinistral displacement and generally strikes parallel to Permo-
Carboniferous bedding. Three groups of fault branch off F_1 into the hanging-wall. The first

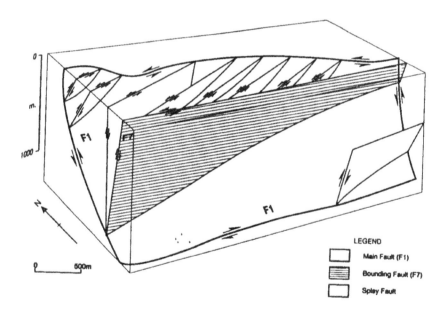

LEGEND

Main Fault (F1)

Bounding Fault (F7)

Splay Fault

Figure 4.Block diagram of the oblique-slip duplex in Jiulonggang coal mine. The structure
formed in approximate E-W dextral shear.

also reverse and sinistral, and dip in the opposite direction to F_1. They intersect F_1. in both the
horizontal and vertical planes at angles of 20° -50° .The third group of branch faults are thrusts
with some strike-slip displacement, which are synthetic to F_1. They occur in the west of the area
at a southward bulge of F_1 (Fig. 4). Faults F_1 and F_7 formed during N-S contraction during the
early or middle Jurassic. The oblique-slip duplex formed during sinistral transpression later in the
Jurassic to Cretaceous [34, 36, 15].

The fault system at Jiulonggang coal mine is a complex network of faults formed by sinistral transpression. The structure can be called an oblique-slip duplex [39]. Several examples of strike-slip duplexes are given by Woodcock and Fischer [39]. Keller et al. give a detailed description of a strike-slip duplex at a contractional overstep along the Carboneras fault, SE Spain [16].

POSITIVE FLOWER STRUCTURE IN LONGWANGDONG MINING DISTRICT

Longwangdong Mining District is Jiangbei, Sichuan Province (Fig. 1b). The structure is a long and narrow NNE-striking anticline at the surface, which is a branch of the Huayingshan anticline. Other NNE-striking right-stepping en echelon anticlines occur in the region. Like the en echelon folds in Nantong coal field, the flower structure at Longwangdong is developed during the Eocene to Neogene Xishan movements.

Folding at Longwangdong is developed above a positive flower structure [12, 1], which consists of contractional and sinistral strike-slip faults (Fig. 5). The structure contains Upper Permian and Lower Triassic coal deposits, and the Xiangguosi as field [17].The oldest rocks exposed are the Upper Triassic coal-bearing Xujiahe Formation, but Silurian rocks occur at depth in the core of the fold.

The fold axis trends N-S in the north and south, but 30° in the south. The anticline is asymmetric, with the west limb being steeper than the east limb. Few faults occur at the surface, but they are common at depth. Four reverse fault zones occur in the west limb, with two in the east limb. These zones strike parallel to the fold axis, and dip towards the hinge. They cut the Upper Palaeozoic to Middle Triassic rocks into long, narrow horses that broaden upwards. The reverse fault with the longest trace lengths occurs in the west limb. It has a trace length which is nearly as long as the fold axis, and it is the furthest fault from the core. Some sinistral displacement occur on the faults, which have a maximum throw of between 640 m and 680 m.

The right-stepping structures in the upper parts of the fold, and the component of sinistral displacement on the reverse faults, show that the positive flower structure was caused by sinistral transpression. The structure probably occurs above a reactivated basement fault.

Examples of similar transpressional structures occur in the east and south of Sichuan Province [31, 41], these often forming important coal mining areas and gas traps. Genna and Debriette describe flower structures in the Ales coal basin, France [8].

FOLDS AND FAULTS IN JINGXING MINING DISTRICT

Jingxing Mining District is at the eastern foot of the Taihang Mountains, Hebei Province (Fig.1a). Deformation at Jingxing is related to sinistral transpression on the Tancheng-Lujiang fault system, which developed during the Yanshan deformation [40]. A series of N-S striking horsts and grabens occur in the Carboniferous and Permian coal-bearing beds along the eastern edge of the Taihang Mountains and on the eastern side of the Zijingguan fault. Jingxing Mining District is in one of t hese horsts (Fig.6). The structure consists of a series anticlines and synclines trending between 30° and 40° , the anticlines having wavelengths of about 500 m.One or two

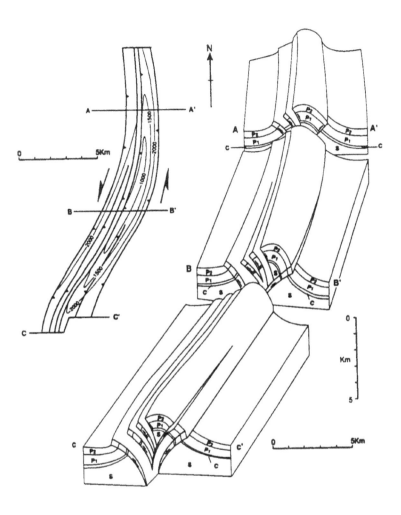

Figure 5. Map and block diagram of the positive flower structure in Longwangdong coal field, Jiangbei, Sichuan Province. See Fig. 1b for a map of the region. The structure formed in approximate NNE-SSW sinistral shear. P_2=Upper Permian, P_1=Lower Permian, C=Carboniferous, S=Silurian.

reverse faults typically occur on one or both limbs of each anticline, these being parallel to the fold axial traces. A series of NW-striking sinistral faults occur perpendicular to the fold axes. There are also many conjugate strike-slip faults striking approximately N-S to WNW-ESE, which complicate the structure [32, 33].

This structure (a ξ-type structure using the terminology of Li [18]) formed by sinistral shear, and accompany with some contraction. The eastern border of the structure was displaced northwards relative to the western border, with sinistral shear and approximately E-W

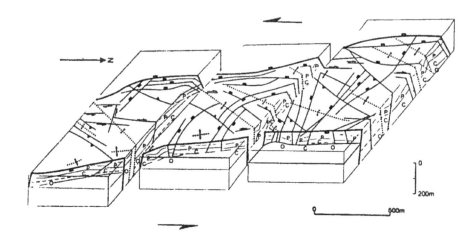

Figure 6. Block diagram of the folds and faults in the western part of the coal mine at Jingxing. The structure formed in approximate N-S sinistral, with some contraction. P=Permian, C=Carboniferous, O=Ordovician.

contraction. Evidence that sinistral simple shear dominated, with contraction being relatively minor, includes: 1)The fold axes and reverse faults strike at nearly 40° to the shear direction. 2) Small secondary NNW-and approximately N-striking anticlines occur in the east of the area. 3) Approximately N-striking vertical faults represent compressional characteristic in the west of the area. The structure at Jingxing was therefore formed mainly by sinistral simple shear, but the latter two facts shows that it has occurred compression at the same time or later stage. This contrasts with the structures at Nantong and Jiulonggang, which were formed mainly by contraction with some shear. This shear is related to the Zijingguan fault in the west , which are dominated by sinistral strike-slip(Fig. 1a)[40, 21].

Many areas have been described which have similar geometries to the Jingxing structure. Examples include the Los Angeles Basin, California [38, Fig. 4], New Zealand ([2], Figs. 5 and 6), and the Buller coal field, New Zealand ([2], Figs. 3 and 8). Each of these areas show folds, thrusts, normal faults and antithetic strike-slip faults developed in a transpressive regime.

SIGMOIDAL STRUCTURE IN BAISHAN COAL MINE

Baishan coal field is at Suixi in northern Anhui Province, in the foreland of the Xuzhou-Suzhou arcuate thrust system [35]. The field lies north of the W-striking Subei fault, west of the Liuqiao fault and east of the Gaohangmiao fault [5, Fig. 1c]. Deformation at Baishan is related to sinistral transpression on the Tancheng-Lujiang fault system, which developed during the Yanshan deformation.

The main structure at Baishan is an approximately N-trending syncline which has a sigmoidal axial trace (Fig. 7). The axis strikes at between 20° and 40° in the north, at 160° in the middle part, and between 15° and 50° in the south. The axial plane dips eastward in the north, westward in the south, and is approximately vertical in the central part. This variation is reflected in bed dips, with the east limb being steepest in the north and gentlest in the south. Bed dips range between 5° and 25°

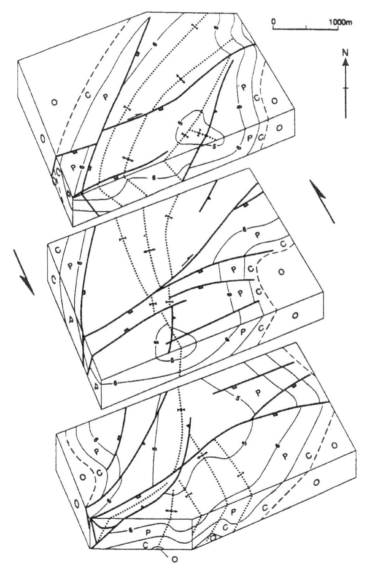

Figure 7. Block diagram of the sigmoidal structure in Baishan coal mine, Suixi, Anhui Province. The structure formed in approximate NW-SE sinistrsl shear. See Fig. 1c for a map of the region.

Secondary structures occur in the core of the syncline. An anticline and two synclines occur in the north, with three synclines and two anticlines occurring in the south. Undulations occur in the hinge of the major syncline, which typically has an anticline in the core, giving it a W-shape. Faults of various sizes and strikes cut the syncline, with about 250 faults having been located in the 22.5 km² mining area. Three major reverse faults occur parallel to the fold axes, indicating they formed in the same stress field that formed the folds. The main faults which cut across the fold are extensional and dextral oblique-slip faults with strikes of between 50° and 70° and with steep dips. These ENE-striking faults are often en echelon, and have a spacing of about 2 km.

The structures in the Baishan district indicate approximate E-W contraction, followed by NW-SE sinistral shear. These stresses caused the sigmoidal syncline and the set of ENE-striking extensional and sinistral oblique-slip faults. Sinistral transpression in this region is related to sinistral displacements on the sinitral Liuqiao and Gaohuangmiao faults (Fig. 1c), these being part of the Tancheng-Lujiang fault system. The contraction from W to E was strongest in the south, because of its proximity to the dextral shear of Subei fault. This sigmoidal transpressional structure represents strong simple shear at the later stage of contraction, which caused anti-clockwise rotation (Fig. 1c, Fig 7).

Similar examples include the Enkou-Doulishan sigmoidal structure in Hunan Province [9], and the Tongguanshan sigmoidal structure in Anhui Province [29]. Other possibly similar examples occur in California [38,Fig.4], and the Midland Valley of Scotland [22, Fig. 8].

CONCLUSIONS

The structures described in this paper are well-constrained in three-dimensions because they occur in coal mines. They are all in areas that have undergone regional transpression. The structures are of economic importance, and provide good examples of the range of different geometries that can be produced by transpression. They vary from being dominated by contraction to being dominated by shear.

The Longquan fault network formed in an environment approaching pure shear. The length of each structure in the Nantong area is large in proportion to the amount of overstep, indicating that contraction was more important than shear, with shear being influenced by the arcuate borders of the zone. Sinistral shear occurs in the north, and dextral shear in the south. The Jiulonggang oblique-slip duplex had approximately equal amounts of shear and contraction, with the vertical and horizontal displacements being approximately equal.

The Longwangdong positive flower structure and the Jingxing structure were formed mainly by shear, with some contraction. The Longwangdong structure was probably controlled by the reactivation of basement faults, while the Jingxing structure was generated by the folding and faulting of sediments at a shallow level without the influence of pre-existing structures. The Baishan sigmoidal structure involved both contraction and shear, with shear being dominant. As the structure evolved, contraction decreased and shear increased.

ACKNOWLEDGEMENTS

Work for this paper was funded by the National Nature Science Foundation of China. Some of the diagrams were drawn by John Abraham. T. Engelder, R. Groshong and Gregory A.Davis and Yadong Zheng are thanked for their comments.

REFERENCES

1. K.T. Biddle and N. Christie-Blick. Glossary-strike-slip deformation, basin formation, and sedimentation. In: *Strike-Slip Deformation, Basin Formation, and Sedimentation.* K. T. Biddle and N. Christie-Blick (Eds). Soc. Econ. Mineral. Spec. Publ. **37**, 375-386 (1985).

2. D.J. Bishop. Neogene deformation in part of the Buller Coalfield, Westland, South Island, New Zealand, *New Zealand J. Geol. Geophys.* **35**, 249-258. (1992).

3. Bureau of Geology and Mineral Resources of Sichuan Province (Eds). *Regional Geology of Sichuan Province.* Geological Society Publishing House, Beijing (in Chinese). (1987).

4. Bureau of Geology and Mineral Resources of Shandong Province (Eds). *Regional Geology of Shandong Province.* Geological Society Publishing House, Beijing (in Chinese). (1987).

5. Bureau of Geology and Mineral Resources of Anhui Province (Eds). *Regional Geology of Anhui Province.* Geological Society Publishing House, Beijing (in Chinese). (1987).

6. P.J. Chen. Time and framework of the huge strike-slip displacement along the Tancheng-Lujiang faults, *Chinese Sci. Bull.* **4**, 289-293. (1988).

7. R.A. Gayer and M. Nemcok. Transpressionally driven rotation in the external Variscides of south-west Britain, *Proc. Ussher Soc.* **8**, 317-320 (1994).

8. A. Genna and P.J. Debriette. Flower structures in the Ales Coal Basin-structural implications, *Comptes Rendus Acad. Sci.* **318**, 977-984 (1994).

9. X.G. Han and Z.H. Li. S-type structure in Enkou-Doulishan District, *Collect. Geomech.* **4**, 120-131 (in Chinese) (1977).

10. T.P. Harding. The Newport-Inglewood trend, California-an example of wrenching style of deformation, *Bull. Am. Assoc. Petrol. Geol.* **57**, 97-116 (1973).

11. T.P. Harding. Tectonic significance and hydrocarbon trapping consequences of sequential folding synchronous with San Andreas faulting, San Joaquin Valley, California, *Bull. Am. Assoc. Petrol. Geol.* **60**, 358-378 (1976).

12. T.P. Harding and J.D. Lowell. Structural styles, their plate-tectonic habits and hydrocarbon traps in petroleum provinces, *Bull. Am. Assoc. Petrol. Geol.* **63**, 1016-1058 (1979).

13. T.P. Harding. Identification of wrench faults using subsurface structural data : criteria and pitfalls, *Bull. Am. Assoc. Petrol. Geol.* **74**, 1590-1609 (1990).

14. W.B. Harland. Tectonic transpression in Caledonian Spitzbergen, *Geol. Mag.* **8**, 27-42 (1971).

15. B.Jiang. Thrust imbricate fan tectonic system of Huainan *Coal Field, Coal Geol. Explor.* **6**, 12-17 (1993).

16. J.V.A. Keller, S.H. Hall, C.J. Dart and K.R. McClay. The geometry and evolution of a transpressional strike-slip system-the Carboneras Fault, SE Spain, *J. Geol. Soc. Lond.* **152**, 339-351 (1995).

17. G. Li, Y. Luo and M. Li (Eds). *Atlas of Gas Fields in China.* Petroleum Press (1987).

18. S. Li (Eds). *Outline of Geomechanics.* Science Press (in Chinese) (1962).

19. S.T. Li, S.G. Yang, C.L. Wu and S.T. Cheng (Eds). Geotectonic background of the Mesozoic and Cenozoic rifting in east China and adjacent regions.In: *Tectonopalaeogeography and Palaeobiogeography of China and Adjacent Regions.* China University of Geosciences Press, Wuhan (in Chinse) (1990).

20. J.D. Lowell. *Structural Styles in Petroleum Exploration.* Oil and Gas Consultants International, Tulsa (1985).

21. X.Y. Ma (Eds). *Lithospheric Dynamics Atlas of China.* China Cartographic Publishing House (1989).

22. A.M. McCoss. Restoration of transpression/transtension by generating the three-dimensional segmented helical loci of deformed lines across structure contour maps,. *J. Struct. Geol.* **10**, 109-120 (1988).

23. R.P. Nickelsen. Sequence of structural stages of the Allegheny Orogeny, at the Bear Valley Strip Mine, Shamokin, Pennsylvania, *Am. J. Sci.* **279**, 225-271 (1979).

24. D.J. Sanderson and W.R.D. Marchini. Transpression, *J. Struct. Geol.*, **6**, 449-458 (1984).

25. Shandong Geological Bureau (Eds). *Notes on the Geological map of Pre-Neogene Bedrock in Shandong Province.* Geological Publishing House, Beijing (in Chinese) (1983).

26. K.A.G. Shiells. The geological structure of north-east Northumberland, *Trans. Roy. Soc. Edinburgh* **18**,449-481 (1964).

27. K.B. Sporli. New Zealand and oblique-slip margins: tectonic development up to and during the Cenozoic. In : *Sedimentation in Oblique-Slip Mobile, Zones,* P. F. Ballance and H. G. Reading (Editors).Spec. Publ. Int. Ass. Sediment **4** , 147-170 (1980).

28. A.G. Sylvester. Strike-slip faults, *Geol. Soc. Am. Bull.* **100**,1666-1703 (1988).

29. Y.A. Tan and D.X. Li. S-type folding and its simulated test: the case study of the Tongguanshan Anticline, *Modern Geology* **1**,112-122(in Chinese) (1987).

30. C.G. Tong (Eds). *Structural Evolution and Oil-Gas Gathering of Sichuan Basin.* Geological Publishing House, Beijing (in Chinese) (1992).

31. C.G. Tong and L.J. Zhang. Regional structural features and traps in south-eastern Sichuan and their relationship to the distribution of gas, *Experimental Petrol. Geol.* **13**,155-165 (in Chinese) (1981).

32. G.L Wang. Predicting structures in mines using geomechanics, *Coal Geol. Explor.* 5 (in Chinese) (1976).

33. G.L. Wang. Forecasting mine structures, *Colllect. Geomech.* **4**,166-180 (in Chinese) (1977).

34. G.L. Wang. Investigation of a disharmonic pitching anticlinorium of sine curve type in Nantong Coal Mine, *J. China Inst. Mining Tech.* **2**,70-82 (in Chinese) (1979).

35. G.L. Wang, B. Jiang, H.Q. Tan and D.C.P. Peacock. The Xuzhou-Suzhou arculate thrust system, *Journal of Southeast Asian Earth Science,* in review (1996)

36. G.L Wang, B. Jiang, Z.B. Xu, D.Z. Liu and S.X. Yan (Eds). *Thrust Nappe, Extensional Gliding Nappe and Gravity Gliding Structures in the Southern Part of North China.* China University of Mining and Technology Press, Xuzhou (in Chinese) (1992).

37. G.L. Wang, B. Jiang, Z. Yu and S. Yan. The system of extensional structures developed in the late Yanshanian orogeny in the west of Shandong Province, China, *Tectonophysics* **238**,217-228 (1994).

38. R.E. Wilcox, T.P. Harding and D.R. Seely. Basic wrench tectonics, *Bull. Am. Assoc. Petrol. Geol.* **57**,74-96 (1973).

39. N.H. Woodcock and M. Fischer. Strike-slip duplexes, *J. Struct. Geol.* **8**,725-735(1986).

40. J.W. Xu, G. Zhu, W.X. Tong, K.R. Cui and Q. Liu. Formation and evolution of the Tancheng-Lujiang wrech fault system: a major shear system to the northwest of the Pacific Ocean, *Tectonophysics* **134**,273-310 (1987).

41. K.S. Yang and K.N. Wang. Strike-slip fault zones and associated flower structures, *Petrol. Seismo-Geol.* **4**,11-27 (in Chinese) (1992).

42. Z. Yang, Y. Cheng and H. Wang (Eds). *The Geology of China.* Clarendon Press, Oxford (1986).

43. A. Yin and S. Nie. An indentation model for the north and south China collision and the development of the Tan-Lu and Homan fault systems, eastern Asia, *Tectonics* **12**,801-813 (1993).

44. Z.W. Zhang (Eds). Structural features and evaluation of the oil-and gas-bearing Sichuan Basin. *In: Symposium on Structural Features of Oil and Gas-Bearing Areas in China.* The Petroleum Industry Press, Beijing (in Chinese) (1989).

Proc. 30ᵗʰ Int'l. Geol. Congr., Vol. 14, pp. 217 - 228
Zheng et al. (Eds)
© VSP 1997

Displacement, Timing and Tectonic Model of the Tan-Lu Fault Zone

GUANG ZHU AND JIAWEI XU

Department of Geology, Hefei University of Technology, Hefei, 230009, China

Abstract

Recent studies further confirm that the maximum horizontal displacement of the Tan-Lu fault zone is over 700 km. Rocks and strata offset or controlled by the Tan-Lu fault zone as well as isotopic dating suggest that large-scale sinistral strike-slip movement in the fault zone took place near the end of Late Jurassic to Early Cretaceous interval, i. e. 110—140 Ma. Many studies demonstrate that the Tan-Lu fault zone is an intracontinental transcurrent fault zone with hundreds of km of displacement. Its development is closely related to oblique motion of the Izanagi Plate in the Pacific region.

Keywords: Tan-Lu fault, displacement, timing, tectonic model, intracontinental transcurrent fault

INTRODUCTION

The NNE-trending Tancheng-Lujiang (Tan-Lu) fault zone is one of the largest continental strike-slip faults in the world. It occurs mainly in Eastern China. In the south, its main part terminates at Guangji County to the north of Yangtze River. It then extends dispersively to Southern China. In the north, the fualt zone continues as the Sungari-Kukansky fault [19] in the Far Eastern Russia and then extends into the Sahalin Gulf of the Okhotsk Sea. The total length of the Tan-Lu fault zone is about 3600 km [22]. Since the fault zone was discovered by one of the authors [23 - 24] and others in 1957, intense research had been carried on the fault zone by many workers. The fault zone has been widely recognized as a major sinistral strike-slip fault zone with hundreds of km of displacement [1, 3, 6, 8, 11, 15, 22 - 30, 37]. This paper presents our recent studies on maximum displacement, timing and tectonic model of the Tan-Lu fault zone.

MAXIMUM DISPLACEMENT

After studies by many workers [22, 36], it now is clear that the orogenic belt between the North China Block (NCB) and South China Block (SCB), the Dabieshan-Jiaonan orogenic belt, was sinistrally displaced by the Tan-Lu fault zone (Fig. 1). This offset indicates about 500 km displacement of the Tan-Lu fault zone. Controversies remain on maximum displacement on the Tan-Lu fault. The authors once suggested that the displacement of the Dabieshan-Jiaonan orogenic belt is not the maximum displacement on the Tan-Lu fault

zone, and the maximum displacement is over 700 km on the basis of correlation between
the West Shandong and North Liaoning massifs on both sides of the Tan-Lu fault zone
[22]. However, some workers [20] argued that offset of the Dabieshan-Jiaonan orogenic
belt represents the maximum displacement. According to our recent studies, more and
more facts show that the West Shandong and North Liaoning massifs have similarities in
many aspects.

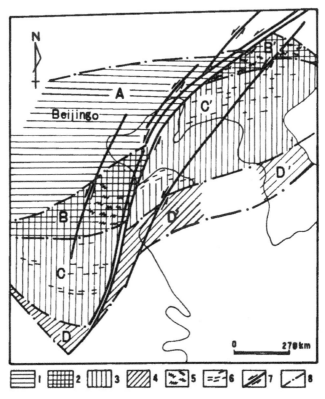

Figure 1. General tectonic map showing sinistral offset by the Tan-Lu fault zone. 1 = Archaean-Middle
Proterozoic basement with Middle Proterozoic cover; 2 = Middle-Late Archaean basement with tourmaline-
bearing rocks, locally with Early Proterozoic basement; 3 = Late Archaean-Early Proterozoic basement; 4 =
Archaean-Proterozoic basement in the orogenic belt; 5 = foliation strikes in Archaean rocks; 6 = foliation strikes
in Proterozoic rocks; 7 = strike-slip faults; 8 = block boundary; A = East Hebei-West Liaoning Massif; B-B′ =
West Shandong-North Liaoning massifs; C-C′ = Huaihe-Jiaoliao (South Liaoning, Jiaodong) massifs,
extending to Nangrim and Pyeongnam areas of Korea; D-D′ = Dabieshan-Jiaonan collision belt, extending to
the Kyonggi Massif of Korea.

The West Shandong and North Liaoning massifs all belong to the NCB, and have extensive
exposures of Archaean basement, e.g. the Taishan Group in the West Shandong Massif
and the Anshan Group, Longgang Group, Jiabigou Group in the North Liaoning Massif.
Early Proterozoic basements are locally preserved in the two massifs. Studies by Shen et
al.[18] demonstrate that Archaean basement in the West Shandong and North Liaoning
massifs all contain greenstone belts (Fig. 2) and their greenstone belts show similarities in

protolith, geochemistry and ages whereas they are different from greenstone belts in West Liaoning, East Hebei and Jiaodong massifs. Archaean granites in the two massifs (Fig.2) also have similar petrology and petrochemistry [16]. On the other hand, basements in the two massifs all contains banded iron formation and tourmaline-bearing dark rocks, such as amphibolite [26]. Gold deposits have been found in the greenstone belts of the West Shandong and North Liaoning massifs [18]. Isotopic dating [26] indicates that protolith ages of the Archaean basement range from 2500 Ma to 3000 Ma. Original tectonic lines, such as foliation strikes, fold axes and ductile shear belts, in the basements of the two massifs all trend WNW-ESE. In the two massifs, covers to the basement are Late Proterozoic Qinbaikou and Palaeozoic strata. Middle Proterozoic cover, the Changcheng System and Jixian System, is absent in the two massifs, which is different from the presence of thick Middle Proterozoic cover in the North Hebei-West Liaoning Massif. The two massifs are all poor areas for Carboniferous to Permian coal [26].

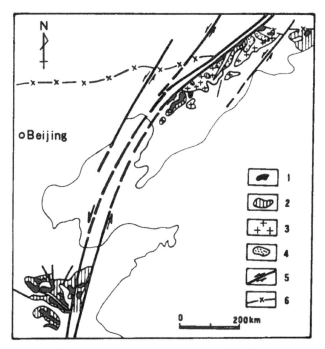

Figure 2. Distribution of Archaean granite-greenstone belts in the West Shandong and North Liaoning massifs (after Shen et al., 1993). 1 = Archaean greenstone belt; 2 = Archaean granitoid in the granite-greenstone belts; 3 = Archaean granitoid in high-grade area (not shown in the West Shandong massif); 4 = supercrustal rocks in high-grade area (not shown in the West Shandong massif); 5 = strike-slip fault; 6 = the northern boundary of the North China Plate.

The above facts proved that the West Shandong Massif west of the Tan-Lu fault zone and North Liaoning Massif east of the fault zone were an integrated massif prior to the offset by the Tan-Lu fault zone. Therefore, maximum sinistral displacement of the Tan-Lu fault zone, shown by the offset of the two massifs, is over 700 km. The sinistral, horizontal displacement of the Tan-Lu fault zone then decreases gradually both to the south and to the north [22].

It was often questioned why sinistral displacement on the northern boundary of the NCB shows not more than 200 km and most of counterpart of the East Hebei-West Liaoning Massif disappears east of the Tan-Lu fault zone (Fig. 2). Studies by Yang [31] and Yao [32] demonstrate that intense thrusting and napping happened in the region around the northern boundary of the NCB east of the Tan-Lu fault zone. It can be seen in Figure 1 that the Tan-Lu fault zone curves to NE-ENE strikes near the boundary region. In this case it is expected that intense transpression should have taken place around the boundary region east of the fault zone during the strike-slip movement. It is proposed therefore that disappearance of most of the East Hebei-West Liaoning Massif as well as obvious decrease of the sinistral displacement around the boundary region east of the fault zone were due to intense thrusting, napping and decollement around the boundary region east of the fault zone during the strike-slip movement.

TIMING OF THE SINISTRAL DISPLACEMENT

Controversies remain also on timing of the sinistral displacement on the Tan-Lu fault zone. Some workers hypothesized that the displacement took place before deposition of the Late Proterozoic Qinbaikou cover in the Shandong part of the Tan-Lu fault zone [7, 35]. This hypothesis was mainly based on the presence of pre-Cambrian ductile shear belts and termination of the basement ductile shear belts at unconformity between basement and cover in the Tan-Lu fault zone. On the basis of the presence of Late Triassic high-pressure rocks, such as blueschist, in the southern part of the fault zone and parallelism of pre-Late Triassic folds with the Tan-Lu fault zone, others suggested Late Triassic (Indosinian) displacement on the fault zone [20, 33]. We had also suggested the Late Triassic displacement in 1980s [22]. However, our recent studies demonstrate that the large-scale sinistral displacement on the Tan-Lu fault zone took place near the end of Late Jurassic to Early Cretaceous.

Our studies [25] also confirmed the presence of the pre-cover ductile shear belts in the Shandong part of the Tan-Lu fault zone. However, the pre-cover shear belts can be traced from inside to outside of the Tan-Lu fault zone. They strike NE inside the fault zone whereas they gradually change into NW strikes outside the fault zone. Moreover, it has been found that the pre-cover shear belts is cut by NNE-striking shear belts developed during strike-slip movement in the Tan-Lu fault zone. It is interpreted therefore that the pre-cover shear belts were developed during solidification of the basement and were then curved into nearly parallel to the Tan-Lu fault zone inside the fault zone. Their termination at the unconformity cannot be used as evidence supporting the pre-cover movement in the Tan-Lu fault zone. Our recent studies [40] show that the Late Proterozoic to Palaeozoic cover in the Shandong part of the Tan-Lu fault zone was clearly involved in the strike-slip deformation in the fault zone. The Tan-Lu strike-slip deformation in the Shandong part is characterized by formation of NNE-striking ductile shear belts in the Archaean basement and shearing folds, sinistral sliding along bedding and unconformity plane in the cover

(Fig. 3). Therefore, it can be concluded that the sinistral displacement of the Tan-Lu fault zone took place after deposition of the Palaeozoic cover.

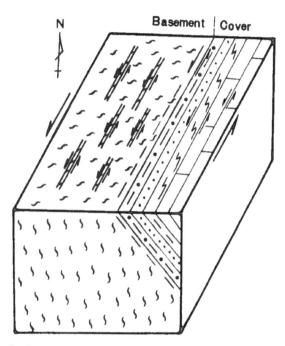

Figure 3. Pattern of strike-slip structures in the Shandong part of the Tan-Lu fault zone, showing ductile shear belts in the basement and shear folds, sliding in the cover.

The southern part of the Tan-Lu fault zone, called the Zhangbalin Belt, occures between the Dabieshan and Jiaonan collision belts (D-D′ in Fig. 1). The eastern side of the fault zone contains blueschist with a $^{40}Ar/^{39}Ar$ phengite age of 245 ± 0.5 Ma (Early Triassic) [12]. Our work demonstrates that the Zhangbalin Belt was a part of the Dabieshan-Jiaonan collision Belt prior to the Tan-Lu fault zone. The high-pressure metamorphism in the Zhangbalin Belt was caused by continent-continent collision between the NCB and SCB rather than the Tan-Lu faulting. It has been found in the field that NNE-striking Tan-Lu faults cut deformation structures associated with the high-Pressure metamorphism in the belt. Many studies [25, 40] have shown that the original foreland folds caused by the collision of the NCB and SCB in the Indosinian movement striked nearly E-W in the SCB, and were then passively curved to NE-NNE strikes during strike-slip movement of the Tan-Lu fault system. Therefore, timing of the Indosinian foreland folding is prior to the sinistral displacement of the Tan-Lu fault zone. These facts suggest that the Tan-Lu sinistral faulting did not take place in the Indosinian movement.

The Hefei basin north of the Dabieshan orogenic belt and the Leiyang basin north of the Jiaonan orogenic belt (Fig. 4) are Early to Late Jurassic foreland molasse basins developed due to uplift of the Dabieshan-Jiaonan collision belt. With offset of the Dabieshan-Jiaonan belt, they were sinistrally displaced by the Tan-Lu fault zone similarly. In the Shandong part of the Tan-Lu fault zone, it can be found that original E-W striking Late Jurassic strata in the western margin of the Leiyang basin are curved to NE-NNE strikes near the Tan-Lu fault zone, indicating strike-slip faulting effect on the basin. To the east of the Tan-Lu fault zone in Anhui region, Late Triassic to Middle Jurassic foreland molasse strata to the south of the Jiaonan collision belt are also curved to NE-NNE near the Tan-Lu fault zone. It is clear therefore that pre-Cretaceous strata have been offset by the Tan-Lu fault zone.

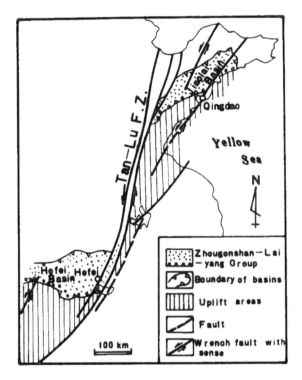

Figure 4. Offset of Jurassic foreland basins by the Tan-Lu fault zone.

Distribution of Early Cretaceous volcanic rocks, called the Qinshan Formation with K-Ar ages between 107 Ma and 113 Ma [8], in the Shandong part of the Tan-Lu fault zone is obviously controlled by the fault zone. Our recent work found that inside the fault zone the volcanic rocks were involved in the left-slip faulting at some places. In the Anhui part of the Tan-Lu fault zone, the Znangbalin Belt, distribution of several granitic bodies (Fig. 5) with an U-Pb zircon age of 128 + 1 Ma [13] are also controlled by the fault zone. It is believed that the Early Cretaceous eruption and intrusion along the Tan-Lu fault zone took place in later stage of the left-slip faulting. Therefore, their distribution were controlled by the fault zone.

Five mylonite whole-rock K-Ar ages between 95 ± 5 Ma and 103 ± 5 Ma (Table 1) were

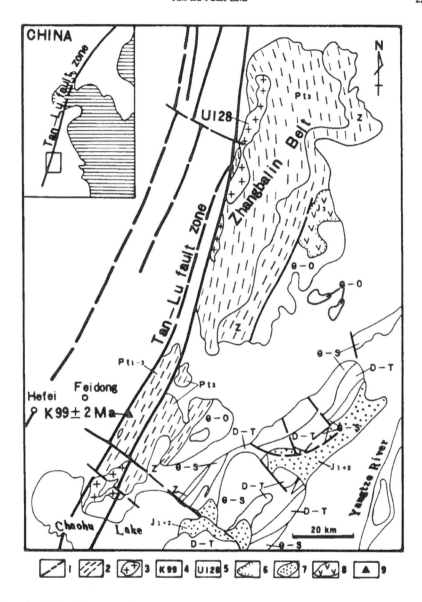

Figure 5. Geological sketch map of the Zhangbalin belt on the eastern side of the southern part of the Tan-Lu fault zone. 1 = fault; 2 = traces of Indosinian stretching lineation; 3 = Yanshanian granite bodies; 4 = K-Ar age; 5 = U-Pb age; 6 = unconformity; 7 = Early to Middle Jurassic basin; 8 = Late Jurassic volcanic basin; 9 = sampling location.

botained from a sinistral, ductile shear belt in the southern part of the Tan-Lu fault zone (Fig. 5) in our recent work. The 5 ages are consistent within error (see Table 1), indicating that the K-Ar systems in the mylonite samples have been equilibrated by a resetting event. It is interpreted that their weighted mean, 99 ± 2 Ma (Early Cretaceous), provides the best estimate for the cooling age of the mylonites developed in the Tan-Lu fault zone. According to studies by Dou et al. [4], a fault rock from the northern part of the Tan-Lu fault zone, the Yilan-Yitong fault, gave ^{40}Ar/^{39}Ar biotite ages between 95.7 ± 2.7 Ma and 100 ± 2.3 Ma. They interpreted the ages as the strike-slip timing of the Tan-Lu fault zone. Using ^{40}Ar/^{39}Ar analysis and multiple diffusion domain model treatments of K-feldspar from the Tan-Lu fault zone, Chen et al. [2] identified a rapid cooling event between the 120 Ma and 125 Ma in the fault zone and interpreted them as the timing approximately contemporaneous with the strike-slip movement.

Table 1. K-Ar whole rock analytical results of mylonites from a ductile shear zone in the southern part of the Tan-Lu fault zone

Sample No.	K%	Rad ^{40}Ar (μg/g)	^{40}Ar/^{40}K	Atm Ar (%)	Age (Ma)
K - 1	3.764	0.02577	0.005739	20.0	96 ± 5
K - 2	3.828	0.02581	0.005652	19.1	95 ± 5
K - 3	3.854	0.02660	0.005785	16.9	97 ± 5
FC - 3	3.294	0.02406	0.006123	23.5	102 ± 5
KC - 3	3.924	0.02890	0.006173	20.6	103 ± 5

Rad ^{40}Ar = radiogenic ^{40}Ar; Atm Ar = atmospheric Ar

We believe therefore that the large-scale sinistral displacement of the Tan-Lu fault zone took place near the end of Late Jurassic to Early Cretaceous, i.e. between 110 Ma and 140 Ma. Afterwards, the fault zone underwent extension in the Late Cretaceous to Paleogene interval, and then compression with minor right-lateral offset since the Neogene [25].

TECTONIC MODELS

In recent years, attention has been devoted to an important role of the Tan-Lu fault zone in East Asian tectonism. However, controversies remain regarding tectonic models of the Tan-Lu fault zone. Four tectonic models of the fault zone have been proposed by different workers (Fig.6), i.e. (1) intracontinental transcurrent fault model [25]; (2) suture line model [39]; (3) transform fault model [9, 21]; (4) collision-related indentation model [33].

In the transcurrent fault model (Fig.6A), the Tan-Lu fault zone is regarded as an major intracontinental transcurrent fault zone with hundreds of km of sinistral displacement [25]. Its development is closely related to oblique motion of the Izanagi Plate in the Pacific region. The suture line model (Fig.6B) is mainly based on the presence of high-pressure rocks, such as blueschist, along the southern part of the Tan-Lu fault zone. In this model, the fault zone is interpreted as a suture line between the NCB and SCB or some terranes [6, 10 – 11, 39]. In the transform fault model (Fig.6C), the fault zone is considered to be

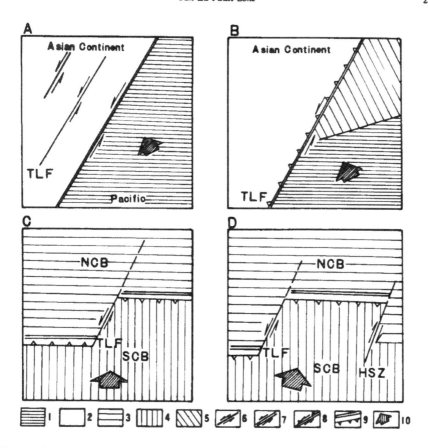

Figure 6. Four tectonic models for the Tan-Lu fault zone. A. Intracontinental transcurrent fault model; B. Suture line model; C. continental transform fault model; D. collision-related indentation model; 1 = oceanic plate; 2 – 5 = continental plates or terranes (2 = Asian continent; 3 = North China Block; 4 = South China Block; 5 = terrane); 6 = strike-slip fault; 7 = boundary transform fault; 8 = suture line or boundary transform fault; 9 = suture line and collision-related orogenic belt; 10 = movement direction of block; NCB = North China Block; SCB = South China Block; TLF = Tan-Lu fault; HSZ = Honam shear zone.

a continental transform fault developed due to opposite A-type subduction along the Dabieshan collision belt and Jiaonan collision belt (Fig. 1). On the basis of left-lateral dislocation of the Tan-Lu fault zone and right-lateral dislocation of the NNE-trending Honam shear zone in South Korea, Yin & Nie [33] proposed that the northern Jiangsu, Jiaonan, and Korean Kyeonggi-Ogcheon massifs between the Tan-Lu fault zone and Honam shear zone constituted a northern-projecting part of an irregular passive margin of the northern SCB before collision of the NCB and SCB (Fig. 6D). The Tan-Lu fault zone is interpreted therefore as one of the irregular margin of the SCB in the collision-related indentation model.

We agree with the intracontinental transcurrent fault model on the basis of the following main facts:

(1) As metioned at first, the Tan-Lu fault zone not only occures between the Dabieshan and Jiaonan collision belts (Fig. 1) but also extends into the NCB and the Far Eastern Russia (the Sikhote-Alin area). This fact is difficult to interpret in the suture line or indentation model.

(2) Many studies [22, 25] have proved that formation of the fault zone was associated with development of a series of NNE-striking, simultaneous sinistral faults, the Tan-Lu wrench fault system which involves the whole East Asian continental margin. This fact suggests that the East Asian continental margin suffered NNE-SSW sinistral shearing during development of the Tan-Lu fault system. The suture line, transform fault or indentation model is difficult to interpret this dynamic setting.

(3) The Tan-Lu fault zone offset not only the Dabieshan-Jiaonan orogenic belt, but also the Hefei-Leiyang foreland molasse basin, Huaihe-Jiaobei, Liaonan blocks, West Shandong-North Liaoning blocks (Fig. 1). The suture line or indentation model cannot interpret the large-scale offset made by the Tan-Lu fault zone.

(4) As mentioned before, the Zhangbalin Belt with high-pressure rocks was a part of the Dabieshan-Jiaonan orogenic belt prior to the Tan-Lu faulting and was then dragged and curved into NNE trend by displacement of the Tan-Lu fault zone (Fig. 1). So the present arrangement of the Zhangbalin belt is caused by the Tan-Lu faulting, it does not represent an original suture line. In this case, the presence of the high-presure rocks in the southern part of the Tan-Lu fault zone, the Zhangbalin Belt, cannot be used as evidence to support the suture line model.

(5) Many studies [9, 17] have suggested that the NCB subducted under the SCB along the Jiaonan Belt to the east of the Tan-Lu fault zone. Magneto-telluric sounding results [14] and seismic P-wave analysis by Zhang [34] also show that the NCB subducted under the SCB in the Dabieshan collision belt west of the Tan-Lu fault zone. So the A-type subduction did not show opposite direction on both sides of the Tan-Lu fault zone. Therefore, it is difficult to interpret the Tan-Lu fault zone as a transform fault.

(6) It was mentioned above that large-scale displacement of the Tan-Lu fault zone took place near the end of Late Jurassic to Early Cretaceous. But more and more evidence, such as timing of high-pressure rocks [12], molasse basins [25], syncollisional intrusions as well as palaeomagnetic data [38], indicate that continent-continent collision of the NCB and SCB happened in the Late Triassic to Middle Jurassic interval. Therefore, sinistral displacement of the Tan-Lu fault zone took place after the collision of the SCB and NCB rather than contemporary with the collision. In this case, only the intracontinental transcurrent fault model can interpret the development of the Tan-Lu fault zone.

The above facts suggest that is is difficult to use the suture line, transform fault or indentation model to interpret the Tan-Lu fault zone. We think that the intracontinental transcurrent fault model gives the best interpretation for the formation of the Tan-Lu fault zone. Development of the Tan-Lu fault zone and its associated Tan-Lu wrench fault system were caused by oblique motion of the Izanagi Plate in the Pacific region [5].

Acknowledgments

We greatly acknowledge the National Science Foundation of China of its financial support of this project (No. 49172130).

INTRODUCTION

1. P. J. Chen. Age and framework of horizontal displacement of the Tan-Lu fault zone, *Chinese Sci. Bull.* **34**, 482 – 487 (1989) (in Chinese).

2. W. J. Chen, T. M. Harrision, M. T. Heizler, R. X. Liu, B. L. Ma and J. L. Li. Studies of MDD $^{40}Ar/^{39}Ar$ thermal chronology on cooling history in the North Jiangsu-Jiaonan tectonic melange belt, *Journal of Petrology* **8**:1, 1 – 17 (1992) (in Chinese).

3. N. G. Deng. Structural system of Cathaysian type of Mesozoic and characteristics of Tancheng-Lujiang fractural system and mechanism of their formation, *Coll. Struct. Geol.* **3**, 33 – 38 (1984) (in Chinese).

4. L. R. Dou, J. G. Song and Y. Wang. Geochronological studies on formation of the northern part of the Tan-Lu fault zone and their significance, *Geological Review* (in press).

5. D. C. Engebretson, A. Cox and R. G. Gordon. Relative motions between oceanic and continental plates in the Pacific Basin, *Geological Society of America*, *Special Paper 206* (1985).

6. Z. J. Fang, M. L. Ding, H. F. Xiang, F. L. Ji and R. C. Li. Basic characteristics of the Tancheng-Lujiang fault zone, *Kexue Tongbao* **31**, 1405 – 1411 (1986).

7. C. J. N. Fletcher, W. R. Fitches, C. C. Rundle and J. A. Evans. Geological and isotopic constraints on the timing of the movement in the Tan-Lu fault zone, northeastern China, *Journal of Southeast Asian Earth Sciences* **11**, 15 – 22 (1995).

8. Z. Y. Guo. Structures, mechanism and history of the middle segment (Yishu Belt) of the Tangcheng-Lujiang fault zone. In: *The Tancheng-Lujiang Wrench Fault System*. J. W. Xu (ed). pp. 77 – 88. John Wiley & Sons Ltd. (1993).

9. K. J. Hsu, Q. Wang, J. Li, D. Zhou and S. Sun. Tectonic evolution of Qinling mountains, China, *Eclogae. Geol. Helve.* **80**, 735 – 752 (1987).

10. D. Jia, Y. M. He, Y. S. Shi and H. F. Lu. Decking history of the East Shandong composite terrane, East China. In: *Terrane Analysis of China and the Pacific Rim*. T J Wiley et al. (eds). pp. 345 – 346 (1990).

11. Q. Y. Lao. The origin and development of the Tancheng-Lujiang fault zone in Precambrian and Palaeozoic, *Coll. Struct. Geol.* **3**, 80 – 93 (1984) (in Chinese).

12. S. G. Li, D. L. Liu and Y. Z. Chen. Formation time of blueschists in central China, *Scienta Geologica Sinica* **28**, 21 – 27 (1993) (in Chinese).

13. X. M. Li, B. X. Li, X. Zhang and T. X. Zhou. Geochronology of Guandian intrusive body in Anhui Province and dynamic metamorphism in Tancheng-Lujiang fracture zone, *Jour. China Uni. Sci. Tec.*, JCUST 85037, supplement, 254 – 261 (1985) (in Chinese).

14. X. X. Li, D. L. Liu, H. J. Wang, X. Z. Shen and A. M. Xue. On the division between North China and Yangtze plates, eastern China. In: *Research on the Structures of Sea and Land Lithosphere and its Evolution*. J L Li (ed). pp. 32 – 45. China Sci. Tech. Press (1992) (in Chinese).

15. J. L. Lin and M. Fuller. Paleomagnetism, North China and South China collision, and the Tan-Lu fault, *Philos. Trans. Roy. Soc.*, *London*, A331 (1991).

16. Q. Lin, F. Y. Wu, S. W. Liu and W. Z. Ge (eds). *Archaean granites in the eastern part of the North China Plate*. Sci. Publ (1992) (in Chinese).

17. A. L. Okay and A. M. C. Sengor. Evidence for intracontinental thrust related to exhumation of the ultra-high-pressure rocks in China, *Geology* **20**, 411 – 414 (1992).

18. B. F. Shen, X. L. Peng, H. Luo and D. B. Mao. Archaean greenstone belts in China, *Acta Geol. Sinica* 67,

208 – 220 (1993) (in Chinese).

19. V. P. Utkin. Wrench faults of Sikhote-Alin and Accretionary and destructive types of fault dislocation in the Asia-Pacific transition zone. In: *The Tancheng-Lujiang Wrench Fault System*. J W Xu (ed). pp. 225 – 237. John Wiley & Sons Ltd. (1993).

20. T. F. Wan and H. Zhu. The maximum sinistral strike-slip and its forming age of the Tancheng-Lujiang fault zone, *Geological Journal of Universities* 2:1, 14 – 27 (1996) (in Chinese).

21. M. P. Watson, A. B. Hayward, D. N. Parkinson and Zh. M. Zhang. Plate tectonic history, basin development and petroleum source rock deposition onshore China, *Marine and Petroleum Geology* A317, 13 – 29 (1983).

22. J. W. Xu, G. Zhu, W. X. Tong, K. R. Cui and Q. Liu. Formation and evolution of the Tancheng-Lujiang wrench fault system: a major shear system to the northwest of the Pacific Ocean, *Tectonophysics* 134, 273 – 310 (1987).

23. J. W. Xu. The horizontal displacement of the Tancheng-Lujiang fault zone and its geological significance. In: *Scientific Papers on Geology for International Exchange*, No. 1, Beijing Geol. Publ. House, pp. 129 – 142 (1980) (in Chinese).

24. J. W. Xu. The Tancheng-Lujiang wrench fault system, *Coll. Struct. Geol.* 3, 18 – 23 (1984) (in Chinese).

25. J. W. Xu. Basic Characteristics and tectonic evolution of the Tancheng-Lujiang fault zone. In: *The Tancheng-Lujiang Wrench Fault System*. J W Xu (ed). pp. 17 – 50. John Wiley & Sons Ltd. (1993).

26. J. W. Xu. Rediscussion on the maximum horizontal displacement of the Tan-Lu fault zone. *Memoirs of Shenyang Inst. of Geol. and Miner. Resources* 3, 43 – 55 (1994) (in Chinese).

27. J. W. Xu, W. X. Tong, G. Zhu, S. F. Lin and G. F. Ma. An outline of the pre-Jurassic tectonic framework in East Asia, *Jour. Southeast Asian Earth Sci.* 3, 19 – 45 (1989).

28. J. W. Xu, W. X. Tong, G. Zhu, S. F. Lin and G. F. Ma. The Meso-Cenozoic geodynamic evolution of the East Asia continental margin. In: *Asian Marine geology*. P X Wang et al. (eds). pp. 113 – 138. China Ocean Press (1990).

29. J. W. Xu and G. F. Ma. Review of ten years of research on the Tancheng-Lujiang fault zone, *Geological Review* 38, 316 – 324 (1992) (in Chinese).

30. X. S. Xu. The horizontal displacement of the Tancheng-Lujiang fault zone, *Coll. Struct. Geol.* 3, 56 – 65 (1984) (in Chinese).

31. M. X. Yang. The Tan-Lu fault zone and its coal-controlling effect in the Lower Liao River region, *Geology of Liaoning* (2), 38 – 47 (1983) (in Chinese).

32. D. Q. Yao. Strike-slip shearing and nappe structures in the eastern part of the northern margin of the Sino-Korean Platform, *Regional Geology of China* (3), 262 – 268 (1989) (in Chinese).

33. A. Yin and S. Y. Nie. An indentation model for the North and South China collision and the development of the Tan-Lu and Honam fault systems, Eastern Asian, *Tectonics* 12, 801 – 813 (1993).

34. J. Q. Zhang. A preliminary study on the character of the deep-seated structures of the southwestern Dabie Mts. area. M. Sc. thesis, Hefei University of Technology (1988) (in Chinese).

35. J. S. Zhang. The basement ductile shear zones of the middle Yishu fault belt, *Seism. Geol.* 5, 11 – 23 (1983) (in Chinese).

36. R. Y. Zhang, T. Hirajima, S. Banno, B. L. Cong and J. G. Liou. Petrology of ultrahigh-pressure rocks from the southern Su-Lu region, Eastern China, *J. Metamorphic Geol.* 13, 659 – 675 (1995).

37. Y. X. Zhang and L. L. Li. The giant strike-slip along the Tancheng-Lujiang fault zone and its influence on the nearby structures, *Coll. Struct. Geol.* 3, 1 – 8 (1984) (in Chinese).

38. X. X. Zhao, and R. C. Coe. Palaeomagnetic constraints on the collision and rotation of North and South China, *Nature* 327, 141 – 144 (1987).

39. T. Z. Zhou, J. Z. Lu and P. H. Huang. Studies on the mechanics of the Tancheng-Lujiang fault zone, *Proceedings of 1st Symposium on the tancheng-Lujiang fault zone*, pp. 94 (1980) (in Chinese).

40. G. Zhu and J. W. Xu. Shear deformation of the cover in the Yishu fault belt and its structural implication, *Journal of Changchun University of Earth Sciences* 25:3, 279 – 285 (1995).

Proc. 30th Itnt'l. Geol. Congr., Vol. 14, PP. 229-250
Zheng et al. (Eds)
© VSP 1997

Evolution of Tan-Lu Strike-Slip Fault System and Its Geological Implications

XIAOFENG WANG [a], Z.J. LI[a], B.L. CHEN[a], L.S. XING[a], X.H. CHEN[a], Q. ZHANG[a], Z.L. CHEN[a], S.W. DONG[b], H.M. WU[c], and G.H. HUO[c]

a Institute of Geomechanics, CAGS, Beijing 100081, China
b Chinese Academy of Geological Sciences, Beijing 100037, China
c Shandong Bureau of Geology and Mineral Resources, Jinan 250013, China

Abstract

The Tan-Lu strike-slip fault system in eastern China extends more than 2000km with a total sinistral offset about 300-100km for its central-southern part and northern part respectively and the main fault and branch faults compose a herringbone structure. The convergence of the North and South China Blocks in late Triassic produced a south-convex belt at the south edge of the Dabie massif in the Qinling-Dabie latitudinal structural system. The Tan-Lu sinistral strike-slip offset accommodated parts of the movement of the west Pacific Ocean plate, and occured mainly during the Yanshanian. The displacement on the main fault was absorbed by N-S shortening in shear zones, thrusts and folded zones of the adjacent areas, e.g., a group of NE-trending shear zones on its east side such as the Sihong-Xiangshui, Wulian-Rongcheng, Dunhua-Mishan and Yilan-Yitong faults. The palaeomagnetism data from Triassic-early Cretaceous strata on both sides of the Tan-Lu zone suggest that there was a 15°-25° anticlockwise rotation in its east side before the Early Cretaceous and not a large sinistral offset between two sides of the Tan-Lu fault. This is concordant with the changing tendency of fold trends in South China Block, from south to north, fold axes turning from ENE via NE to NNE. Crust shortening increased northward from 18% through 38% to 70%. The offset of the Tan-Lu system in its western side was absorbed by shortening of a series of south-convex arc-structures, e.g., the Luxi and the Dabie arcs; the former consists of a series of listric imbricate south-thrusting faults which were translated into detachment faults at a later time.The mechanism of the Tan-Lu faulting might be a segmental progressive northward migration. There are four stages of stress evolution in the Tan-Lu fault system : sinistral transpressional ductile shear (T_3J_{1-2}); sinitral brittle transtensional shear (J_3,K_1); extension (K_2,E); dextral transpressional and transtensional shear (N to now).

Keywords:strike-slip system, multilayer gliding model, palaeomagnetic analysis, thermochronology, tectonic stress field

INTRODUCTION

The Tan-Lu fault zone is a major structural belt in eastern China with a NNE - trend and extends more than 2400 km. The formation and evolution of the Tan-Lu fault zone controlled the formation and distribution of sedimentary facies, magmatic activities, metamorphism, ore deposits and oil and gas resources in eastern China. It is also a seismic belt at present.

The investigation of the Tan-Lu fault zone has been carried out for nearly a century [19]. The fault zone has attracted great attention of geologists all over the world since it was named as " the Tancheng-Lujiang deep fracture zone " by the Aeromagnetic Survey Team No. 904 of the Geological Ministry of China in 1957. There are five tectonic scenarios of the features, evolution and dynamic mechanism of the fault zone. As the first researcher of the Tan-Lu fault zone, Xu [23-26] suggested the maximum strike-slip displacement over 740km mainly achieved in the Late Jurassic to the Early Cretaceous (140-110Ma). Zhou considered it as a suture and margin transform fault [21]. Li [9] and Okay [18] considered it as a transform fault. Yin and Nie [29] proposed an indentation model for the development of the Tan-Lu fault system. Deng [5] supposed that it belongs to the late Neocathaysian structural system formed in the Late Yanshanian.

GEOLOGIC SETTING

Matching of the North and South China Blocks and Initiation of the Tan-Lu faulting
The background of the initiating of the Tan-Lu fault system was the matching of the North and South China Blocks. The South China Block wedged into underneath the North China Block, which caused the gradual formation of an arcuate matching boundary protruding slightly southwards. Along the matching belt there are ultra-high pressure high temperature metamorphic belt in the depths and high pressure low temperature metamorphic belt in epizone with eclogites and blueschists as their typical rocks respectively, with which were associated felsic mylonitic belts. The Tan-Lu strike-slip fautlting was synchronous with indentation of South China Block into North China Block occurred in the Triassic and Jurassic. The main lines of evidence include :

(1) Sedimentological data : The timing of final disappearance of marine sedimentation along the northern margin of the South China Block in Qinling region [29] concludes that the matching between the North and South China Blocks occurred in the Late Triassic.

(2) Ages of deformation along the matching belt : The age range of deformation and metamorphism constrained by U-Pb, ^{40}Ar / ^{39}Ar, Rb-Sr and Sm-Nd dating techniques is between 250 and 180Ma [1, 3, 4, 11-14, 16, 17, 20, 32, 34, 36].

(3) Paleomagnetic data : The existing paleomagnetic data from North and South China Blocks seem to suggest that the age of their final amalgamation did not occur until the Late Jurassic [7, 22, 29, 35] because paleomagnetic poles older than Cretaceous do not coincide.

Framework and Tectonic Style of Tan-Lu Left-lateral Strike-Slip Fault System
The Tan-Lu fault zone and its adjacent fault zones compose a herringbone tectonic system (Fig. 1). On the east side, strike-slip displacement on the NNE trending major fault zone was mainly accommodated by generally N-S shortening and resulted in a series of left lateral shearing belts in NE and ENE directions and thrusting belts including the Sihong-Xiangshui, Wulian-Rongcheng and Dunhua-Mishan zones from south to north in order . On the west side, it was absorbed by a series of southward convex arcuate belts including the West Shandong and Dabie Arcs and others from north to south. The West Shandong Arc consists of a series of listric and imbricate faults, which were south thrusting faults in early time and then changed into detachment faults. The tectonic style of the Tan-Lu fault zone as a whole reflects that the terranes have moved northwards on its east side and southwards on its west side.

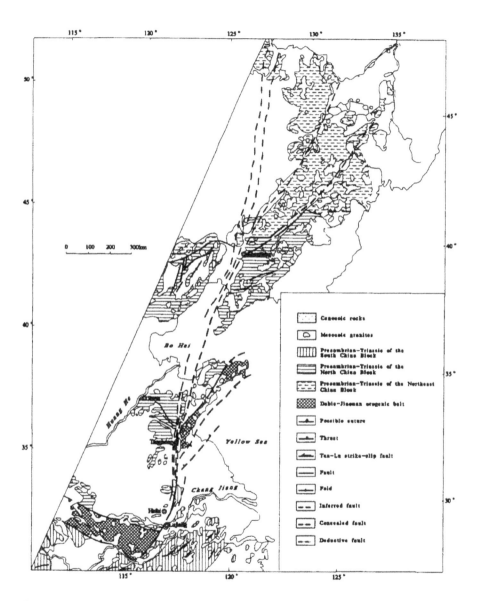

Figure 1. Tectonic map of Tan-Lu Fault.

Indicator of sinistral translation in Tan-Lu fault zone--correlation of Dabie and Sulu (Jiangsu-Shandong) orogenic belts.

Based on correlation of the basic structural elements, deformation and metamorphism and their ages, and deep-seated geological processes of the Dabie and Sulu orogenic belts on the sides of the Tan-Lu fault zone, it is believed that they were the same orogenic belt developed due to the

matching of the North and South China Blocks, and offset later by the Tan-Lu shear translation.
The main lines of evidence include :

(1)The Dabie orogenic belt is comparable with the Sulu orogenic belt in structure elements
(Table.1, Fig. 2).

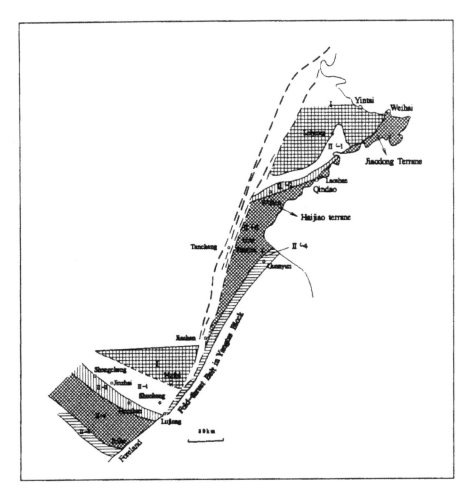

Figure 2. Comparison of the structure between Dabie and Sulu orogen. I , North China Block; II -1, Mesozoic Hefei
Basin; II -2, North Huaiyang belts; II -3, Complex zone in Dabie orogen; II -4, Shusong belts; II '-1, Laiyang Basin;
II '-2, Wulian belts; II '-3, complex in Jiaonan belts; II '-4, Lujian-Jiashan-Guanyun belts.

(2)High and ultrahigh pressure metamorphic complexes in Dabie and Sulu areas are basically
comparable in terms of metamoprphism and protolith composition. The main expressions are as
follows:(a) the metamorphic complexes in both areas consist of Late Archaean-Early Proterozoic
crystalline basement and Mid-Proterozoic cover rocks, which were extensively   reworked by
early Indosinian high and ultrahigh-pressure   metamorphic events; (b) the cover rocks belong to

bimodal spilite-quartz keratophyre formations formed in the same structural environment, and at their base developed phosphatic rocks experienced a high-pressure low-temperature metamorphism ; (c) all the gneiss complexes consist of granitic gneiss and epidermic rocks and underwent regional progressive metamorphism from epidote amphibolite, amphibolite facies to amphibole-gneiss facies ; (d) eclogite widely occurs and is characterized by the presence of coesite. The compositions of protolith and the P-T-*t* path of the retrograde metamophism are roughly comparable (Figs. 3 , 4). The above shows that the metamorphic complexes in Dabie and Sulu area belong to the same geologic body.

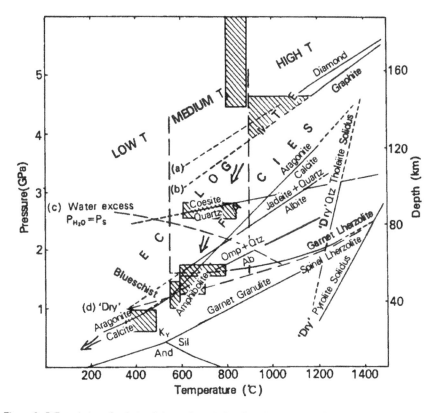

Figure 3. P-T evolution of eclogite facies rocks and blueschists in Qinling-Dabie high and ultrahigh pressure metamorphic belt of central China. Arrows represent retrograde trends. Ps-T petrogenetic grid is after [2], and simplified. Data are from Zhang *et al.*[33], Xu *et al.*[27] and Zhang and Liou [31].

(3) It is shown obviously from deep structural features that South China Block subducted northward under North China Block and then transformed into large detachment fault in the Late Indosinian-Early Yenshanian (Fig. 5) and that North China Block driven by the Tan-Lu left lateral strike-slip fault subducted southward under the Dabie Orogen (opposite subduction) in the Eearly Jurassic (Fig. 6). The above evidence indicates the Tan-Lu left lateral strike-slip

kinematics.

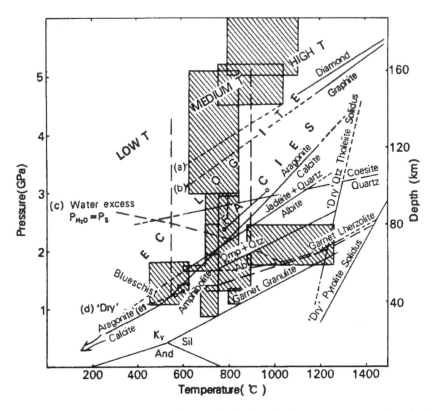

Figure 4. P-T evolution of eclogite and eclogite facies rocks in the Su-Lu ultrahigh pressure metamorphic belt of East China. Arrows represent retrograde trends. Ps-T petrogenetic grid is after Carswell [2] and simplified. Data are from Enami and Zhang [6], Yin [30], Fan *et al.* [9] and Yang *et al.* [28], Lines which link two P-T points shows (retrograde) evolution trends of some rocks in this metamorphic belt.

Table 1. Correlation of Dabie and Sulu orogenic belts

Structure Unit	Dabie Orogenic Belt	Sulu Orogenic Belt	Main Characteristics
I	Hefei Basin	Laiyang Basin	Molasse basins at back margin of Mesozoic basins
II	Northern huaiyang belt (Fuziling and groups)	Wulian Belt Meishan (Wulian Group)	Stratigraphic composition are comparable
III	Dabie Complex Belt	Jiaonan complex Belt	Main body of orogenic belt, high-and ultrahigh pressure metamorphic zone
IV	Susong Belt	Lujiang-Zhangbaling-Guanyun Belt	Front marginal belt of orogenic belt, which consists of blueschist and quartz schist equivalent to those (protolithes) in Zhangbaling Group

The displacement on the Tan-Lu fault zone derived from the distribution of Dabie and Sulu high and ultra-high pressure metamorphic zones is 560km. If it were so, the Dabie-Sulu Belt would have extended in an E-W strike before being translated. Palaeomagnetic studies, however, showed that the South China Block on the eastern side of the Tan-Lu fault zone had rotated counterclockwise 15°-25° during the Late Jurassic. Therefore, after restoration, the Dabie-Sulu zone may show an arcuate form and the sinistral translation of the Tan-Lu fault may be less than 560km.

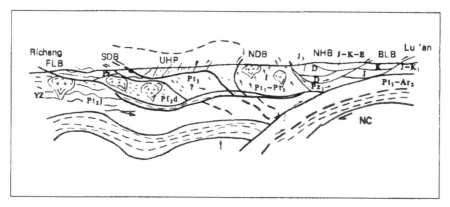

Figure 5. Crustal structure of Dabie orogen. FLB, foreland belt; UHP, Ultra-high-pressure belt; NDB, North Dabie block; NHB, Beihuaiyang Caledonian compressional belt; BLB: Hinterland Basin; Pz, foreland fold belt: Pt_{2j}, subducting basement of South China Block; Pt_{2d}, subducting Dongling basement; Pt_1-Pt_3, metamorphic rock of Mount. Dabie; D, Devonian-Silurian Flysch; Pz_1, Caledonian compressional belt; J-K-E, Heifei Basin; Pt_1-Ar_2, high-conductive layer in North China Block. Arrows represent direction of movement.

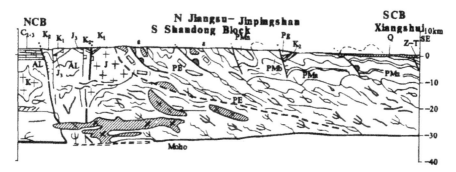

Figure 6. The structural map of south segment of Xiangshui-Mandula geotraverse (from Ma *et al*.[15])

MULTILAYER GLIDING MODEL FOR TAN-LU FAULT SYSTEM

Division of deformation and metamorphism domains

The development of the Tan-Lu strike-slip fault system resulted in three deformation and metamorphism domains at different levels in the fault zone and its adjacent areas. The first is a ductile-brittle domain comprising folds and thrusts in the Sinian-Paleozoic cover rocks at depth less than 10-15 km. The second is a ductile domain mainly composed of mylonites and schistic mylonites of blueschist or greenschist facies in Zhangbaling group at depth of 15-25km. The third, a deep level ductile domain, is composed of high and ultra-high pressure metamorphic rocks represented by eclogites at depth up to 80-100km.

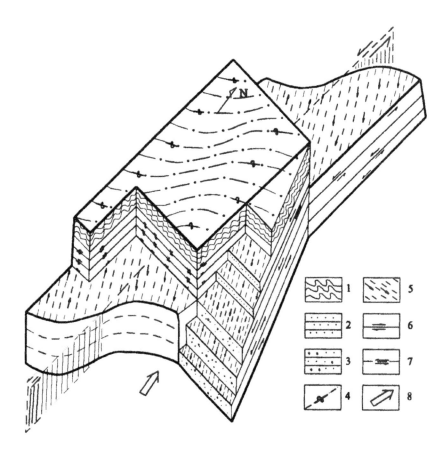

Figure 7. Multi-layers model showing the initial left-lateral shearing of Tan-Lu fault. 1, Covering layers and folds; 2, Brittile-ductile layers; 3, Ultra ductile layers; 4, Fold trends and inclination; 5, A-type lineation and shear sense of upper layer; 6, Intrallayer tangential shear and shear sense; 7, Sense of deep horizontal shear and possible location of brittle shear; 8, Direction showing the deep mantle transportation.

Characteristics of the domains

The middle-south segment of the Tan-Lu fault zone is taken here as an example to illustrate the three dimentional strain image for the development of the Tan-Lu fault system (Fig. 7) and the propagation style of the major intracontinental strike-slip fault system from deep to shallow. As mentioned above, the NNE trending Tan-Lu left-lateral strike-slip faulting initiated first at the apex of the Dabie arc due to the irregular geometer of the matching boundary between the North and South Blocks. Because of the anisotropy of the media between cover and basement, an echelon fold system, such as the NE-ENE trending folds in Sinian-Palaeozoic strata in the South China Block east of the Tan-Lu fault zone, was generated in the covering strata and inclined to the deep-level NNE trending Tan-Lu sinistral ductile shear zone in an acute angle. The mantle fold trends vary gradually from ENE to NNE as approaching the major fault of the Tan-Lu system, until sub- parallel to the major fault in Zhangbaling-Lujiang area, and form a fan-shaped pattern on map with the south edge of the Dabie terrane as its convergening centre. Near the major fault zone, the axial surfaces of these folds are subvertical or at high dip angles to ESE. With distance increasing to the major fault zone, the dip angles of fold axial surfaces are becoming more and more gentle. In sections crosscutting the major strike-slip fault, the axial surfaces around the major fault also show a fan-shaped pattern. From the Yangtse River to south Wulian-Rongcheng tectonic belt east of the Tan-Lu fault zone, the crust shortening and contraction ratio caused by mantle folding in a SSE direction has been estimated here are 240-300km and 40-45% respectively by means of balanced section construction (Fig. 8). This is compatible with sinistral displacement at deep levels.

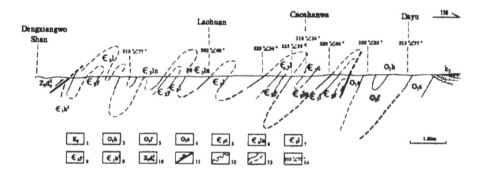

Figure 8. Cross section from Denxiangwo Shan to Dayu,Quanjiao County,Anhui Province. 1, Upper Cretaceous ; 2, Lower Ordovician Honghuayuan Fm.; 3, Lower Ordovician Fexiang Fm.; 4, Lower Ordovician Ochong Fm.; 5, Upper Cambrian Cheahuitong Fm.; 6, Upper Cambrian Langyashan Fm.; 7, Upper Cambrian Longpan Fm.; 8, Middle Cambrian Yangliugang Fm.; 9, Lower Cambrian Huanglishu Fm.; 10, Upper Sinian Dengying Fm.; 11, Thrust fault; 12, Unconformity; 13, Boundary; 14, Attitude.

In adjacent regions of the Tan-Lu fault zone, including Dabie Shan, Zhangbaling and Su-Lu areas, penetrative SSE-dipping foliations at low to moderate angles, and SE-SSE plunging stretching lineations imply the presence of large-scale multilayer detachment faults and the plastic rock flow in the middle and lower crust (Fig. 9). Indicators of shear sense, such as S-C fabrics and fold-vergence also illustrates top-to- the SSE shearing on the detachment surface. The shear strains are ranged from 1.98 to 6.71, average 3.35 (Fig. 10). Thus 230-270km of shearing displacement in SSE is derived for the upper layer in South China Block east of the Tan-Lu fault zone.

In short, either the 240-300km of shortening of the covering rocks or the 230-270km of shearing displacement of the upper detachment layers in a SSE direction east of the Tan-Lu fault zone, was the result of gradual deformation and shearing flow of the rocks at various levels due to the left-lateral strike-slip movement of the Tan-Lu fault system in a NNE at middle and deep levels.

These two values, the crust shortening of covering rocks and the shearing displacement of middle-deep levels, are roughly equal and balanced with the left-lateral displacement of the Tan-Lu fault zone. It is therefore suggested that the development of the left-lateral strike-slip movement was associated with the rock flow and progressive deformation at various levels of lithosphere, where detachment surfaces were formed. Materials at lower levels flew and migrated in the same direction of the left-lateral horizontal displacement, and those in upper layers escaped in reverse. Ductile shearing in the same direction took place in upper layers at middle-deep levels, while echelon folds and thrusts in NE were displaying in the cap rocks at shallow levels. This is the three dimensional strain image for the development of the Tan-Lu left-lateral strike-slip fault system, and the maximum cumulative displacement of c. 300km has been derived hence for the middle-south segment of the Tan-Lu fault zone.

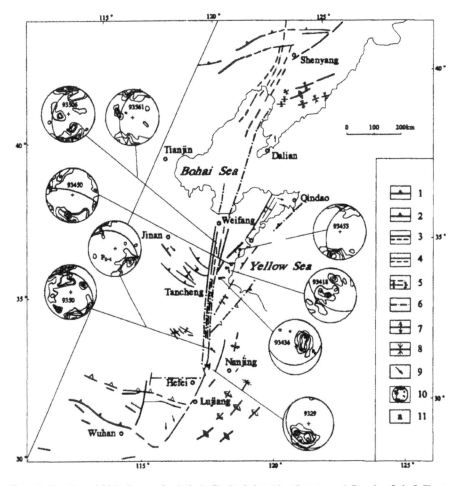

Figure 9. Lineation and fabric diagram of rocks in the Tan-Lu fault and its adjacent area. 1, Boundary fault; 2, Thrust fault; 3, Major fault; 4, Fault; 5, Fault inferred from physical exploration; 6, Fault inferred from remote sense; 7, Anticline; 8, Syncline; 9, A-type lineation; 10, C-fabric diagram of quartz.

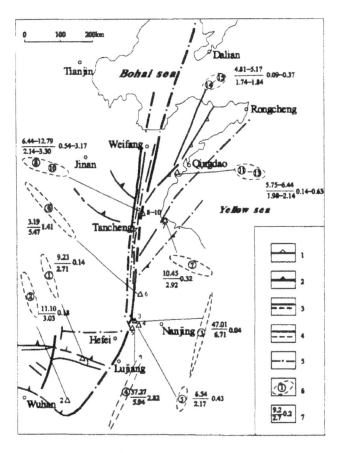

Figure 10. Strain measurements of the Tan-Lu fault system and its adjacent area. 1, Boundary fault; 2, Thrust fault; 3, Major fault; 4, Fault; 5, Burried fault; 6, x/z showing ellipticity ; 7, Flinn parameter.

FAULTING OF THE TAN-LU LEFT-LATERAL STRIKE-SLIP SYSTEM AND ROTATION OF ITS ADJACENT TERRANES

The paleomagnetic analysis of about 150 samples from Middle and Upper Triassic, Jurassic and Lower Cretaceous strata both sides of the Tan-Lu fault zone suggests that, there is not considerable left-lateral displacement on the fault at least since the Middle Jurassic, and a counterclockwise rotation of 15°-25° of South China Block east of the Tan-Lu occured mainly in the Late Jurassic before the Early Cretaceous (Table 2, Figs. 11, 12) [22]. This is consistent with the switch in trends of mantle folds from ENE to NE to NNE with the approach of the Tan-Lu fault zone. In addition, the comparison of the apparent polar wandering paths (APWP) of the North and South China Blocks revealed that two blocks might have been linked with or closing to each other at the east end since the Late Permian [7, 22, 35], and that there was an angle of 60°-70° existing westwards beween their boundaries. Since the counterclockwise rotation of the North China Block and the clockwise rotation of the South China Block, they came into contact at the apex of the southwards convex Dabie arc during late Indosinian, and the   Tan - Lu   fault

Table 2. Rotational angles of sampled regions as opposed to APWP of the South China Block

age	western side	Eastern side
K₁	Huoshan -7°,Mengyin 3°	Zhucheng -8°,Zhejiang -4°
		Zhucheng -6°
J₃	Huoshan 4°, Mengyin -5°	Lujiang -13°,Zongyang -27°
		Zhejiang -2°, Zhejiang -21°
J₂	Feixi 5°,Mengyin 4°	Anqing -22°
T₃		Anqing -25°
T₂		Anqing -36°

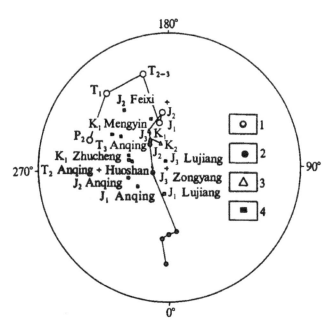

Figure 11. The Apparent Polar Wander Path (APWP) of the south and north China since Upper Permian and paleomagnetic pole position of sampled regions between both sides of the Tan-Lu fault zone. 1, South China Block; 2, North China Block; 3, China Block; 4, Location of sample.

zone formed the boundary between them. In the Jurassic the counterclockwise rotation of the South China Block east of the Tan-Lu fault zone resulted from the compression of the North and South China Blocks at the apex of Dabie arc and the northwestward push of the Pacific Plate on the Tan-Lu fault. In the Early to the Late Cretaceous, the various China Blocks adjusted together with the Eurasian Plate, rotated slightly clockwise, and reached their present positions. Thus, the stress state might changed reversely due to the clockwise rotation of the various China blocks since the Cretaceous. The discovery of the counterclockwise rotation of the partial South China Block east of the Tan-Lu fault zone, is of importance not only to understand of movement of the blocks in eastern China adjacent to the west Pacific Plate since the Mesozoic, but also to the restore of the original azimuths of the Pre-Cretaceous features.

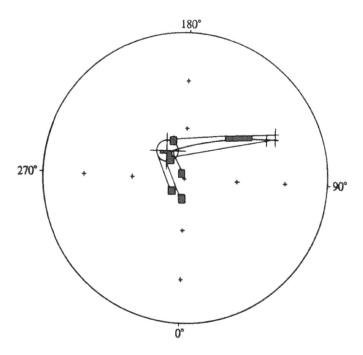

Figure 12. Paleomagnetic pole position of sample regions from the eastern side of Tan-Lu fault zone as opposed to the western side of it in Late Jurassic (after rotated correction)

TIMING OF TAN-LU SINISTRAL STRIKE-SLIP FAULTING

It is suggested that the main time of sinistral translation along the central-southern segment of the Tan-Lu fault is Early to Late Jurassic according to the following lines of evidence :

(1) Evidence from the analysis of the deformation. Megascopically, the Tan-Lu fault zone cut Dabie-Sulu high and ultrahigh pressure metamorphic zone that formed in the Late Triassic, so it may be later than Triassic. There are two sets of lineations in the schistose mica and quartz mylonites in the Zhangbaling area : the SSE-trend stretching a-lineation that resulted from matching of the North and South China Blocks, yields a $^{40}Ar/^{39}Ar$ age 212.6±0.4Ma for phengite ; the NNE-striking a-lineation that were formed in company with the Tan-Lu sinistral strike-slip fault and superimposed on the previous set. This relationship also shows that the main activation period of the Tan-Lu sinistral strike-slip fracture is later than Triassic.

(2) Stratigraphic evidence. Early Jurassic piedmont molasse formations on both sides of the southern segment of the Tan-Lu fault are comparable and the Tan-Lu fault clearly controlled the

sedimentary thickness.

(3) Palaeomagnetic evidence. As above mentioned, no large-scale sinistral translation has occurred on the Tan-Lu since the Late Jurassic. Before Cretaceous, mainly in Late Jurassic, the massif on the east rotated anticlockwise 15°-25°. Therefore, sinistral translation on the Tan-Lu probably occurred in the Early-Middle Jurassic.

(4) Evidence of sedimentary petrology. By the Late Triassic, Lower Yangtze Basin consisting of the Huangmaqing Group had become a trans-compressional basin (Fig. 13). Heavy concentrations of apatite and lepidomorphite in the Huangmaqing Group in Lujiang were derived from the Susong Group at the southern margin of the Dabie Mountains, and heavy concentrations of blue corundum in the Huangmaqing Group in Yueshan of Huaining were derived from blue corundum kyanite-schist of the Susong Group. All these are expressions of syndepositional sinistral strike-slip movement, showing initation of the fault as early as Middle-Late Triassic.

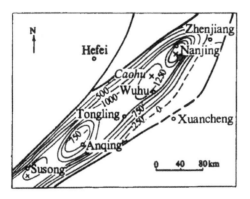

Figure 13. Depositional isochore map of Huangmaqing Fm. along Yangtze River (from Li *et al.* [10]).

(5) Evidence of isotopic chronology. Dating data (including U/Pb, $^{40}Ar/^{39}Ar$, Sm/Nd, K/Ar, ESR methods, etc) from dynamometamorphic rocks along the Tan-Lu fault and the Dabie-Sulu high and ultra-high pressure metamorphic rocks were plotted in a histogram (Figs. 14, 15), in which appeared several peak values, including 230-220Ma, 220-210Ma, 150-140Ma and 110-90Ma. Data from eclogites, concentrated 230-220Ma, approximately reflect the age of high and ultra-high pressure metamorphic zone and the matching of North and South China Blocks ; data from blueschist and schistose mylonites concentrated on 210-220Ma, approximately reflect the time span of transform from matching of the two blocks to the Tan-Lu sinistral transferring ; and data from mylonites concentrated on 150-140Ma, showing the activation time of sinistral transferring on the Tan-Lu fault system. Therefore, the Tan-Lu sinistral strike-slip movement occurred mainly in the Jurassic.

In Sanligang of Lujiang schistose mica and quartz mylonites are exposed, where two sets of mylonitic foliation are developed ; the NE-striking and the NNE-striking sets, the latter clearly cuts the former. Phengites from schistose mylonites in them also show clear differences in texture. In the NE-striking foliation, they are predominated by 2M type with b_0 of 9.012-9.039Å ; while in the NNE-striking foliation, phengite is predominated by 3T type with b_0 of 9.045-9.058Å. $^{40}Ar/^{39}Ar$ dating of phengite yields plateau ages (209.0±6.6Ma and 163Ma) (Figs. 16, 17). They reflect the enclosed ages of two thermal events respectively : matching of the North China Block and the South China Blocks and initiation of the Tan-Lu fault.

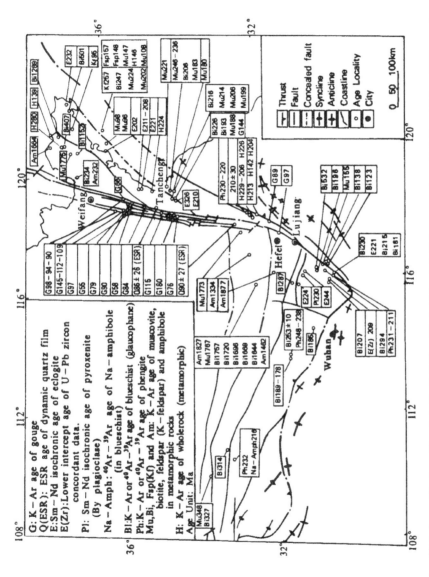

Figure 14. Distribution of isotopic ages of metamorphic rocks in Qinglin-Dabie and Su-Lu high and ultrahigh pressure metamorphic belts, and K-Ar and ESR ages of gouges and dynamic quartz film along Tan-Lu fault zone and its adjacent faults, showing the formation age of Qinglin-Dabie and Su-Lu orogen and active epoch of brittle Tan-Lu fault. QDB, Qinglin-Dabie high and ultrahigh pressure metamorphic belt; SLB, Su-Lu high and ultrahigh pressure metamorphic belt, including Zhangbaling area of Anhui Province [1, 3, 4, 11-14, 16, 17, 20, 32, 34, 36].

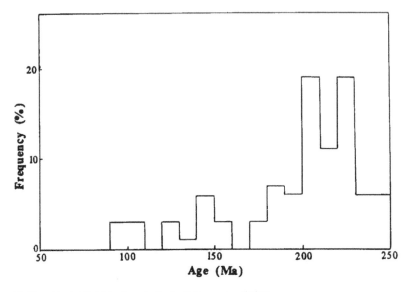

Figure 15. Histogram showing data of ages in Tan-Lu fault zone since 250Ma.

PROGRESSIVELY SEGMENTATIONAL GROWING AND PROPAGATING MECHANISM FOR TAN-LU FAULT SYSTEM FROM SOUTH TO NORTH

According to the differences in the structural composition and evolution history of the Tan-Lu fault zone, it is divided into three segments, the northern, middle and southern ones. The fault zone is generally characterized by the progressively segmentational growing and propagating from south to north.

(1) The initiational and developmental ages of the left-lateral horizontal displacement get more and more young from south to north. The left-lateral ductile shearing displacement of the south segment in Dabie area initiated in the late Indo-Chinese epoch with its main action in the Early to Middle Jurassic. The north segment, Dunmi fault zone, initiated and developed mainly in the Early Cretaceous. The Tan-Lu fault zone therefore was not linked up from south to north until the Early Cretaceous [21].

(2) The left-lateral horizontal displacements get more and more little from south to north. The overall displacement of the Tan-Lu fault zone concentrated mainly on its middle-south segment in Jiangsu, Shandong and Anhui provinces with the cumulative magnitude of 300km, and was partially absorbed by the N-S shortening of the adjacent terranes. On the middle-south segment, the N-S crust shortening of 40-45% or of 240-300km took place mainly from late Indo-Chinese epoch to early-middle Jurassic. The cumulative displacement achieved 100-150km on the north segment in Liaonin province, and it might be mainly absorbed by the N-S crust shortening of 27% or of 100-150km, which resulted from the E-W trend folding of the P-T strata in the south Liaonin area north to the Su-Lu region. The shortening in Liaonin province might happen mainly in late Indo-Chinese epoch to late Jurassic, and ended at 152.9±0.4Ma, which was later than the middle-south segment, has been obtained from an $^{40}Ar/^{39}Ar$ closure age of K-feldspar measured from mylonites in detachment fault lying over the E-W trending folds. Evidence above shows a

trend of decreasing in scale from south to north.

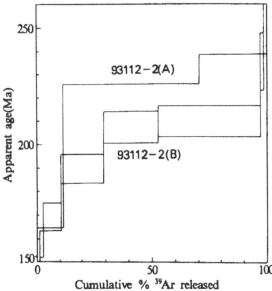

(3) In terms of time span of the development of syntectonic sedimentary basins, it is proved that, the south segment of the Tan-Lu fault zone had controlled the Huangmaqing Formation basin in the Middle Triassic [10], the middle-south segment had controlled the Middle to Lower Jurassic basin, and that the middle segment has controlled the Cretaceous basin of volcanic and volcanic clastic rocks. The development ages of the syntectonic sedimentary basins decrease from south to north.

Figure 16. ^{40}Ar/^{39}Ar spectra of muscovite from Sanligang, Lujiang, Anhui Province. (A) in T.M. Harrison laboratory, Department of Earth and Space Sciences, UCLA, U.S.A.; (B) in ^{40}Ar/^{39}Ar Laboratory, Institute of Geology, CAS, China.

(4) The volcanism controlled by the Tan-Lu faulting decrease in age from south to north, and the ending ages of these volcanisms get younger in the same direction. The Cenozoic volcanism related to the north segment showed higher frequency and wider distribution than that to the south and middle segments.

DEVELOPMENT OF THE TAN-LU FAULT SYSTEM AND EVOLUTION OF THE

TECTONIC STRESS FIELD

According to the quatitative analyses on the tectonic deformation in Tan-Lu Fault system, we divided the evolution of the Tan-Lu Fault system into following six stages (see Figs. 18, 19).

Preceding the Tan-Lu stage (later P_2-T_2)
During the Late Permian to the Middle Triassic, the Paleo-Tethys between the North and South China Blocks began to shrink and to close, and the South China Block wedged into the North China Block in a SSE-NNW direction, forming the Dabie-Sulu high to ultra high-pressure metamorphic belt. Tectonic stress field in this stage is σ_1 in a NNW-SSE direction with differential stress (σ_1-σ_3) > 100MPa.

Initiating stage(T_2-T_3)
During this stage, the convergence between the South China Block and the North China Block

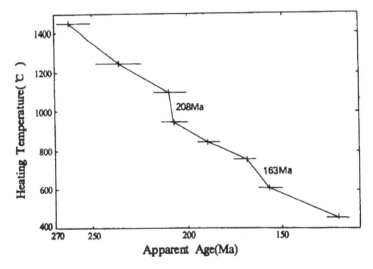

Figure 17. Curve showing the relation between the heating temperature and ages of muscovite from Sanligang, Lujiang, Anhui Province.

transformed into the Tan-Lu left-lateral ductile shear belt. Under the NNW-SSE principal compressive stress (differential stress was 50-110MPa), arcualte convergence boundary slightly protruding southwards formed as a result of heterogeneous medium. Taking the arcuate protruding parts as the starting point, the Tan-Lu left-lateral ductile shear took place in the middle-lower crust and the left-lateral ductile strike-slip fault belt formed in the basement. As a result, the NE trending echelon fold system formed in the cover and the N-S crust-shortening of 300km happened. In the basement east of the Tan-Lu fault system, top-to-the SSE ductile shear detachment belts developed with the cumulative displacements probably as large as 300km.

Left-lateral strike-slip faulting stage (J₁.₂-J₃)

During this stage West Pacific kinetic regime began to replace the paleo-Tethys one and the σ_1 direction switched from NNW-SSE to NW-SE (the stress was about 160-190MPa). Meanwhile, the Tan-Lu fault extended from middle-deep to shallow levels, and transformed from left-lateral ductile strike-slip fault to left-lateral brittle strike-slip faulting. While the Tan-Lu left lateral strike-slip movement controlled the formation and the distribution of Jurassic sediments in the central southern part of the fault system, they could lead to the uplift and exhumation of the Dabie-Sulu high-pressure to ultra high-pressure metamorphic belts along the both sides of the fault system. Therefore, the left-lateral strike-slip faulting in the central-southern part of the fault system was shaped during the Jurassic. During the Late Jurassic, the eastern block of the fault system rotated 15°-25° anticlockwise.

Extension stage (K₁-K₂)

From the end of late Jurassic to the beginning of Early Cretaceous, as a result of the Pacific oceanic crust subducted under the Eurasian continental crust, the shallow level crust in East China became in the extensional state and endowed the Tan-Lu fault system therefore some extension characteristics. The Tan-Lu fault system continued to extend to the surface and to the deep and the fault system became larger. The Tan-Lu fault system controlled extensive volcanism and plutonism of Late Jurassic and Early Cretaceous ages. Faulted basins formed along the fault system in both flanks of the fault system and left-lateral strike-slip with a displacement up to

100-150km took place in the Dunmi fault zone. We believe that during the Early Cretaceous the Tan-Lu fault system may be lined through from south to north and the frameworks of the fault

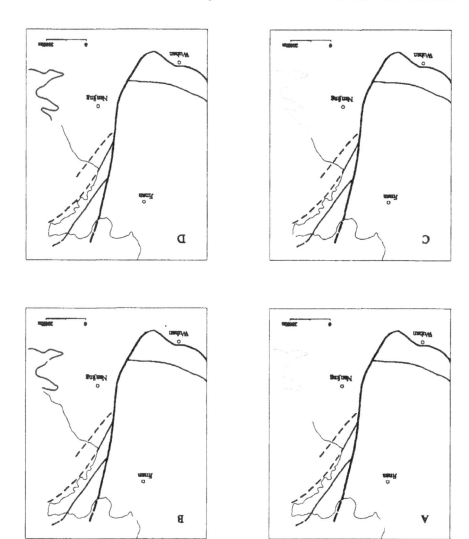

Figure 18. Changes of the direction of the principal tectonic stress during brittle-deformed episode of Tan-Lu fault and its adjacent area. A, Left-lateral strike-slip shearing episode ($J_{1.2}$-J_3); B, Extensional episode (K_1-K_2); C, Compression to right-lateral strike-slip episode (K_2-E); D, Present (N-Q).
\

system was formed. Moreover, the extension in the Tan-Lu fault system also resulted in the rapid rise and the whole ascent and the orogenesis of the Dabie-Sulu belt, at the same time the Hefei basin and the Laiyang basin formed. Tectonic stress field in this stage is characteristic of σ_3 in near E-W direction with stress of 130-150MPa.

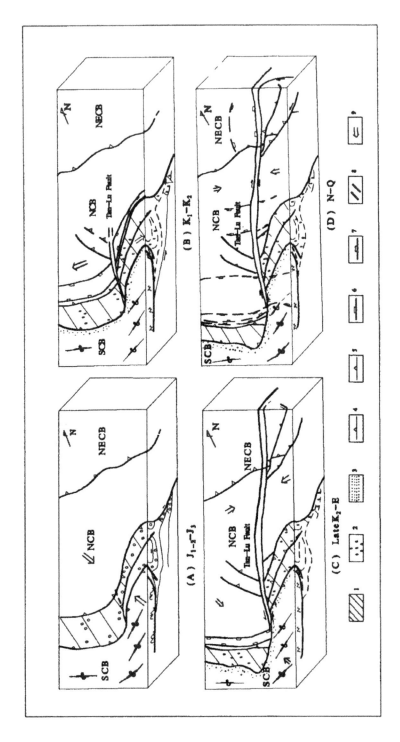

Figure 19. Tectonic model of the evolution of Tan-Lu fault system in brittle-deformed episode. 1, Ultrahigh pressure metamorphic belt; 2, Eclogite; 3, Blueschist belt; 4, Boundary fault; 5, Thrust fault; 6, Mesozoic basin; 7, Cenozoic basin; 8, Fault; 9, Direction of principal stress.

Right-lateral strike-shift compressive stage (later K_2-E)

At the end of Late Cretaceous, the subduction direction of the Pacific Plate relative to Eurasian continent switched from NW to NNW and during the Eocene the continent of Indian underthrust into the continent of Asian. These resulted in the variations of the tectonic stress field in the Tan-Lu fault system and with changed from near E-W extension to near E-W compression with of the stress of 90-130MPa. In the northern part of the fault system therefore the compression with right strike-shift was the dominant factors, and the southern part of the fault system is characterized by compression and left strike-slide. At the shallow levels, the Cretaceous sequences formed a large-scale compressional fold system. When the Tan-Lu fault system continued to develop into the upper mantle, the mantle-source enclaves were distributed along the fault. The right lateral strike-shift of the Tan-Lu fault system also resulted in the formation of the rotational shear structres in western Shandong west of the Tan-Lu fault system. In addition, the generation and the development of a series of Tertiary sedimentary basins and gas reservoirs are also controlled by the Tan-Lu fault system.

Neoid (late) active stage (N-Q)

The tectonic stress field in Tan-Lu fault system since the Neogene is essentially the same as that in paleogene, i. e., near the E-W compression. At the shallow levels Cenozoic folds and thrust faults developed, the Quaternary sediments are overlaid by the older sequences as of thrusting. The Tan-Lu fault system penetrated through Moho into the upper mantle and acted as an ascending conduct of upper mantle materials, with the result that Cenozoic basalts and mantle-source enclaves distributed along the fault system. Along the Tan-Lu fault system, doming and flexure might appear in the lower crust and at the upper levels of the asthenosphere, and the low-velocity zone in the middle crust was distributed concentratelly. It is inferred that the Tan-Lu fault system is now a seismic belt.

ACKNOWLEDGMENTS

Thanks are due to prof. Dong Faxian of Institute of Geomechanics, CAGS and to prof. An Yin and prof. T.M. Harrison of Department of Earth and Space Sciences, UCLA, U.S.A.

REFERNCES

1. L. Ames, G.R. Tilton and G. Zhou. Timing of collision of the Sino-Korean and Yangtze Cratons : U—Pb zircon dating of coesite-bearing eclogites, *Geology*. 21, 339 — 342(1993).
2. D.A. Carswell. *Eclogite Facies Rocks*. Blackie, Glasgow London (1990).
3. W.J. Chen, Q. Li, D. M. Li and X. Wang. Geochronological implications of K/Ar isotope system of fault gouge-a preliminary study, *Phys. Chem. Earth*. 17, 1723(1989).
4. Y. Z. Chen, S.G. Li, B.L. Cong, R.Y. Zhang and Z.Q. Zhang. The formational time and genesis of Jiaonan eclogites-evidence from Sm-Nd isotopic geochemistry and chronology, *Chinese Science Bulletin*. 37, 2169 (1992) (in Chinese).
5. N.G. Deng. Structural system of Cathaysian type of Mesozoic and characteristics of Tancheng — Lujiang fractural system and mechanism of their formation, *Coll. Struct. Geol*. 3, 33 — 38 (1984) (in Chinese with English abstract).
6. M. Enami and Q. Zang. Quartz pseudomorphs after coesite from Shandong province, east China, *Am. Mineral*. 75, 381 — 386 (1990).
7. R.J. Enkin, Z. Yang, Y. Chen and V. Courtillot. Paleomagnetic constraints on the geodynamic history of the major blocks of China from the Permian to the Present, *J.Geophy. Res*. 97, 13953-13989 (1992).
8. Q. Fan, R. Liu and B. Ma. A preliminary study of high-pressure metamorphic ultramafic rocks in North Jiangsu and South Shandong, China, *Acta Petrologica Sinica*. 8(1), 90 — 95 (1992) (in Chinese).
9. C.Y. Li. Geologic Map of Asia, with explanatory text. Geological Publishing House, Beijing (1984) (in Chinese

with English abstract).

10. P.J. Li and B.D. Xia. Transpressional Basin-a case study of Mid-Late Triassic basin around Yangtze River, Lower Yangtze, *Scientia Geologica Sinica*. 30, 130 — 138 (1995) (in Chinese with English abstract).

11. S.G. Li, N. Ge, D. Liu, Z. Zhang, X. Yie, S. Zhen and C. Peng. Sm-Nd age of c-group eclogites in northern Dabie Mountains and its tectonic significance, *Chinese Science Bulletin*. 34, 522 — 525 (1989) (in Chinese).

12. S. G. Li, S.R. Hart, S.G. Zheng, D.L. Liu, G.W. Zhang and A.L. Guo. Timing of collision between the North and South China Blocks-The Sm-Nd isotopic age evidence, *Sci. Sin.(Ser. B)*,32, 1393 — 1400 (1989) (in Chinese).

13. S.G. Li *et al*. Sm-Nd age of coesite-bearing eclogites in southern Dabie Mountains, *Chinese Science Bulletin*. 37, 346 — 349 (1992) (in Chinese).

14. S.G. Li, *et al*. Timing of blueschists in central China. *Scientia Geologica Sinica*. 28, 21 — 27 (1993) (in Chinese with English abstract).

15. X.Y. Ma, C. Q. Liu and G. D. Liu (chief editors). Geoscience Transect from Xiangshui, Jiasngsu to Mandula, Neimenggu (1:1000000), with Explanatory Notes, Geological Publishing House. Beijing (1991) (in Chinese).

16. M. Mattauer *et al*. Tectonics of the Qinling belt: Build-up and evolution of eastern Asia, *Nature*. 317, 496 — 500 (1985).

17. B.G. Niu, Y.L. Fu, G. Z. Liu and J.S. Ren. $^{40}Ar/^{39}Ar$ ages of blueschists in northern Hubei and their geological significance, *Chinese Science Bulletin*. 38, 1309 — 1313 (1993) (in Chinese).

18. A.I. Okay and A.M.C. Sengor. Coesite from the Dabie Shan eclogites, central China, *European Journal of Mineralogy*. 1, 595 — 598 (1989).

19. F.V. Richthofen. The geological structures of Shandong (Jiaozhou) and mineral deposit, *J. Appl. Geol.* March (1898) (in German).

20. X. Wang, J.G. Liou, and H.K. Mao. Coesite-bearing eclogites from the Dabie Mountains in central China, *Geology*. 17, 1085—1088 (1989).

21. X.F. Wang, Z.G. Li, Q. Zhong, B.L. Chen, X.H. Chen and Z.L. Zhen. *Tecnonic feature of the middle part of the Tan-Lu fault zone*. In *30th IGC Field Trip Guide T316*. Geological Publishing House, Beijing (1996).

22. L.S. Xing, Z.J. Li, X.F. Wang and B.L. Chen. Counterclockwise rotation of South China Block east of the Tan-Lu fault zone, *Journal of Geomechanics*. 1(3), 31 — 37 (1995) (in Chinese).

23. J.W. Xu. The Tancheng-Lujiang wrench fault system, *Coll. Struct. Geol.* 3, 18 — 33 (1984) (in Chinese with English abstract).

24. J.W. Xu, K.R. Cui, Q. Liu, W.X. Tong and G. Zhu. Mesozoic sinistral transcurrent faulting along the continetal margin in eastern Asia, *Mar. Geol. Quat. Geol.* 5(2), 51 — 64 (1985) (in Chinese with English abstract).

25. J.W. Xu and G.F. Ma. Review of ten years (1981 — 1991) of research on the Tancheng-Lujiang fault zone, *Geol. Rev.* 38(4), 316 — 324 (1992) (in Chinese with English abstract).

26. J.W. Xu(ed). *The Tancheng-Lujiang Wrench Fault System*. John Wiley Sons, New York (1993).

27. S.T. Xu, W. Su, Y. Liu, L. Jiang, S. Ji, A.I. Okay and A.M.C. Sengor. Diamond in the high pressure metamorphic rocks in the eastern part of Dabie Mountains, *Chinese Science Bulletin*. 17, 1318 — 1321 (1991) (in Chinese).

28. J. Yang, G. Godard, J.K. Kienast, Y. Lu and J. Sun. Ultrahigh-pressure (60kbar) magnesite-bearing garnet peridotites from north-eastern Jiangsu, China, *J. Geology*. 101, 541 — 554 (1993).

29. A. Yin and S.Y. Nie. An indentation model for the North and South China collision and the development of the Tan-Lu and Honam fault systems, eastern Asia, *Tectonics*. 1294, 801 — 813 (1993).

30. Y. Yin. Characteristics and petrogenesis of the eclogites in Shandong and Jiangsu, *Acta Petrologica et Mineralogica*. 10 (1), 11 — 20 (1991) (in Chinese with English abstract).

31. R.Y. Zhang and J.G. Liou. Significance of magnesite paragenesis in ultrahigh-pressure metamorphic rocks, *Am. Mineral*. 79, 397 — 400 (1994).

32. R.Y. Zhang, B.L. Cong and J.G. Liou. Su-Lu ultrahigh pressure metamorphic terrane and its genesis, *Acta Petrologica Sinica*. 9(3), 211 — 226 (1993) (in Chinese with English abstract).

33. S.Y. Zhang, Z.G. Zhou and Y.R. Xing. The characteristics of blueschist belt on the north margin of Yangtze platform, *J. Changchun Univ. Earth Sci.* 4, 53 — 70 (1987) (in Chinese with English abstract).

34. S.Y. Zhang, K. Hu, X.C. Liu and L.Y. Qiao. The Characteristics of Proterozoic blueschist-whiteschist-eclogite in central China. a trinity of ancient intercontinental collapsion-collision zone, *Journal of Changchun University of Earth Science*. Special Issue of Blueschist Belt in Hubei and Anhui Provinces Aps. 152 — 160 (1989) (in Chinese with English abstract).

35. X. Zhao and R.S. Coe. Paleomagnetic constrains on the collision and rotation of North and South China, *Nature*. 327, 141 — 144 (1987).

36. G.Z. Zhou *et al. Research on the blueschist belt in northern Hubei, China*, Geological Publishing House, Beijing (1991) (in Chinese with English abstract).

Part 5

CONTRACTIONAL TECTONICS

Proc. 30th Int'l. Geol. Congr. Vol. 14, pp. 253-262
Zheng et al. (Eds)
© VSP 1997

Evidence for Growth Fault-Bend Folds in the Tarim Basin and Its Implications for Fault-Slip Rates in the Mesozoic and Cenozoic

HUAFU LU, DONG JIA, CHUMING CHEN, DONGSHENG CAI, SHIMIN WU, GUOQIANG WANG, LINGZHI GUO,YANGSHEN SHI

Department of Earth Sciences, Nanjing University, Nanjing 210093, China

Abstract

Growth fault bend folds were discovered in the Tarim basin for the first time. The growth fault bend fold is an excellent indicator for timing the emplacement of thrust fault. Typical growth triangle and growth strata are well exposed on surface and identified in seismic profiles. They are of $V_s/V_u>1$ type, $V_s/V_u=1$ type and $V_s/V_u<1$ type (V_s : sedimentary rate; V_u : uplift rate), and occurred during the Permian, Triassic, Jurassic and Cenozoic. In light of the growth triangle, the average movement rate of the thrust is estimated as 0.06 mm/a during Jurassic. The slow thrust emplacement rate is favorable for oil and gas accumulation.

Keywords: growth fault bend fold, growth triangle, fault timing, thrust movement rate, Tarim basin

INTRODUCTION

The processes of tectonic deformation take certain time duration. Traditionally, geologists take the span between the youngest strata age below the unconformity surface and the oldest strata age upon the unconformity surface. Using this conception, the unconformity usually provides a loss and an even wrong answer of the timing of deformation. The duration of the deformation is not necessary the same as that of erosion. Many structures are formed under the water beneath the erosional surface. An unconformity represents the erosion events which are mainly related to the crustal uplift. The geologists tend to find the shorter span of the strata ages beneath and above the unconformity surface, which means the deformation event timing is precise. That implies the idea of fast deformation episode. But in fact, many deformation events take long time duration.

Suppe et al.[11] proposed a model of growth fault-bend folds, which provided a powerful timing method for structural deformation. We report here the finding of growth fault-bend folds in the Tarim basin for the first time. The critical point for the formation of growth fault-bend fold is that the total or part of a fault-bend fold develops above a blind flat-ramp thrust occurring during deposition. Under this situation, several sedimentary features appear, which can be used as excellent timing and kinematic indicators of structural deformation (Fig. 1). In Suppe's model [10], the growth strata exhibit various thickness in different parts of the growth fault-bend fold, while the pre-The growth triangle is confined by the active axial surface and growth (inactive) axial surface (Fig. 1). The quotient of growth triangle base length(L_0) over the span of growth strata(T_t) is the average slip rate of the thrust, $V=L_0/T_t$. The quotient of length difference between the bottom and top surface of any horizon in the triangle(Δl_i) over the span of the horizon(Δt_i) is the average slip rate during the depositional duration of the horizon, $V_i = \Delta l_i / \Delta t_i$ (Fig.2).

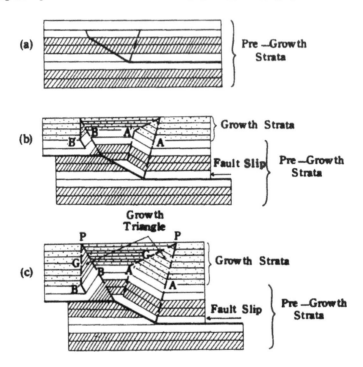

Figure 1. Model of growth fault-bend fold. Symbols: A and B, active axial surfaces; A' and B', inactive axial surfaces (simplified as the active surface and inactive surface respectively in the following features); G, growth (inactive) axial surface; P, point of intersection between G and A, B respectively. After Suppe et al. [11] and Shaw [7, 8].

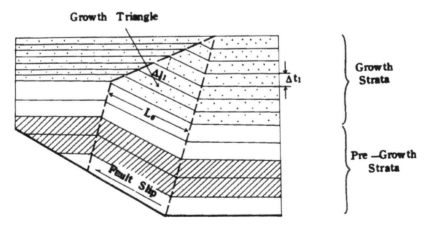

Figure 2. Growth triangle and average slip rate on the footwall ramp. Average slip rate (V) is the ratio of the base length of strata in growth triangle (L_0) and the span of growth strata (T_1). After Shaw [7, 8].

In light of internal structures of the growth strata, they are divided into three types by the ratio of sedimentary rate (V_s) and uplifting rate (V_u) of fault-bend anticline: $V_s/V_u>1$, $V_s/V_u=1$ and $V_s/V_u<1$ [1,11] (Fig. 3). The sections of Fig. 3 a and Fig. 1 are of $V_s/V_u>1$ type. For the rest two types there is no deposition on the top of the anticline, and the onlap of growth strata occurs on the frontal limb of the anticline, which is named the growth onlap wedge. While the growth truncation wedge occurs on the rear limb of the anticline where the growth strata are truncated. The polarity of the growth onlap wedge and growth truncation wedge provides a good kinematic indicator of movement direction of blind stepped thrust below the growth fault-bend fold.

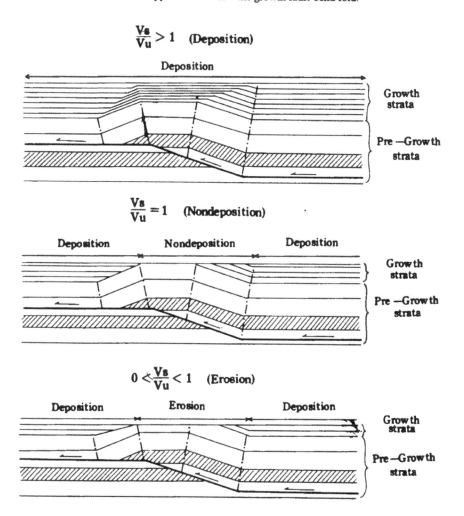

Figure 3. Different growth strata structures under different V_s/V_u ratios. Symbols: V_s, sedimentary rate; V_u, uplift rate. After Suppe et al. [11].

The above mentioned timing and kinematic indicators are well exposed in both surface sections and seismic profiles in the Tarim basin, which helped us to improve our understanding to the tectonics of the basin.

GROWTH FAULT-BEND FOLDS IN THE TARIM BASIN

The Tarim basin is located directly south of the rejuvenated Tianshan Paleozoic collisional orogenic belt (Fig. 4a).

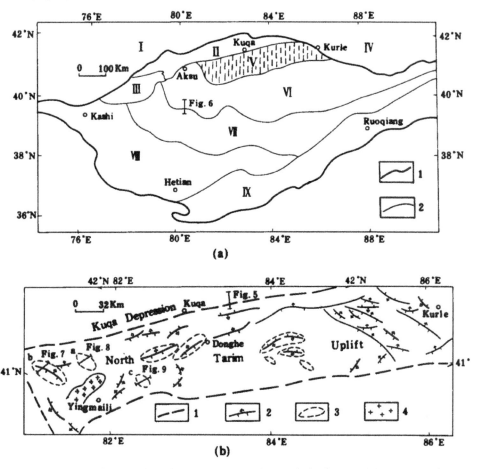

Figure 4. (a). Tectonic units of Tarim basin and Tianshan Paleozoic collisional orogenic belt. Symbols: 1, basin boundary ; 2, boundary of tectonic units;. I , South Tianshan Paleozoic orogenic belt; II , Kuqa rejuvenated foreland deformation belt; III , Kalpin rejuvenated foreland basement deformation belt; IV, Keluketake rejuvenated foreland basement deformation belt; V, North Tarim Uplift; VI, Manjaer depression; VII, Central Uplift (left part is Bachu Uplift); VIII, Southwest depression; IX, Southeast depression. (b).Sketch map of the structures in the North Tarim Uplift, which is located in the dashed area on Figure 4a. Symbols: 1, boundary of the North Tarim. Uplift ; 2, thrust faults; 3, anticlines; 4, granitoid bodies; a, Yingmai No. 8 anticline; b, Nanka well No. 1 anticline; c, Yingmai No. 1 anticline. Modified from Jia[3].

The growth fault-bend folds which we discovered are scattered in the Kuqa Cenozoic rejuvenation foreland deformation belt[5,6], the North Tarim Post-Paleozoic uplift, and the Bachu Cenozoic uplift[3](Fig. 4b).

We observed a series of growth fault-bend folds in the Tarim basin in both surface sections and seismic profiles, This is the first discovery of this kind of structures in China. They developed in Late Paleozoic-Early Mesozoic time and Cenozoic time, and belong to the type of $V_s/V_u>1$ and types of $V_s/V_u \leqslant 1$. The Yaken anticline and the anticline on the eastern end of the Bachu uplift in SN-87-400 line are formed in Cenozoic and are of $V_s/V_u=1$ type and $V_s/V_u>1$ type respectively. The Yinmai No. 8 anticline and Yingmai No. 1 anticline in North Tarim Uplift formed during Late Paleozoic-Early Mesozoic and are of $V_s/V_u<1$ type. Nanka well No. 1 in the western termination of the North Tarim Uplift formed during the Triassic and Jurassic periods and is of $V_s/V_u>1$ type.

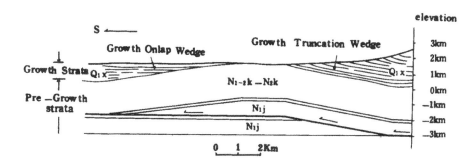

Figure 5. Yaken Quaternary growth fault-bend fold. For location see Figure 4. The section is constructed according to the deduction by the surface section survey, the geometric features of the deformed strata, the thickness of subsurface strata of $N_{1-2k}~N_{2k}$ and N_{1j}, and the gypsum bearing horizon of Jidike formation(N_{1j})[3].

The Yaken anticline is located at the southern margin of the Kuqa rejuvenated foreland deformation belt which exposed on the south flank of the northern Tarim basin for 40 km long in the E-W direction. The pre-growth strata in the Yaken fault-bend anticline cropping out on surface are sandstones and conglomerates of Late Pliocene Kuqa formation (N_{2k}). The sandstones, mudstones, gypsum and salt beds of the Late Miocene-Early Pliocene Kangcun formation (N_{1-2k}) and the Early Miocene Jidike formation (N_{1j}) are inferred being involved in the anticline formation in the subsurface. The growth strata is black gravel of the Early Quaternary Xiyu formation. The frontal limb (southern limb) made up of Kuqa formation dips to the south with 8°, and the rear limb (north limb) dips to north with 10°. The Xiyu formation exhibits the growth onlap wedge against the top of Kuqa formation of frontal limb of the anticline, while the Xiyu formation exhibits the growth truncation wedge which in fact is the growth triangle in a special case on the rear limb of the anticline. On the flat top of the anticline there are only a few meters of the Xiyu formation, which approaches zero in the macrostructure. Therefore the anticline is formed in Quaternary and is of $V_s/V_u=1$ type (Fig. 5). Taking the base length of growth triangle as 6 Km and span of Xiyu formation as 2 Ma, the slip rate of the blind stepped southward thrust is roughly estimated as 3mm/a.

The anticline in the seismic profile on the eastern end of the Bachu uplift is a growth fault-bend fold. On its rear limb, the growth triangle is well-developed (Fig. 6). The inactive axial surface dips to the south, and it starts from the bottom of the Cambrian, passes through the total Paleozoic and Mesozoic strata, and ends in the Lower Tertiary strata, which are of pre-growth strata exhibiting constant thichness on both sides of the inactive axial surface. The active axial surface runs parallel to the inactive axial surface and extends into the middle part of the Tertiary strata. The growth strata is inferred as Oligocene and Miocene in age which represents the duration of thrusting. The growth axial surface dips gently to the south, above which the growth strata are 2/3 thinner than those beneath that. The footwall ramp of the blind thrust is in the Lower Cambrian strata and dips to the north. Both the growth strata and the footwall ramp suggest that the thrust moved from north to south. The growth fault-bend fold anticline is of $V_e/V_u>1$ type.

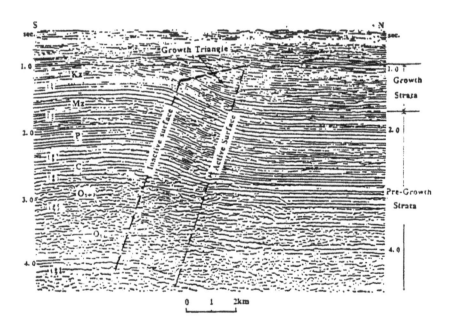

Figure 6. Growth triangle of a growth fault-bend fold on the eastern end of Bachu uplift. For location see Figure 4. Seismic profile from the Tarim Petroleum Exploration and Development Bureau(TPEDB).

The Nanka well No. 1 growth fault-bend anticline is shown in seismic profile (Fig. 7). The growth triangle appears at the central part of Fig. 7. The inactive and active axial surfaces dip NE. The blind thrust moves northeastward. The inactive axial surface terminates at the bottom of Jurassic strata, and the active axial surface extends to the top of Jurassic strata. The growth axial surface dips to northeast gently. The thickness of Jurassic strata in the growth triangle is two times thicker than that above the growth axial surface. The growth strata of Jurassic reveal that a blind northeastward thrusting occurred in Jurassic period, and the growth fault-bend fold is of $V_e/V_u>1$ type. The Triassic strata between the active and inactive axial surfaces are truncated by the bottom surface of Jurassic, forming the wedge feature. The strata before Triassic exhibit constant thickness as the pre-growth strata. That suggests a $V_s/V_u<1$ type growth fault-bend folding and a northeastward thrusting occurred in Triassic time. In light of Jurassic strata in triangle, the slip rate of thrusting is estimated as 0.06 mm/a (3165 meter/(208-145)10^6a=0.0597 mm/a).

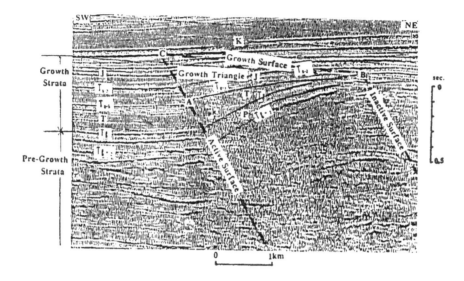

Figure 7. Nanka No. 1 growth fault-bend fold. For location see Figure 4. Seismic profile from TPEDB.

The growth onlap wedge of the Yingmai No. 8 anticline occurs in seismic profile shown in Fig. 8. The anticline is trending northwest. Inactive and active axial surfaces dip to northeast. Pre-growth strata are Paleozoic. Horizontal growth strata of Triassic and Jurassic lay on the southwest dipping top surface of Paleozoic strata formed a typical growth onlap wedge. The thickness of Pre-growth Paleozoic strata between inactive and active surfaces is thicker than that on top of anticline, which implies that there was not only nothing deposited on the top of anticline during Triassic and Jurassic but also local erosion occurred there. Therefore the anticline is of $V_s/V_u<1$ type. The growth onlap wedge thins on northeastward suggesting a southwestward blind thrusting occurred beneath the Paleozoic strata.

The Yingmai No. 1 anticline is a northwest trending fault-bend fold. The footwall ramp of stepped thrust cuts through Sinian and Lower Cambrian strata, and dips to southwest. The pre-growth strata are Sinian, Cambrian, Ordovician and Silurian, and the growth strata are Devonian, Carboniferous, and Permian. On the southwest limb of the anticline, the rear limb, the growth strata were eroded and covered by the Triassic strata, forming a growth truncation wedge. On the northeast limb of the anticline, the frontal limb, the growth strata overlap on the top of Silurian strata forming a growth onlap wedge (Fig. 9). indicating a $V_s/V_u<1$ type growth fault bend anticline. That implies the northwest trending fault thrusts northeastward during the Late Paleozoic.

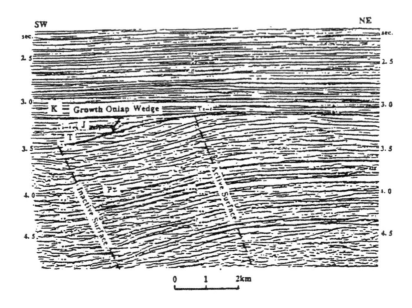

Figure 8. Onlap sedimentary structure on the rear limb, southwest limb of Yingmai No. 8 growth fault-bend fold. For location see Figure 4. Seismic profile from TPEDB.

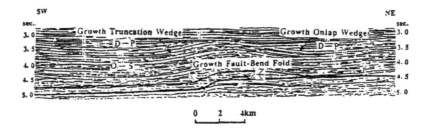

Figure 9. Yingmai No. 1 growth fault-bend anticline of $V_s/V_u<1$ type. For location see Figure 4. Seismic profile from TPEDB.

DISCUSSION AND CONCLUSIONS

1. Growth fault-bend folds are discovered for the first time on surface and seismic profiles in the north flank of the Tarim basin. They are of $V_s/V_u>1$, and $V_s/V_u=1$ and $V_s/V_u<1$ types.

2. Part of the growth fault-bend folds, the Yaken anticline and the Bachu uplift eastern end anticline and their related thrusts occurred in the Cenozoic and the thrust slip rate using the Yaken anticline as example is estimated as 3mm/a. The rest of the growth fault-bend folds and related thrusts (the Nanka well No. 1, Yingmai No. 8, Yingmai No. 1 anticlines) developed during the Late Paleozoic and Early Mesozoic, and the thrusting slip rate using the Nanka well No. 1 anticline as example is estimated as 0.06 mm/a.

3. It appears that the Cenozoic thrusting slip rate is more than a hundred times higher than that of Paleozoic-Early Mesozoic thrusting, indicating the differences of structural deformation strength between the main processes of structural emplacement of rejuvenated foreland deformation belt[5] and the intracraton strike slip deformation after main emplacement processes of the foreland deformation[6,9]. During the Cenozoic, the remote tectonic effect of India-Asia collision[2] resulted in the rejuvenation of the Permian-Triassic foreland basin to form the Kuqa Cenozoic rejuvenated foreland deformation belt[5]. That represents a main intracratonal orogenic processes implying A-type subduction. Therefore it has relatively higher thrust slip rate. The Permian, Triassic and Jurassic slow thrust slip rate are resulted from a second order stress field by the strike slip movement along the E-W trending boundary fault of the North Tarim Uplift after the Tarim-Kazakstan collision and its related foreland deformation event in the Carboniferous and Permian[6,9]. The emplacement of this kind of thrusting and fault related folds lasted for several tens of million years. The anticlines resulted from the slow deformation processes are favorable for the accumulation of petroleum and nature gas.

Acknowledgments

This contribution is a part of results of the Eighth Five-Year Plan Research Program, Tarim Petroleum Province Tectonics, entitled "Study on the Structural Features of West Segment of North Margin, Kalpin and Yingmaili Areas of Tarim Plate". We thank China National Petroleum Corporation, Tarim Petroleum Exploration and Development Bureau(TPEDB) and Dr. C. Jia for their help in our study and for their permission of using seismic profiles in this paper. We thank Professor J. Suppe for his discussion on the growth fault-bend fold materials in this paper and his encouraging us in this study. We also thank Professor Y. Zheng and Professor A. Yin for their careful review and valuable suggestions to our manuscript.

REFERENCES

1. D.A. Medwedeff. Growth fault-bend folding at Southeast Lost Hills, San Jouquin Valley, California. Amer. Assoc. Petrol. Geol. Bull. 73, 54-67(1989).
2. L. Guo, H. Lu and Y. Shi. Two kinds of remote structural effects of the India-Tibet collision. In: *Symposium of Modern Geology (Volume I)*. Q. Li, J. Dai, R. Liu and J. Li (Eds).pp.1-8. Nanjing University Press(in Chinese with English abstract)(1992).
3. C. Jia. Tectonic evolution of Tarim basin. In: *Symposium of Modern Geology (Volume I)*. Q. Li, J. Dai, R. Liu and J. Li (Eds). p.22. Nanjing University Press(in Chinese with English abstract)(1992).
4. H. Lu , D.G. Howell, D. Jia, D. Cai, S. Wu, C. Chen, Y. Shi, Z.C. Valin and L. Guo. Kalpin transpression tectonics, Northwestern Tarim Basin, Western China. Inter. Geol. Rev. 36, 975-981(1994a).
5. H. Lu, D.G. Howell, D. Jia, D. Cai, S. Wu, C. Chen, Y. Shi and Z.C. Valin. Rejuvenation of the Kuqa foreland basin, Northern flank of the Tarim Basin, Northwest China. Inter. Geol. Rev. 36, 1151-1158(1994b).
6. H. Lu, D. Jia, D. Cai, S. Wu, C, Chen and Y. Shi. Plate tectonic evolution of the Tarim and west Tianshan. In: *Progress in Research on the Petroleum Geology of the Tarim basin.*. X. Tong, D. Liang, and C. Jia (Eds). pp. 235-245. Science Press of China(in Chinese with English abstract)(1996).
7. J.H. Shaw. Active blind thrust faulting and strike-slip fault-bend folding in California., Doctoral thesis,

Princeton University(1993).

8. J. H. Shaw and J. Suppe. Active faulting and growth folding in the eastern Santa Barbara Channel, California. Geol. Soc. Amer. Bull. **106**, 607-626(1994).

9. Y. Shi, H. Lu, D. Jia, D. Cai, S. Wu, C. Chen, D.G. Howell and Z.C. Valin. Paleozoic plate tectonic evolution of the Tarim and Western Tianshan regions, Western China. Inter. Geol. Rev. **36**, 1058-1066(1994).

10. J. Suppe. Geometry and kinematic of fault-bend folding, Amer. Jour. Sci. **283**:9, 684-721(1983).

11. J. Suppe, G.T. Chou and S.C. Hook. Rates of folding and faulting determined from growth strata, In: *Thrust Tectonics*. K.R. McKlay(Ed).pp. 105-121. Chapman Hall publisher(1992).

Proc.30⁰ Int'l. Geol. Congr.,Vol.14, pp. 263-274
Zheng et al.(Eds)
© VSP 1997

Determination of Contractional Fault Systems in the Southern Margin of the Ordos Basin, China, by Multisource Data Processing

YONGJIE TAN

Coal Geological Survey of China (CGSC), 40 Fanyang Road, Zhuozhou, Hebei, 072752 China

J. L. VAN GENDEREN

International Institute for Aerospace Survey and Earth Sciences (ITC), 7500 AA Enschede, The Netherlands

Abstract

In this paper, multisource geoscientific data processing are used to study the reverse faults in the southern margin of Ordos basin, China, such data include NOAA data, the data of Bouguer gravity anomaly and aeromagnetic anomaly, and the multisource geophysical processing methods include upward continuations, vertical derivative and gradient image, and synthetic image processing method - remote sensing image processing and geophysical image processing under I^2S600. On the basis of these results and with geological investigation in the field, the authors have determined the existence of an EW striking contractional system, a NE striking contractional system and a NW striking contractional system. These three systems reveal the evolution of geological structures along the southern margin of Ordos basis and in the north side of Qinling orogenic belt.

Keywords: multisource data, data processing, contractional system, deformation, Ordos basin, China

INTRODUCTION

Ordos basin covers the northern part of Shaanxi Province, China, and small areas of other provinces. It is a famous energy resource basin in China and in the world for it contains large coal resources, petroleum, gas and other sedimentary mineral deposits.

Ordos basin is a depositional sequence in the Mesozoic and partly reformed in the Cenozoic. It is located on the northern side of the Qinling orogenic belt and the eastern side of the Helanshan Mountains - Liupanshan Mountains belt (Fig.1). It is, therefore, at the intersection of two boundaries that divide China's geology into eastern - western tectonic provinces and northern - southern tectonic provinces.

The study area for this paper is located at the southern margin of Ordos basin. Because this area is largely covered by the Mesozoic sediments, vegetation and is over printed by later deformation. Knowledge of its structure, by traditional geological surveys, has not been very successful. This situation negatively affects the geologists' ability to understand better the region's coal resources and other deposits. For this reason, multisource geoscientific data processing has been used to study the structural features of this area. The source data include TM data, NOAA data, Bouguer gravity anomaly (1:200000), aeromagnetic anomaly

(1:200000) and seismic data. From interpretation of the processed results and geological investigations in the field, a number of reverse faults have been determined and grouped into three contractional fault systems. The three systems define the geological structures of the area and are the basis for studying the distribution of coal resources in this part of the Ordos basin.

STRATIGRAPHIC SEQUENCES IN STRUCTURAL EVOLUTION AND IN GEOPHYSICS

STRUCTURAL/ STRATIGRAPHIC SEQUENCES

The structural/ stratigraphic sequences represent a series of strata that are formed in a limited region during one structural stage. According to the structural histogram and strata present in this area [7], six structural/ stratigraphic sequences have been determined. They are the basis for interpreting faults and thrusts from remote sensing images and multisource images.

The structural/ stratigraphic sequence of Archeozoic and Lower Proterozoic eras (SS1) is composed of hypometamorphic rocks and constitutes the metamorphic basement of this geological basin. The structural/ stratigraphic sequence of Middle and Upper Proterozoic Era (SS2) is composed of epi-metamorphic rock, sandstone and limestone. The total thickness in this sequence is relatively thin. The structural/ stratigraphic sequence of the

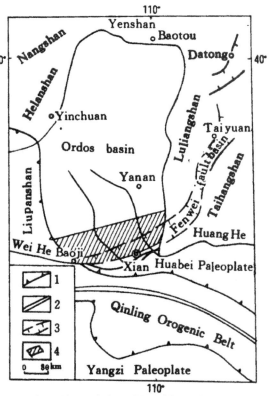

Fig.1 The geologic position of the study area
1-The thrust at the margin of orogenic belt; 2-The boundary of paleoplate collision; 3-The margin of fault basin; 4-study area

Cambrian and Ordovician periods (SS3) is composed of limestone. It is the basement of Carboniferous and Permian coal series, Huabei coal basin which includes the Ordos basin. Hence the top of this sequence is an important geological surface. The structural/ stratigraphic sequence of the Carboniferous, Permian and Triassic Periods (SS4) is composed of thin limestone, sandstone and coal seams. Furthermore, it is basement of the Jurassic coal series, Ordos coal basin. Then the top of the sequence is also an important geological surface. The structural/ stratigraphic sequence of the Jurassic and Lower

Cretaceous Periods (SS5) is composed of sandstone, shale and coal seams and is the main part of Ordos basin. The structural/ stratigraphic sequence of the Cenozoic Era (SS6) is composed of clastic rocks and detrital sediments.

THE STRATIGRAPHIC SEQUENCES IN GEOPHYSIC FEATURES

Geophysic Features and Geophysic Stratigraphic Sequences

The geophysical features of rocks and strata are the basis for geophysical data processing. Table 1 is a statistical summary of geophysical features of the rocks in this area [1,2,3].

Table 1. The geophysic sequences and geophysical features of strata and rocks

strata and rock	geophysics	density g/cm³	magnetic susceptibility 10^{-5}SI	residual magnetization 10^{-5}SI	seismic velocity (km/S)
cover-ing strata	Q: loess, sand	1.67	60	26	1.7~
	R: clay rock, mudstone	2.14			3.8
	Mz-Pz$_2$: sandstone	2.53			5.5~
	Pz$_1$-Pt$_2$: limestone	2.71			6.1
base-ment	Pt$_1$-Ar: metamorphic rock of basement, Pt: metamorphic rock in Qinling Belt	2.73~ 2.90	899	267	6.25

Density stratigraphic sequences: From Table 1, it is obvious that there are five density stratigraphic zones: Quaternary, Tertiary, Mesozoic Erathem - Upper Paleozoic Erathem, Lower Paleozoic Erathem - Middle Proterozoic Erathem, and metamorphic basement. Of the four interfaces, between them, the oldest three interfaces are important for the interpretation of geological structures.

Magnetic stratigraphic sequences: From Table 1, it is obvious that there are two magnetic zones: covering strata and metamorphic basement rock. Therefore, only one magnetic interface.

The reflective surfaces of seismic wave: There are five reflective surfaces for us in this area [5]: the top of Triassic System (T_5), the top of Lower Paleozoic Erathem(T_9), the top of Upper Proterozoic Erathem (T_{12}), the top of middle Proterozoic Erathem (T_{14}) and Lower Proterozoic Erathem (T_8).

Spectral Reflectance of Stratigraphic Sequences

Table 2 is the result of spectral reflectance measurements on rocks from the northwestern part of the Ordos Basin [4]. The strata and rocks in that area are almost the same as in our study area, and are, therefore, applied to this remote sensing image interpretation and image processing.

Table 2. Spectral reflectance of main rock units, northwestern Ordos Basin

band spectrum(μm)	TM1 0.45-0.52	TM2 0.52-0.60	TM3 0.62-0.69	TM4 0.76-0.90
Archean metamorphic rock	40.29	46.40	52.96	56.13
lower Paleozoic limestone	31.71	37.37	39.58	42.39
Carboniferous-Permian coal series	23.92	27.91	32.45	34.96
Jurassic coal series	19.71	19.51	19.37	22.51
Quaternary sediments	26.64	30.09	32.73	39.52

MULTISOURCE GEOSCIENTIFIC DATA AND THEIR PROCESSING

TM DATA

Six scenes of TM data (bands 2, 3, 4) were used to form one color mosaic image under the computer image processing system, and then the mosaic image was processed through enhancement function; the result is shown in Plate 1. From this image, some faults can be interpreted, but because some portions of them are not obvious, certain some interested subareas of the TM images were also processed further.

In order to show the results of the multisource data processing, we take the determination of fault F_4 as an example. F_4 is the Wufengshan - Caiershan reverse fault (see Fig.5), which extends in an E-W direction through the middle part of the study area. Its central section is exposed at the surface, where the northern wall is composed of Carboniferous-Triassic sandstone(SS4); and the southern wall is composed of the Cambrian and Ordovician limestone (SS3). The fault's eastern section is covered by loess and sand; the western section is covered by Jurassic sandstone.

Plate 1. The mosaic TM image (2,3,4) of the study region

The spatial resolution of our TM data is 30m, not very good for the interpretation of regional faults. Therefore, a new method (TTOM) was developed in this research. The average value of a window (3 pixels × 3 pixels) is put on the center point of the window, and the window is shifted one pixel by one pixel to produce a new image. This new image is then processed by the following enhancement of computer system. Plate 2 is the result of TM in the central part of the study area with this processing way (TM347, TTOM, Histogram-equalization). The middle portion of fault F_4 appears clearly on it.

Plate 2. The processed TM image of the mid-section of reverse fault F_4

The middle western section of F_4 is not obvious in TM234 mosaic image because it is covered by a thin layer of soil and vegetation. Another processing method was developed in this research, named relative gradient processing (RG). A new image was obtained from the following calculation:

$$g(i,j) = (d(i-1,j) - d(i,j)) / d(i,j) \qquad \text{or}$$
$$g(i,j) = (d(i+1,j) - d(i-1,j)) / d(i,j)$$

Plate 3 is a processed image from TM347, TTOM, RG and Scale. It shows the existence of F_4 in its middle western section. Plate 4 is obtained in the same area with almost the same processing: TM347, TTOM, Exponent. The middle western section of F_4 obviously exists on it.

Plate 3. The processed TM image of the middle western section of reverse fault F_4

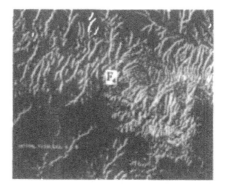

Plate 4. The processed TM image of the middle western section of reverse fault F_4

NOAA DATA

NOAA data images have a lower spatial resolution 1100m, but they are very useful in

detecting differences in vegetation and moisture at the surface or in shallow soil. NOAA and
TM data are overlain to form a new
image that can increase the precision
of fault determination, even for those
faults are shallowly buried by
sedimentary rocks. Plate 5 is a
processed image that comes from the
superposition of NOAA and TM data
under the control points. It clearly
shows the existence of middle
western section of fault F_4. Fig.2, the
interpretation map from the mosaic
TM image, NOAA image and some
processed subarea images, shows the
main structures of the area.

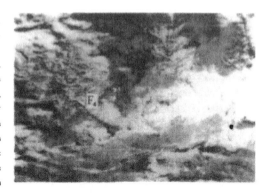

Plate 5. The composition of NOAA and TM data

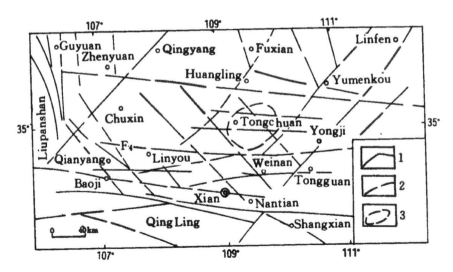

Fig. 2. The interpretation map of TM image and NOAA image
1-fault; 2-lineament; 3-annular structure

THE DATA OF BOUGUER GRAVITY ANOMALY

As mentioned above, there are four density interfaces in this geological area. This is the
basis for using the data of Bouguer gravity anomaly to determine fault locations. For most
of the area's larger faults, the two walls do not have the same density. So gravity data can
be used to determined the faults' buried traces, even when buried under a cover of sediment
or hidden by vegetation.

The 1:200000 map of Bouguer gravity anomaly was digitized by computer, then the digital data is transferred into raster form with 1cm×11cm window, and then is processed with the method of first derivative processing at vertical direction or gradient processing at 0° direction. The initial result is transferred into image processing system S600 and is processed with the functions of S600, such as Convolution interpolation enlargement, False color, etc. Plate 6 is the image from the processing of first vertical derivative of Bouguer gravity anomaly. It shows

Plate 6. The first vertical derivative of Bouguer anomaly

the existence of fault F_4 in its western section despite its cover of Jurassic-Lower Cretaceous strata. A seismic prospecting profile (Fig.3) also confirms the existence of F_4 in the western part of this area.

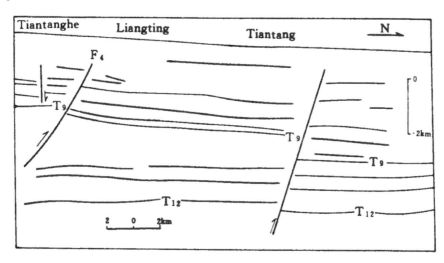

Fig. 3. The structural profile of seismic reflective surfaces along Tiantanghe River, Linyou
T_9 - about the top of Pz_1; T_{12} - about the bottom of Pz_1

In order to get a good result of interpretation from processed image, the above image (Plate 6) is changed with the following formula:

$$g = (-p \cdot \cos\theta \cdot \cos\varphi - a \cdot \sin\theta \cdot \cos\varphi + \sin\varphi) / \sqrt{1 + a^2 + p^2}$$

among it: $p = \partial z/\partial x$; $a = \partial z/\partial y$; θ=azimuth to look at; φ=the dip to look at

then the image g is multiplied by the three bands of its false color image (Plate 6) respectively. A resulting image is a color composite picture with false stereoscopic appearance. This processing can be finished by a series of commands under S600. The result is shown in Plate 7.

The combined Bouguer gravity anomaly and TM images yield not only the existence of faults, but also their position at the surface. Plate 8 is a composite of band 2 of plate 7, TM 2 and band 1 of Plate 7. Fig.4 is the interpretation of Plate 7 and Plate 8. It shows the structural sketch of this area.

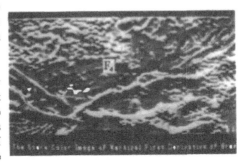

Plate 7. The stereoscopic sensation image processed withthe first vertical derivative of Bouguer gravity anomaly

AEROMAGNETIC ANOMALY DATA

The aeromagnetic anomaly data (1:500000 and 1:200000) were transferred into computer in the same way as Bouguer gravity anomaly was. The data is calculated with different distances of upward continuations. The result only presents the faults which cut the metamorphic basement. Plate 9 is the result with this method (5km upward), it shows that fault F_4 also cuts the basement of this area.

THE DETERMINATION OF CONTRACTIONAL FAULT SYSTEMS

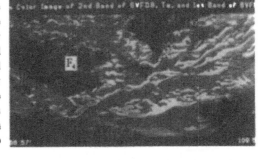

Plate 8. A composite composition of TM and Bouguer gravity anomaly image

On the basis of the data processing techniques discussed above, a field investigation was made along a series of geological profiles to check and to confirm whether the predicted faults do exist and to define their nature. Fig.5 is a structural geological map derived from the interpretation of processed multisource geoscientific images, the result of investigation in the field, and seismic prospecting data and other data. It shows the geological structural framework of the study area.

EW STRIKING CONTRACTIONAL SYSTEM

E-W striking reverse faults are the dominant structures in the area and are distributed in its north part. They are accompanied by asymmetric folds and other contractional structures.

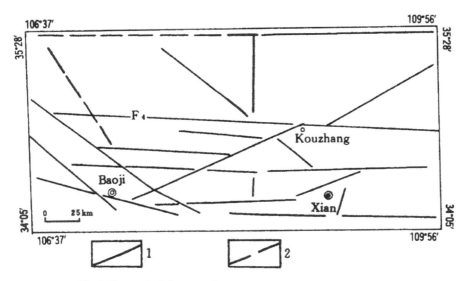

Fig. 4. The synthetic interpretation map of multisource image processed
in the southern margin of Ordos basin
1-lineament; 2-inferred fault

Among them, there are eight main faults: Tianfesi-Caojiaping reverse fault (F$_1$), Jiaojie-Zhuanghegou reverse fault (F$_2$), Zhangjiazui reverse fault (F$_3$), Wufengshan-Caiershan reverse fault (F$_4$), Lidipe-Didian reverse fault (F$_5$), Chenglou reverse fault (F$_6$), Zaomiao-Dukanggou reverse fault (F$_7$) and Yunmengshan-Aitiecun reverse fault (F$_8$).

Some reverse faults of this system are covered by Jurassic strata, which means that this system formed before the Jurassic, i.e., in the

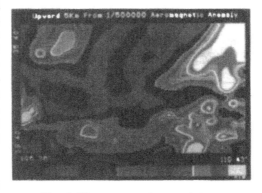

Plate 9. The aeromagnetic anomaly image
with 5km upward

deformation from the end of Ordovician to the end of Triassic. The other study [5] also reveals that the structures of this system appears in an imbricate arrangement from south to north direction, the intensity of deformation decreases from south to north, and the age of deformation is younger toward the north. Given that the Qinling orogenic belt was formed in the same - the end of Ordovician to the end of Triassic [6,7]. The formation of this southern Ordos system was controlled by the generation and evolution of Qinling orogen.

NE STRIKING CONTRACTIONAL SYSTEM

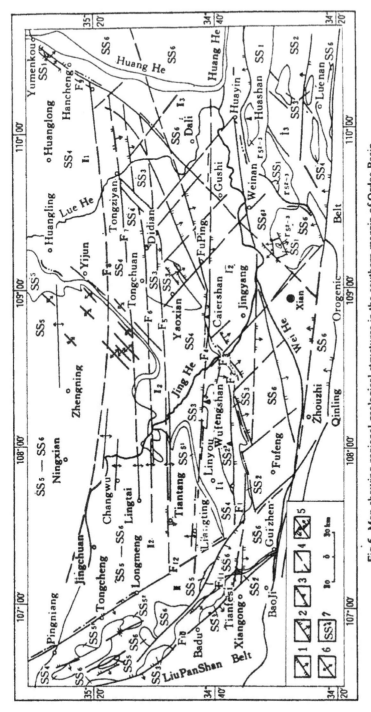

Fig.5. Map showing regional geological structures in the southern margin of Ordos Basin

1-reverse fault; 2-normal fault; 3-ductile thrust; 4-inferred fault or buried fault; 5-transferring fault; 6-anticline; 7-the boundary of structural/ stratigraphic sequences; SS_6-the structural/ stratigraphic sequence of Cenozoic; SS_5 -the structural/ stratigraphic sequence of Jurassic - Lower Cretaceous; SS_4-the structural/ stratigraphic sequence of Carboniferous - Triassic; SS_3 -the structural/ stratigraphic sequence of Cambrian - Ordovician; SS_2-the structural/ stratigraphic sequence of Middle and Upper Proterozoic; SS_1-the structural/ stratigraphic sequence of Archeozoic - Lower Proterozoic

Reverse faults and asymmetric folds which strike northeasterly are found in the eastern part of the study area. The main fault is the Yumenkou-Yingshan reverse fault (F_9).

This system was formed in the Jurassic-Cretaceous deformation because the asymmetric folds are locally covered by the lower Cretaceous strata. This system has features which indicate that deformation spread northwestwards. The formation of this system was controlled by the generation and evolution of the Luliangshan tectonic belt [5].

NW STRIKING CONTRACTIONAL SYSTEM

This system is distributed in the southwestern part of the study area, and is composed of reverse faults and asymmetric folds. The main faults are Badu-Guizhen reverse fault (F_{10}), Qianyanghe reverse fault (F_{11}) and Tongcheng-Longmen reverse fault (F_{12}).

The formation of this system ended at the end of Cretaceous to the end of Eogene because there is an unconformity under the bottom of lower Cretaceous System and the strata of lower Cretaceous system is pushed and put over the lower Tertiary. The research also shows that the formation of this system is controlled by the generation and evolution of Liupanshan tectonic belt [5]

CONCLUSIONS

The conclusion of this research are as follows:

1. TM data, NOAA data and the processing methods developed are very effective in determining the location of faults in the research area, so it is also an important work for the same study.

2. Bouguer gravity anomaly and magnetic anomaly data have also been used successfully to determine the location of some sections of buried faults.

3. The geological structures of the study area consist of three contractional fault (and fold) systems that have controlled the geological evolution in the southern margin of Ordos Basin.

Acknowledgments

The authors would like to thank the Aerophotogrammetry and Remote Sensing of China Coal (ARSC) for providing the geophysical, geological and remotely sensed data used for this study. Thanks are also due to Prof. Huang Kexing, Prof. Gao Wentai and Prof. Guan Haiyan for their assistance in this research.

REFERENCES

1.X. Li and Z. Wang, 1991, On the Controlling process of structures for the ground water in the Northern Part of Weihe River, China: Project Report, Xi'an Geological Institute, Xi'an, China, p.12-13.
2.Z. Pan, 1989, On the Prediction of Minerals and the Lithosphere with geophysical data in Qinling-Bashan Region, China: Project Report, Xi'an Geological Institute, Xi'an, China, p.7-10.
3.Shaanxi Geological Survey, 1990, The Appreciation of Crust Stability and Geological hazards in the

Region of Xi'an, Project Report, Shaanxi Geological Survey, Xi'an, China, p.36-38.

4. Y. Tan, 1992, Landsat TM Image Analysis for the Thrust System of Helanshan Mountains, the Northwest of China, p.35-39 in International Archives of Photogrammetry and Remote Sensing, Volume XXIX, Part B7, Washington, D.C., USA.

5. Y. Tan, 1994, The Geological structural features and their controlling process for Coal Series in the Southern Margin of Ordos Basin, China: Journal of China Coal Geology, Vol.6, No.3, p.78-83.

6. G. Zhang, 1988, The Formation of Qinling Orogenic Belt and its Evolution: Publishing House of Northwest University, Xi'an, China, p.1-16.

7. G. Zhang, 1991, The evolution, Texture and Composition of the Lithosphere in Qinling Orogenic Belt, China: Journal of Northwest University, Vol.21, No.2, p.77-86.

PART 6

CONTINENTAL DEFORMATION AND GEOMECHANICS

Proc. 30th Int'l. Geol. Congr. , Vol. 14, pp. 277-292
Zheng et al. (Eds)
© VSP 1997

On the Mesozoic and Cenozoic Intracontinental Orogenesis of the Yanshan Area, China

Shengqin Cui and Zhenhan Wu

Institute of Geomechanics, CAGS, 11 Minzu Xieyuan Nanlu, Haidian District, Beijing, China, 100081

Abstract

The Yanshan Range in northern China is a Meso-Cenozoic intracontinental or intraplate orogenic belt formed on Precambrian cratonic basement. Its orogenesis differs from geosynclinal folding, marginal subduction and continental collision. Here we mainly discuss the geophysical properties and crustal structures, structural deformation, tectonic evolution, orogenic processes and orogenic mechanism of the Yanshan intracontinental orogenic belt, which are important for continental geodynamics.

Keywords: intracontinental orogenesis, tectonic evolution, Yanshan Range, Mesozoic, Cenozoic

INTRODUCTION

Orogenises and orogenic mechanism caused by crustal movement is a fundamental problem of modern geoscience. The term " orogeny " appeared in the middle of last century. Its meaning and content have developed with the whole geoscience, and constantly formed new hotspots and new arguments. Bewteen 1857-1960s, most geologists took the orogenesis as the results of the closure of geosynclines [18, 23]. In early 1960s, the theory of " Plate Tectonics " was founded, and the orogenesis has been linked with subduction and collision of plates afterwards [6, 12-16, 22, 30]. The orogenesis is often related to the subduction of plates in the marginal area to form the marginal orogeny and the collision between continents or plates to form the intercontinental or interplate orogeny [6, 14, 15]. There is growing recognition that the continental crust does not behave rigidly, but deforms over broad regions (e.g. Molnar, 1988 [26]). Some studies have described Cenozoic intracontinental deformation in central Asia and attributed to the identation of India into Asia [1, 2, 27, 32], and a few studies discussed the Mesozoic intracontinental deformations found in the Andes [20, 21], the Laramy orogeny and the Sevier orogenic belt of the western United States [8, 9], the Sino-Mongolian border area [33, 34] and the Beijing area of China [17, 18]. Most geologists, however, seem not to realise or deny the existence of intracontinental or intraplate orogenesis and its related orogenic belts occurred within the craton. Some geologists admitted of the intraplate orogenesis while confused with the orogenic mechanism.

The Yanshan Range is located in the northern part of the North China Plain and, geologically, lies in the north part of the Sino-Korea platform or the North China block of the east Asia (Fig. 1). Here developed the renowned Meso-Neoproterozoic stratigraphic section (formerly called the " Sinian System of northern China " or the " Sinian Suberathem ")[17, 19], has the oldest continental crust of China, is the place where the well-known Yanshanian Movement was named [9, 29, 31], and served as the " cradle " for the geological sciences of China. The Yanshan Range is also an intracontinental orogenic belt related to the Mesozoic and Cenozoic crustal movements on the Luliangian cratonic basement finally formed in 1800-1900 Ma [18, 23, 29]. During the long development stage of Meso-Neoproterozoic to Paleozoic (1800-250 Ma) aulacogen and cratonic stable cover, no any obvious tectonism occurred in this area [7]. In Mesozoic-Cenozoic Era, the main part of the Yanshan tectonic belt was as far as 1400-1800 km away from the west Pacific ocean-continental boundary, but strong orogenesis happened during the Indosinian, Yanshanian and Himalayan periods [7-10]. Its orogenic mechanism was neither geosynclinal closing nor marginal subduction and interplate collision, but typical intracontinental orogenesis based on the palaeocraton basement. So it is very important to study the Meso-Cenozoic Yanshan intracontinental orogenic belt for understanding the intraplate geodynamics and continental geodynamics. Further more, the Yanshan orogenic belt is also an important tectonic-magmatic-metallogenic belt of the Mesozoic and an active tectonic belt. It is of important significance to study the belt in detail for the exploration of mineral resources and the reduction of geological hazards.

GENERAL FEATURES OF YANSHAN INTRACONTINENTAL OROGENIC BELT

Structural Features Illustrated by the Remote Sensing Image

The 1:200000 coloured TM remote sensing image and the 1:500000 MSS remote sensing image illustrated well the structural features near the surface of the Yanshan orogenic belt. Two kinds of structures, the linear structures and the circular structures, are identified by the interpretation of the remote sensing images (Fig. 2). The linear structures, corresponding mainly to the Mesozoic and Cenozoic faults, trend in many different directions such as NE, NNE, E-W, NW and N-S, similar to the directions of the characteristic lines of the magnetic field. The circular structures, corresponded mainly to the Mesozoic and Cenozoic magmatic complexes and volcanic craters of different magnitude, distribute all over the whole Yanshan area and have evident relation with the deep crustal geological processes and magmatic activities.

Geophysical Features and Three Dimentional Structural Framework

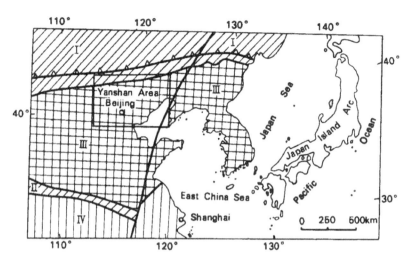

Figure 1. Location and tectonic setting of Yanshan intracontinental orogenic belt. I, Mongolia Paleozoic Geosyncline. II, Qinling Paleozoic-Triassic Geosyncline. III, Sino-Korea Platform with its basement finally formed in 1900-1800 Ma. IV, Yangzi Platform with its basement finally formed in 800 Ma.

Figure 2. Sketch map of the linear and the circular structures of the Yanshan orogenic belt interpreted from the MSS and the TM remote sensing images, illustrating the directions of faults and the distribution of the circular structures related to the Mesozoic magmatic intrusions. The asterisks mark the location of the Mesozoic volcanic craters.

The regional gradient zone of Bouger gravity of the Yanshan orogenic belt mainly trends in north-northeast and east-west,   and reflects the fact that the two main groups of structures -- the NNE and E-W structures on the crustal surface and their compounding correspound to geologic and tectonic setting at depth. The Moho depth and the lithospheric thickness of the Yanshan orogenic belt illustrate the similar features as the deep gravity anomaly of the Yanshan orogenic belt. The crust thickness of the Yanshan orogenic belt decreases gradually from 42 km in the northwest via 32 km in the southeast to 30-31 km in its southeast border area -- the Bohai Gulf basin where an arc-related mantle upwelling zone is developed [24].The thickness of the lithosphere of the Yanshan orogenic belt decreases gradually from 160 km in the northwest via 100 km in the southeast to 60-80 km in its east and south border area -- the Bohai Gulf basin and the North China Plain (basin) [24]. Another notable fact is that the compounding of the E-W and NNE structures in the crust gradually changes to the evident N-S structures at the depth of the lithospheric mantle in the Yanshan orogenic belt.

The heat flow in the Yanshan and its nearby area correlates closely to the gravity and the thickness of the crust and the lithosphere. The heat flow in Yanshan orogenic belt is often less than 1.0 HFU except its west area where Cenozoic rift basins are present with heat flow reaching 1.2-1.3 HFU. In the southeast border area of the Yanshan orogenic belt--the Bohai Gulf basin and the North China Plain (basin), the heat flow reaches to 1.8-2.3 HFU [24].

The crust of the Yanshan orogenic belt is divided into three layers: the upper crust, the middle crust and the lower crust. In its south part, the upper crust mainly consists of sedimentary rocks and green schist facies metamorphic rocks with thickness ranging from 0 to 15 km, the middle crust is mainly consisted of rock series of amphibolite facies and ranged vertically from 15 to 25 km, and the lower crust is mainly consisted of granulite facies and ranged vertically from 25-35 km. The upper section of the middle crust preserved liquid in its fracture, formed the layers of high conductivity and low velocity. Three layers of low velocity existed at different depth of the lithosphere in the middle Yanshan orogenic belt.

Mesozoic Regional Unconformities and Cenozoic Erosional Surfaces

Six periods of Mesozoic regional unconformities or disconformities and two periods of Cenozoic erosional surfaces were identified in the Yanshan orogenic belt (Fig. 3) [7-10]. The first period of regional unconformity formed at the end of Middle Triassic, and marked the begining of the Mesozoic intracontinental orogeny. The second period of regional unconformity formed at the end of the Late Triassic. Two Jurassic regional unconformities occurred at the end of Early Jurossic and Late Jurossic respectively, followed by two Cretaceous regional unconformities occurred at the end of Early Cretaceous and Late Cretaceous respectively. The second, the fourth and the sixth Mesozoic unconformities distributed widely in the Yanshan orogenic belt. The two periods of Cenozoic erosional surfaces formed in Early Eogene and Late Miocene respectively in the Yanshan and its adjacent areas.

Geologic Time		Age (Ma)	Formation	Orogenic Episode	Orogenic Stage
Genozoic			Clastic rocks		
		65		IV	
Cretaceous	K_2		Basalt with Clastic rocks		Late Yanshanian Stage
Cretaceous	K_2		Clastic rocks		
		95		III	
Cretaceous	K_1		Coal – bearing rocks		
Cretaceous	K_1		Shale,sandstone		
Cretaceous	K_1		Andesite, tuff		
Cretaceous	K_1		Sandstone, oil shale		
Cretaceous	K_1		Andesite, Basalt		
Cretaceous	K_1		Rhyolite, andesite		
Cretaceous	K_1		Rhyolite, trachyte		
		135		II	
Jurassic	J_3		Conglomerate, Sandstone		Early Yanshanian Stage
Jurassic	J_2		Andesite, Clastic rocks		
Jurassic	J_2		Tuffaceous clastic rocks		
				I	
Jurassic	J_1		Breccia, Sandstone		
Jurassic	J_1		Coal – bearing rocks		
Jurassic	J_1		Basalt, andesite		
		205		II	
Triassic	T_3		Conglomerate, Sandstone		Indosinian Stage
Triassic	T_3		Clastic rocks		
Triassic	T_3		Coal – bearing rocks		
				I	
Triassic	T_2		Clastic rocks, tuff		
Triassic	T_2		Conglomerate, sandstone		
Triassic	T_1		Clastic rocks		
		250			
	P_2		Conglomerate, sandstone		

Figure 3. Mesozoic sedimentary-volcanic formations and unconformities of the Yanshan orogenic belt.

Meso-Cenozoic Sedimentary Formations and Volcanic Activities

In the Mesozoic, the NE-NNE and E-W trending intracontinental basins were separated by the parallel uplifts developed in the Yanshan area, and deposited many periods of sedimentary-volcanic rock series (Fig. 3). Many periods of Mesozoic sedimentary cycles were identified. Most sedimentary cycles began with fine clastic rocks as the shale, mudstone, coal, siltstone and fine sandstone, and ended by coarse clastic rocks as the conglomerate, coarse sandstone, breccia similar to molasse formation. All the Mesozoic sedimentary rocks formed in the continent, and no Mesozoic marine sedimentary rocks occurred in this area. In south of the Yanshan orogenic belt and its surrounding areas, Cenozoic rift bains developed and deposited lucastrine sediments of thousands meters thick.

Very strong Mesozoic intracontinental volcanic activities happened in the Yanshan orogenic belt [3]. The Mesozoic volcanic rocks distributed in the whole Yanshan orogenic belt and its nearby areas, and the Mesozoic volcanic cycle began with Late Triassic-Early Jurassic basalt, followed by Late Jurassic andesite, and ended with the Early Cretaceous rhyolite, trachyte, some andesite and basalt. The Cenozoic volcanic activity became much weaker, only Tertiary and Quaternary basalt were found in some localties of the north Yanshan orogenic belt.

Mesozoic Magmatic Intrusion and Metallogenesis

During the least three periods of magmatic intrusion developed in the Yanshan orogenic belt, including the Late Permian-Triassic intrusion, the Jurassic intrusion and the Cretaceous intrusion, formed more than 300 larger Mesozoic magmatic complexes. Mesozoic magmatic rocks include granite, granodiorite, diorite, monzonite. syenite and alkali rocks with gabbro and ultrabasic rocks.

More than 100 gold, silver, copper, lead, zinc and molybdenum deposits on larger scales are distributed in the Yanshan orogenic belt. Most of them formed in the Mesozoic (Fig. 4) as a result of Mesozoic intracontinental orogenesis and related to Mesozoic tectonic and magmatic activities. Their spatial distribution was controlled evidently by the E-W, NE, NNE and N-W Mesozoic structural-magmatic zones and their compounds.

TECTONIC EVOLUTION OF THE YANSHAN INTRACONTINENTAL OROGENIC BELT

The Formation and Evolution of the Cratonic Basement

The cratonic basement of the Yanshan orogenic belt formed in Archean to Paleo-proterozoic stage, and was characterized by small blocks composed of high-grade metamorphic granulite-gneiss and relatively active granite-greenstone belts in bettween before 2500 Ma. An old continental nucleus formed under the influence of many tectono-thermal events in the latest Archean, followed by the activity of Paleo-Proterozoic rifting in the north and the east Yanshan areas and the strong tectonism manifested by strong structural deformation, magmatic activities and regional metamorphism occurred in the Yanshan and its adjacent area at the end of Paleo-Proterozoic Era with the peak of the outstanding tectono-thermal event in 1900-1800 Ma [23, 29]. The epoch-marking crustal movement related to the final formation of the united Sino-Korea craton with the Yanshan area as the middle section of its north part, was the tectono-thermal event occurred in 1900-1800 Ma, which was called by J.S. Lee as the "Luliangian Revolution" in geological history [23].

Formation of Aulacogen-Cratonic Covering Strata

After the formaton of the metamorphic basement, an aulacogen developed in the Meso--Neoproterozoic with the maximum thickness of sedimentary rocks of 10 km, and was

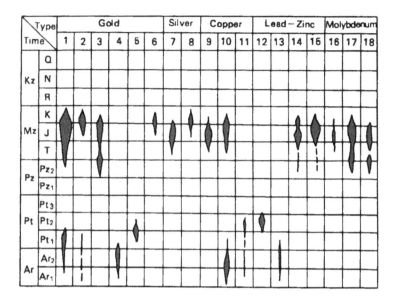

Figure 4. Metallogenic periods of the Yanshan orogenic belt. 1, hydrothermal quartz-vein type gold deposit. 2, volcanic-rock type gold deposit. 3, altered-rock type gold deposit. 4, metamorphic genetic gold deposit. 5, palaeoconglomerate type gold deposit. 6, porphyry type gold deposit. 7, silver deposit related to magmatic intrusion. 8, silver deposit related to volcanic eruption. 9, skarn type copper deposit. 10, hydrothermal quartz-vein type copper deposit. 11, volcanic-rock type copper deposit. 12, strata-controlled lead-zinc deposit. 13, metamorphic type lead-zinc deposit. 14, skarn type lead-zinc deposit. 15, hydrothermal quartz-vein type lead-zinc deposit. 16, skarn type molybdenum deposit. 17, porphyry type molybdenum deposit. 18, quartz-vein type molybdenum deposit.

characterized by the fault-bound and noncompansative depression with intermidiate-basic submarine volcanic eruption and continental magmatic intrusion in its early stage, and two periods of uplifting occurred in 1000 Ma and 800-600 Ma respectively [7, 11]. After the closure of the aulacogen, the stable Cambrian-middle Ordovician and Carboniferous-Permian cratonic deposits, mainly the clastic-carbonate strata of shallow marine facies, were not affected by tectono-magmatic activities until 250 Ma [6, 19, 29]. The long time interval of Mesoproterozoic to Paleozoic stage (1800-250 Ma) in the Yanshan area can be viewed as the preorogenic period of Meso-Cenozoic intracontinental orogeny [11].

Mesozoic-Cenozoic Orogenesis and Its Tectonic Evolution
After the formation of the 1800-250 Ma aulacogen and stable cratonic stage, the Yanshan area entered into a new stage of tectonic activities marked by tectonic deformation, many periods of magmatic intrusion and volcanic eruption, and the widespread metallogeny [7-10, 15]. Different period formed different tectonic frame-

work in the Mesozoic-Cenozoic stage (Fig. 5-7), resulting in the formation of the Yanshan intracontinental orogenic belt.

Permian to Triassic Period--the primary orogenic period. This episode of tectonism produced wide range of deformation features such as the solid-state plastic flow, ductile shearing, thrust faulting, complicated tight folding and schistosity in the Yanshan area. It was accompanied by magmatic intrusions such as granite, granodiorite, diorite and some minor diabase, gabbrophyre and alkaline rocks. Many magmatic complexes distributed in ductile shear zones as lenses. The E-W structures dominated in the "primary-Yanshan" orogenic belt with some NE strutures distributed in the east part (Fig. 5).

Jurassic Period--the early main orogenic period. This event produced strong depression and sedimentation, volcanic eruption, magmatic intrusion, thrusting and folding in the Yanshan area. By comparing to the Permian-Triassic period, the Jurassic tectonism became much stronger. More than 100 larger magmatic complexes emplacemented, with granite, diorite, granodiorite and mozonite as their major rock types. Early Jurassic basalt and Late Jurassic andesite are distributed over all of the Jurassic basins and their surrounding regions. The Jurassic NE and E-W thrust faults and folds dominated in the early "Meso-Yanshan" orogenic belt, and controlled the distribution of Jurassic basins, magmatic complexes, volcanic eruptions and the related ore deposites (Fig. 6).

Cretaceous Period--the late main orogenic period. It created widely distributed volcanic rocks as rhyolite, basalt, andesite and alkali rock. There existed more than 100 magmatic rock complexes consisting of granite, granodiorite, diorite, monzonite and alkali rocks. In this period strong tectonic deformation such as the folding, thrusting and syndepositional faulting occurred. The NNE-trending structures are dominated together with some minor E-W, NE and NW structures in the late "Meso-Yanshan" orogenic belt, and controlled the distribution of Cretaceous basins, magmatic complexes, volcanic eruptions and the related mineralizations (Fig. 7).

Cenozoic Era--the second orogenic period. It is expressed by the strong rifting in the south Yanshan and its adjacent areas [10, 25], such as the Fenwei rift system with its northern part--the Yanqing basin, the Huailai basin and the Weixian basin etc. in the southwest Yanshan orogenic belt. The rifting event also created the North China Plain-

Figure 5. Permian-Triassic palaeotectonic map of the Yanshan area. 1, the Paleozoic geosynclinal enviroment. 2, thickness contour of Carboniferous-Permian coal-bearing strata. 3, Late Triassic. 4, Early-Middle Triassic. 5, sedimentary rocks. 6, andesite. 7, basalt. 8-20, Permian-Triassic magmatic rocks and structures. 8, granite. 9, granodiorite. 10, diorite. 11, syenite. 12, basic vein. 13, ultrabasic rocks and gabbro. 14, thrust fault. 15, ductile shear zone. 16, solid state plastic flow. 17, anticline. 18, syncline. 19, anticlinorium. 20, synclinorium. 21, isotopic ages. 22, unconformity. 23, disconformity. The latitute and longitude of this figuure can be seen in figure 1.

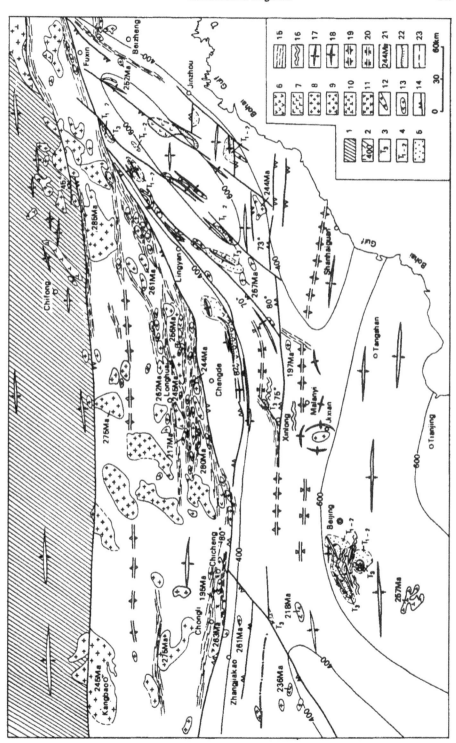

-a huge intracontinental rift basin in the south of the Yanshan orogenic belt, the Bohai Gulf-Lower Liaohe rift basin in the southeast of the Yanshan orogenic belt, the depression zone in front of the Yinshan Mts. in the west of the Yanshan orogenic belt, and the Songliao Plain--another huge intracontinental rift basin in the northeast of the Yanshan orogenic belt. While these rift basins depressed, the "Neo-Yanshan" orogenic belt--the present Yanshan Range rapidly uplifted, resulting in the Tertiary and Quaternary basic volcanic eruption and the formation of present basin and range tectonic features mainly controlled by the NE-NNE, NW and E-W extensional faults.

MECHANISM OF MESO-CENOZOIC INTRACONTINENTAL OROGENESIS

Thrusting and Folding of the Two Groups of Structural System

Two groups of tectonic systems--the EW and NE-NNE systems consisting of folds, thrust faults and schistosity formed in the Mesozoic and dominated in the "Primary-Yanshan" and the "Meso-Yanshan" orogenic belts (Fig. 5-7). Their related thrusting and folding were caused by the N-S and NW-WNW compressions, which controlled the formation and evolution of uplifts, basins, magmatic activities and Mesozoic intracontinental orogenesis. The E-W and the NE-NNE trending thrust-nappe structural systems mainly consisted of en echelon or parallel thrust faults, folds and schistosity with thrusting in the same direction or in the opposite directions, formed in different periods in different regional stress fields. Generally, the E-W tectonic systems formed in the Permian-Triassic and the Jurassic were caused by strong N-S compression and are marked by the evident N-S horizontal shortening (Fig. 5-6), and some of them were reactivized in the Cretaceous and the Cenozoic (Fig. 7), while the NE-NNE structural systems mainly formed in Jurassic-Cretaceous Periods under the condition of strong NW-WNW compression , and are marked by the evident NW horizontal shortening (Fig. 6-7), and many of them reactivized in the Cenozoic. These two groups of structural systems and their compounding formed the basic tectonic framework of the Yanshan intracontinental orogenic belt.

Detachment and Decoupling at Different Depth

The Luliangian unconformity, the Meso-Neoproterozoic sedimentary rock series, and

Figure 6. Jurassic palaeotectonic map of the Yanshan area. 1, Late Jurassic. 2, Middle Jurassic. 3, Early Jurassic. 4, Mesoproterozoic-Triassic rocks. 5, Archean-Paleoproterozoic basement rocks. 6, coarse clastic rocks. 7, fine clastic rocks and coal bearing strata. 8, andesite. 9, basalt. 10-23, Jurassic structures. 10, NE thrust fault system. 11, E-W thrust fault system. 12, normal fault. 13, ε-type structural system. 14, N-S thrust fault. 15, N-S normal fault. 16, NW normal fault. 17, NE compressional-shear fault. 18, syndepositional fault. 19, nappe structure. 20, anticline. 21, syncline. 22, circular fault system. 23, schist zone. 24, Mesozoic fault. 25, burried fault. 26-30, Jurassic magmatic rocks. 26, granite. 27, diorite. 28, syenite. 29, monzonite. 30, gabbro. 31, unconformity. 32, disconformity. 33, dip and strike of strata. 34, dip angle of fault. 35, isotopic age. 36, volcanic crater. The latitute and longitude of this figure can be seen in figure 1.

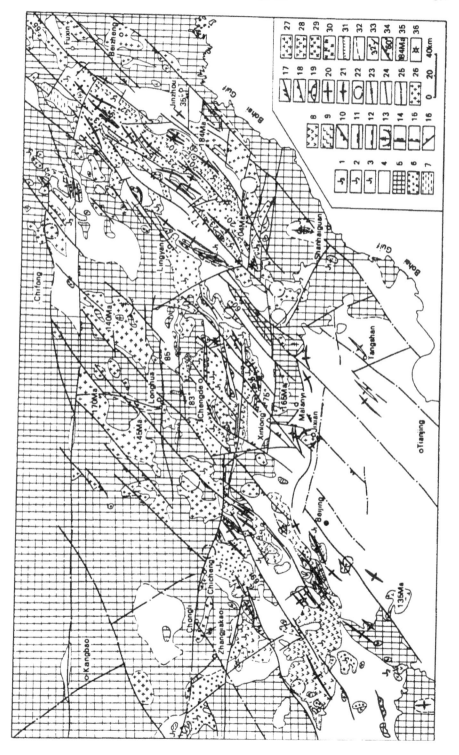

the Late Paleozoic and Mesozoic coal-bearing strata, often became weak layers under the condition of certain temperature, pressure and tectonic stress, and Mesozoic detachment and decoupling systemes developed along them at different depth of the crust. The Mesozoic and Cenozoic detachment and decoupling also formed in deep crust along the low velocity zone as demonstrated by numerous geological and geophysical data. The detachment and decoupling played an important role to the development of the Yanshan orogenic belt and served as one kind of important intracontinental orogenic mechanism in the Yanshan and its adjacent areas.

Magmatic Diapirism and Thermal Upwelling

More than 300 Mesozoic magmatic complexes formed in the Yanshan orogenic belt, and most of them intruded actively in the compressional stages, and caused local melting, plastic flow, folding, ductile shearing, radial and circular fracturing, and contact metamorphism, forming the diapiric tectonic systems. These local tectonic systems, such as the circular systems and the metamorphic core complexes [12], were compounded and disturbed the regional tectonic systems, and complicated the tectonic frameworks of the intracontinental orogenic period.

Fault Depression-Uplifting of the Secondary Orogenic Period

The Mesozoic palaeo-basin and range and the related palaeo-mountains suffered erosion and the region gradually formed the penaplaination. In the early Himalayan stage (Eogene), the Cenozoic rifting began in the surrounding areas of the Yanshan orogeic belt [10, 25], while the uplifting of the Yanshan Range strengthened and lead to the formation of coarse clastic rocks in some places of the Yanshan area as in the southwest of Beijing. In late Himalayan stage (Neogene-Quaternary), the rifting continued in the southeast and southwest of the Yanshan orogenic belt and extended to south part of the Yanshan area [10, 25], formed the Nankou depression in the south of Beijing and the Yanqing rift basin in the northwest of Beijing, while the Yanshan Range entered into another period of rapid uplifting, and lead the early plaination formed in Early Eogene uplift to 1000-1500 m. All these depression and uplifting were controlled by the Cenozoic extensional fault activities, and finally resulted in the formation of present basin and range tectonics and present mountaineous geomorphic landscapes.

Figure 7. Cretaceous palaeotectonic map of the Yanshan area. 1, Late Cretaceous. 2, Early cretaceous. 3, Mesoproterozoic-Jurassic rocks. 4, Archean-Paleoproterozoic basement rocks. 5, coarse clastic rocks. 6, shale. 7, clastic rocks. 8, coal-bearing formation. 9, rhyolite. 10, andesite. 11, basalt. 12, trachyte. 13-29, Cretaceous magmatic rocks and structures. 13, granite. 14, diorite. 15, syenite. 16, gabbro. 17, E-W thrust fault system. 18, E-W normal fault. 19, NE thrust fault. 20, NE compressional-shear fault. 21, N-S normal fault. 22, rotational structural system. 23, rotational circular structural system. 24, NW normal fault. 25, shear fault. 26, anticline. 27, syncline. 28, circular fault system. 29, NNE thrust fault system. 30, Mesozoic fault. 31, burried fault. 32, Cretaceous metamorphic core complex. 33, unconformity. 34, disconformity. 35, dip and strike of strata. 36, dip angle of fault. 37, isotopic age. 38, volcanic crater. The latitute and longitude of this figure can be seen in figure 1.

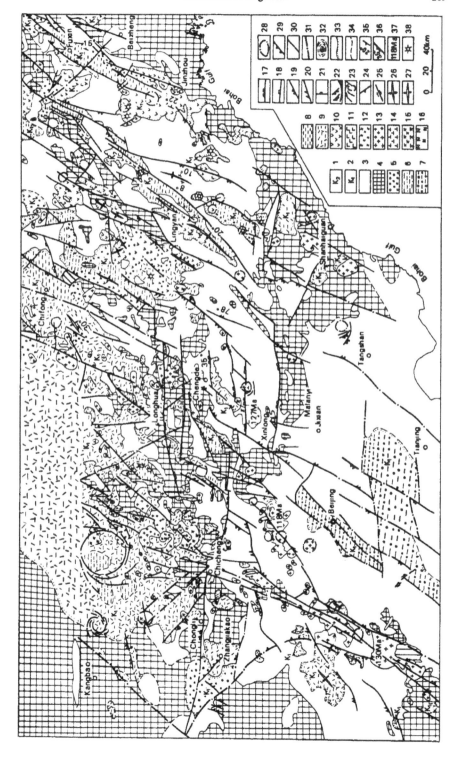

CONCLUTION AND DISCUSSION

The Meso-Cenozoic Yanshan intracontinental orogenic beit is a complicated orogenic belt formed on a Precambrian cratonic basement as the result of the compounding of tectonic systems of different directions, deformation of different types, and intracontinental orogenesis of different periods. It differes from the marginal subductional orogenic belts and the interplate collisional orgenic belts in tectonic setting, sedimentation, volcano-sedimentary formation, magmatic activity, deformational intensity and tectonic features, and orogenic mechanism.
The geodynamic cause of the Yanshan intracontinental orogenesis can not be successfully explained by the plate tectonics and need further research [11, 15]. Some factors as the rotation of the earth, thermodynamic conditions and decoupling in deep crust and mantle, gravity, magmatic upwelling, and intraplate stress, seem related to Meso-Cenozoic Yanshan intracontinental orogenesis.

As a new type of orogenesis, the intracontinental orogenesis are not limited in the Yanshan area, but developed in many ancient cratonic regions, such as the east Eurasia and the Africa continents, with the Yanshan-type orogenic belt as the typical example [11]. And the Yanshan intracontinental orogenic belt will provide a good window for understanding the intraplate or intracontinental geodynamic processes.

Acknowledgements
The authers thank the Ministry of Geology and Mineral Resources of China and the National Science Fund of China for the finacial support. Acads. Dianqing Sun and Qingxian Chen provided good suggestions for our research project. Doctor An Yin and Professor Yadong Zheng reviewed this paper in detail and made many good suggestions. Discussions about continental rifting and faults of east Asia with Acads. N.A. Logatchev and S.I. Sherman were helpful to our research. Thanks also are due to the Institute of Geomechanics for its support of field and laboratory work in 1992-1996.

REFERENCES

1. M. B. Allen, B.F. Windley, C. Zhang and J. Gao. Evolution of the Turfan Basin, Chinese Central Asia, *Tectonics* 12, 889-896 (1993).

2. J.P. Avouac, P. Tapponnier, M. Bai, H. You and G. Wang. Active thrusting and folding along the northern Tian Shan and late Cenozoic rotation of the Tarim relative to Dzungaria and Kazakstan, *Jour. Geophys. Res.* 98 (B4), 6755-6840 (1993).

3. Beijing Bureau of Geology and Mineral Resources. *Regional geology of Beijing Municipality* (in Chinese with English summary). Geological Memoirs. Ser.1 (27). Geological Publishing House, China. pp.598 (1991).

4. B.C. Burchfiel, D.S. Covan and G.A. Davis. Tectonic overview of the Cordilleran orogen in the western United States. In: *The Cordilleran Orogen*. B.C. Burchfiel, P.W. Lipman and M.

C. Zoback (Eds.). 407-479. Boulder, Colorado, *Geol. Soc. Am. N. Am. Geol.* G-3 (1992).

5. B.C. Burchiel and G.A. Davis. Natural and control of Cordilleran orogenesis, western United States; extension of an earlier synthesis. *Am. J. Sci.* 275-A, 363-396 (1975).

6. K.C. Condie. *Plate tectonics and crustal evolution*, second edition. Pergamon Press. pp.188-215 (1982).

7. S. Cui, Z. Yang and Y. Zhao et al. Studies of palaeostructural systems in Yanshan and its adjacent areas, *ACTA Geologica Sinica* 52 (2) (1979).

8. S. Cui and J. Li. On Indosinian tectonic movement of peri-Pacific belt in China, *ACTA Geologica Sinica* 57(1) (1983).

9. Shengqin Cui, Jinrong Li and Yue Zhao. On the Yanshanian movement of the peri - Pacific tectonic belt in China and its adjacent areas, *Scientific papers for the 27th IGC.* Geological Publishing House, China. pp. 221-234 (1985).

10. Shengqin Cui and Jinrong Li. Himalayan tectonic evolution in East Asian peri-Pacific region, *ACTA Geologica Sinica* 3, 233-246 (1990).

11. Shengqin Cui, Ganguo Wu, Zhenhan Wu and Yinsheng Ma. *Structural features and stratigraphy of the Ming Tombs-Badaling area, Beijing--30th IGC field trip guide T218.* Geological Publishing House, China (1996).

12. Gregory A. Davis, Xianglin Qian and Yadong Zheng. T*he Huaorou (Shuiyu) ductile shear zone, Yunmengshan Mts., Beijing -- 30th IGC field trip guide T209.* Geological Publishing House, China (1996).

13. G.A. Davis, X.L. Qian, Y. Yu, C. Wang, G.E. Gehrels, M. Shafiquallah and J.E. Fryxell. Mesozoic deformation and plutonism in the Yunmeng Shan: a metamorphic Core Complex north of Beijing, China. In: *The Tectonic Evolution of Asia.* A. Yin and M. Harrison (Eds.). 253-280. Cambridge University Press, Cambridge (1996).

14. J.F. Dewey and J.M. Bird. Mountain belts and the new global tectonics, *Jour. Geophy. Res.* 75, 2625-2647 (1970).

15. J.F. Dewey and B. Horsfield. Plate tectonics, orogeny and continental growth, *Nature* 225, 521-525 (1970).

16. J.F. Dewey and K. Burke. Tibetan, Variscan, and Precambrian reactivation: products of the continental collision, *Jour. Geol.* 81, 683-692 (1973).

17. A.W. Grabau. The Sinian System, *Bull. Geol. Soc. China* 1, 1-4 (1922).

18. T.K. Huang. On major tectonic forms of China, *Geol. Memoirs. Ser.* A(20), 165 (1945).

19. C.S. Kao, Y.H. Hsiung and P. Kao. Preliminary notes on the Sinian stratigraphy of North China, *Bull. Geol. Soc. China* 13, 243-276 (1934).

20. T.E. Jordan, B.L. Isacks, R.W. Allmendinger, J.A. Brener, V.Z. Ramos and C. Ando. Andean tectonics related to geometry of subducted Nazca plate, *Bull. Geol. Soc. Am.* 94, 341-631 (1983).

21. T.E. Jordan, B.L. Isack, V.S. Ramos and R.N. Allmendinger. Mountain building in the central Andes, *Episodes* 3, 20-26 (1983).

22. G. Kimura, M. Takahashi and M. Kono. Mesozoic collision-extrusion tectonics in eastern Asia, *Tectonophysics* 181, 15-23 (1990).

23. J.S. Lee. *Geology of China.* Thomas Murby and Co.(1939).

24. Xingyuan Ma, Guodon Liu and Jian Su. The structure and dynamics of the continental lithosphere in north-northeast China. *Annales Geophysicae* 2, 611-620 (1984).

25. Xingyuan Ma and Daning Wu. Cenozoic extensional tectonics in China. *Tectonophysics* 133, 243-255 (1987).

26. P. Molnar. Continental tectonics in the aftermath of plate tectonics, *Nature* 335, 131-137 (1988).

27. P. Molnar. Quaternary climate change and the formation of river terrace across growing anticlines on the north flank of the Tian Shan, China, *J. Geol.* 100, 583-602 (1994).

28. P. Molnar and P. Tapponnier. Cenozoic tectonics of Asia: effects of a continental collision, *Science* 189. 418-426 (1975).

29. Dianqing Sun and Shengqin Cui. On major tectonic movement of China, *Scientific papers on geology for international exchange, 1, for the 26th IGC.* Geological Publishing House, China. pp. 15-26 (1980).

30. B.F. Windly. *The evolving continents.* John Wiley (Sons (1977).

31. W.H. Wong. Crustal movement and igneous activities in eastern China since Mesozoic time, *Bull. Geol. Soc. China* 6(1) (1927).

32. A. Yin, S.Y. Nie, T.M. Fillipone, T..M. Harrison, X.L. Qian and M.S. Li. A kinematic model for the Tertiary development of the Tian Shan, Altyn Taigh, and Chapman Fault System in central Asia during the Indo-Asian collision, *Geol. Soc. Am. Abst. with Program* 25, 121 (19 91).

33. Y. Zheng, S. Wang and Y. Wang. An enormous thrust nappe and extensional metamorphic Core Complex newly discovered in Sino-Mongolian boundary area, *Science in China* (Ser. B) 34, 1145-1152 (1991).

34. Y. Zheng, Qian Zhang, Y. Wang, R. Liu, S.G. Wang, G. Zuo, S.Z. Wang, B. Lkaasuren, G. Badarch and Z. Badamgarav. Great Jurassic thrust sheets in Beishan (North Mountain)-Gobi areas of China and southern Mongolia, *Jour. Struct. Geol.* 18, 1111-1126 (1996).

Proc. 30th Int'l. Geol. Congr. , Vol. 14, pp. 293-303
Zheng et al. (Eds)

Structural Control on Concentration and Dispersion of Ore-Forming Elements

GANGUO WU, DA ZHANG

Institute of Geomechanics, CAGS, Beijing, 100081, China

Abstract

Tectonic and ore-forming process is product of crustal movement. The tectonic force may influence the conditions under which chemical reactions take place, bringing about the migration of chemical constituents, lead to stress corrosion, cause strain incompatibility between host rocks and country rocks. The kinetic energy caused by tectonic force can change into other kinds of energy, therefore may change the chemical potential in the ore-forming process.Mang cases indicate that structures of syn- and pre- mineralization may influence physical and chemical conditions of mineralization.In the area involved in the formation of tectonic system, the trend of material flowage and chemical reaction is from higher chemical potential to lower one. Ore-forming fluid move along structural planes or rock pores to deposit there.In the folds resulting from buckling and bending,the elements with bigger ionic radius and the more reactive elements with smaller ionic radius are migrated by their own means. Under the coaction of vertical buoyancy and lateral tensile force, the tectonic denudation occurres in the upper part of fold; At the same time, the deep materials uplift because of the chimney effect. The friction and the dislocation results in shear force, producing shearing crack system, while the tension crack appears in the uplifting mass. If ore-forming fluid appears, it will be deposited in the crack,forming vein deposits. In the compressive and compresso-shear faults, from their main fault plane to the both sides, the stability series of ore-forming elements is in accordance with the series of the elemints'ionic radius from small to big. Therefore,in these kinds of faults mentioned above, ore deposits of high sulfur and low oxygen are easily formed.In the tensil and tensil-shear faults, the more reactive oxyphile elements and high valence ion are enriched, and it is easy to form ore deposits prodominated by higher oxidation state minerals.

Keywords: tectonic system, structures of ore-controlling, ore-forming fluid

INTRODUCTION

Tectonic deformation caused by a tectonic movement is generally not equally and evenly distributed. The deformation is usually expressed as a linear high strain zone combining with associated low-strain blocks [11]. Walshe found the relationship between crustal-scale structures and metallogenic provinces in the Tasman Fold belt system in Australia. He referred that at the regional scale, the distribution of some of the major mineral provinces appears to reflect interplay of pre-mineralization structure [20]. Kutina also stressed the importance of global latitudinal structures in controlling distribution of metal deposits. He investigated the role of deep structure of the lithosphere in metallogeny and suggested that major clusters of metallic ore deposits of different types and ages have originated above mantle-rooted discontinuties, especially where such discontinuities intersect with rift structures at the convergent plate margins. For example, there exists a broad belt of magnetic highs in western United States, and major clusters of ore deposits have developed in this belt. In China, the Yinshan-Tianshan orogenic belt extends latitudinally through northern China between about 40° 30' and 42° 30'N. In central

Europe, there exists a similar feature.Glen found that, in the upper and middle crust, thrusts with different types are associated with various mineral deposits [9]. In the presence of mineral sources, the strongly deformed parts often provide physical and chemical conditions favourable for ore-formation. Tectonic control on mineralization implies that the physical and chemical conditions in some parts of a tectonic system favour the formation, focusing and accumulation of ore-forming material [19]. These problems are discussed in this paper.

TECTONIC INFLUENCE ON PHYSICAL AND CHEMICAL CONDITIONS FOR MINERALIZATION

Structures that control the emplacement of ore body are generally of two types: pre-mineralization structures and syn-mineralization structures. The mineral deposits include several types: some syn-tectonic deposits with mineralization related to deformation, and this kind of deposit formed during or after deformation. Other deposits with mineralization predate deformation [7,8]. The pre-mineralization structures physically provides the space for ore emplacement or the passageway for ore fluid, while the syn-mineralization structures develop during the mineralization under the same system directly affecting the physical and chemical conditions leading to mineralization.

The Effect of Pre-Mineralization Structures
The space provided by the pre-mineralization structures is often located in a specific place of the structure [1,2,4], e.g., a structural plane or a structural zone of some width or a columnar fragmented zone of fault intersections. For example, thrust-associated deposits include those localized on thrusts, above or below thrusts, and in anticlines above inferred blind thrusts [7]. Some mineralization is hosted by quartz veins, which grew in fault dilatant zones and associated fractures. The magnitude of a structural plane or zone [5], their textural features, depth and their mechanical properties may cause a great difference in ore textures and the contents of ore-forming elements and mineral assemblages [6].

During the formation of folds, due to the difference of physical and mechanical properties and stress conditions, deformation of rock, shape of ore body,scale and ore texture and structure of orebody are different. They even cause the ore composition difference. For example,an anticline resulting from bending, in the extensional area above neutral plane, there develops a fan-shaped extensional fracture system vertical to layers and forms shear joint system bedding and oblique to layers. If they are filled by succeeding veins, big vein will occur at the crest of anticline while small vein belt or net vein will occur on the limbs,structural characteristic of most strata-bound ore deposits is that the fold dissected or steep-dipping faults become structures as passage way for ore fluid, while the flat faults developed from the shearing joints are host structures.

The change of geometric shape of textural plane provides various physical space for ore-formation [5]. For instance,the change of fault occurrence can produce space with different scale. Reverse faults may be slightly waved-shape, and the ore-forming fluid may move to the extended area to form ore. That is because when moving, the ore-forming fluid runs into the expanding space. Its pressure decreases and temperaturebecomes lower and the fluid velocity slows down. Thus on the one hand,the micrograin of ore drafting in the fluid precipitates because of decreasing fluid speed and kinetic energy, on the other hand,the chemical reaction develops toward lower temperature and pressure. and then the minerals got out to become ore. So there often forms an ore shoot at the place of expansion of tectonic plane. A good model about precipitation during fault slip is from the Wattle Gully fault zone reported by Cox et.al. [4]. Prior to fault slip, fluid pressure promotes fluid infiltration of the host rocks. The sudden fluid pressure decrease associated with fault slip interrupts this flow pattern by reversing the hydraulic head. The ore-controlling of fault intersection has the same mechanism, a couple of shear zones appear to be lattice-type intersection net, when ore-forming condition exists, ore-node arranged at equal distance is formed.

If faults have no reactivation history and remain the primary structural feature, faults with different mechanic properties have different textural characteristics and they have various ore-controlling characteristics. Extensional and transextensional fractures are useful places for ore liquid residing and ore material precipitating, then produce ore shoot; Shear fractures control thin plate-shaped orebody and compressional fractures are not easily filled by orebody because their fault planes are tight, if fault mud is well-developed, due to its good closure, it can become geochemical obstacle that limits ore-forming fluid to flow along one plate of fault above which orebody is emplaced. It displays ore-controlling of single-plate fault. For example, one mercury ore in Guizhou province, orebody is located in the lower plate of thrust fault but doesn't penetrate the fault.

Tectonic Influence on Physical and Chemical Conditions for Mineralization
The tectonics and mineralization are both the results of crustal movement. They are complicated processes and are difficult to model by experiments. Concerning the relation between them, following points can be drawn:

It can affect the pressure condition of chemical reaction [12]. Because the tectonic stress superimposes on hydrostatic pressure and acts on the rock mass in a close system, causing the change of forces among the rock mass. it will affect one of the main condition of chemical reactions --pressure, and prompts the exchange of material and chemical reaction.

It can provide energy for a change of material composition and affect the temperature of chemical reaction [27]. The formation of structures is a process during which energy releases and is provided to other geological process, especially to metallogenic. Under tectonic dynamics. the accumulation of energy not only causes the movement of rockmass, but also forces the fluid to flow. It indicates that. under stress process. the friction resulting from the movement between the rockmass produces thermal energy (frictional heat) causing the change of the mineral composition and rock near fault plane.

For example, ferruginous membrane on the plane of fault is produced by temperature and pressure which approach the condition for forming specularite during the movement of fault, and then causes the mineral disintegrate, the ferry separated off and arranged orentially on the plane of fault. Some experiments indicate that there exists the thermal effect during strain softening. Totally,the accumulation of thermal energy change the condition of T and P and then prompts the metallogensis.

The continuous stress causes the migration of chemical composition [22]. For example, pressure solution presents that, under the continuous stress action, a part of chemical composition of rock in the place of stress concentration may mirgrate. In pressure shadows, SiO_2 or $CaCO_3$ are emirgrated from compressional area and precipitate in extensional area and arranged parallelly to the principal extensional stress. So pressure solution is a cooling migration of minerals in rock under stress process.

Pressure corrosion [22]. Stress process produces microcrack in rock and if there is fluid, the original texture at the pointed end of crack is broken with the chemical change and causes the crack to expand. At this time, stress corrosion occurs. At first, the corrosion is slight, but once it occurs, it will enlarge in a certain scale owing to chain reaction.

The differences of pressure [28], temperature and accumulation of energy among the structures with various textures cause the difference of other physical and chemical condition(such as reduction-oxidation potential).

On the one hand the tectonic stress of mineralization can form new tectonic feature [21], on the other hand, at the same time, it also make the pre-mineralization tectonic feature react and to be transformed, and then the former structures are apposited to the tectonic system of metallogenic epoch and affect the mineralic physical and chemical conditions as a whole.

Tectonic evolution controls that of deposits [13]. The structures and deposits may be changed after forming, sometimes, the change strength is weak, on the other times, it's tenacious. The change may be physical and shows that the scale and shape of deposits are reformed and broken but texture and composition remain constant place. The change may be both physical and chemical and shows that mineral composition and the content of mineral elements and change, then mostly forms new ore deposits or makes the old deposits whether poor or rich, meanwhile, the ore texture is reconstructed.

The above features suggest that, under special conditions, the kinetic energy resulting from tectonic stresses can change into other forms of energy. Consequently a change in chemical potential during metallogenic process may occur. In the area where the formation of tectonic system is involved, the trend of material flowage and chemical reactions are from higher chemical potential to lower one, as a whole, tectonic-material systems towards the lowest energy. In other words, tectonics and metallogensis is a process during which the crustal material liberates energy by means of mechanic movement. Material adjusting called by Yang Kaiqing is an indication of this process. It is studied that, under stress process, mineral firstly appears contraction of volume and discocation of lattice [27]. Volume contraction leads accumualtion of particles closely.

After that process, the closer texture appears. At this time, in mineral, the atomic and ionic radius are shortened and elelectric nebula is overlapped, therefore, inner energy of mineral crystal increases greatly, so the crystal has high energy. Presently, mineral has not only deformation and restoration energy, but also higher chemical potential, then possibly causes chemical reaction that does not take place in the other conditions [22,27]. Ore-forming fluid is migrated towards the direction of decreasing pressure, temperature and dissipating energy. under the same tectonic stress process, the law and direction of concentrition and disperse of different chemical elements may be different, but the general mode of concentrition and disperse accords with tectonic stress field, so it is concluded that tectonic stress field and geochemical field are universal.

FOLD CONTROLS ON THE CONCENTRATION AND DISPERSION OF ORE-FORMING ELEMENTS

When fold is formed, the stress regularly distributes in different places of fold and changes with its development. This kind of change often prompts variation of physical and chemical condition everywhere in rock layer, thus drives chemical elements to migrate and reset. Later on, the correlation between the distribution of fold stress of fold and distribution of chemical elements appears.

In the folds resulting from buckling, the rule of distribution of stress is that, outside of neurtral plane is extensional area versus inside of that is compressional area, meanwhile, the fold limbs where anticline transits to syncline is the shear area. General rule of distribution of chemical elements is that the elements with bigger ionic radius and the more reactive elements with smaller ionic radius always   migrate to the extensional area, whereas the elements with small ionic radius or the less reactive elements always stay in the original place or migrate to the limbs. So as a whole, in fold, the elements with bigger ionic radius mostly host in the extensional area whereas the elements with smaller ionic radius often host in the compressional area. The result of example about single-layer bending of alloy plate indicates that, when the metal plate consists of two metal elements with unequal atomic radius bends, the metal elements with bigger atomic radius are enriched in extensional area outside of neutral plane while those with smaller atomic radius are enriched in compressional area inside of neutral plane. Other examples of this kind of phenomena also occur in geology. Cheletnicheco (1977) found out when acid-basic rock is controlled by fold, acid rock locates over the basic rock in the core of fold, on the contrary in syncline and neutral rock mostly lies between them. This suggests that, under lateral compressional force process, accompanying with the formation of buckling fold, intrusive magma has   geochemical deferentiation due to the stress controlling, at this time,elements with bigger ionic radius are migrated towards extensional area while the smaller ones are enriched in compressional area, and then it is appeared that magma of various constituent distributes orderingly and vertically in fold. Direction of material movement of buckling fold is from two limbs to the recess, then thin limb and thick-top simular fold is formed. But to the rock layer with obscure plastic flowage, there will produce prostration because of slip between layers and produce reverse thrust due to detachment of layers and shearing in limbs in order to adapt the stress and movement of material. The signature place of the above structures is enrichment area of revelent chemical elements. For example, the prostration of the crest of anticline is often filled by emigrated material from other places, so it's useful site of ore-controlling of fold (Fig. 1).

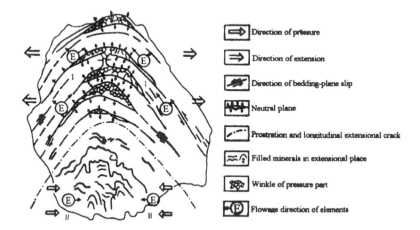

Figure 1. The Prostration in the Crest of Anticline Filled by Emgrated Material from Two Limbs(Modified after Yang Kaiqing, 1993).

Starostin studied that after folding of bedding ore-layers, the contents of ore-forming elements Cu, Fe, Sb are different in various structural places [16](Fig. 2). The high density area of the above elements exists in extensional area of fold(such as at the top of anticline) and in shear area of limbs while the lower density area is in the compressional area. That suggests that various stress can make chemical elements active and migrated and different tectonic spaces provide the location in which different chemical elements are enriched.Some stratabound ore deposits are reconstructed by post fold, and then cause some characteristic tectonic parts (for instance the break of fold) to be in oxidation enviroment of open system, so original deposit is changed into secondary oxidation deposit and then ore shoot is formed. For example, $MnCO_3$ ore layer in kuangxi province is poor ore which is unable to be used by industry, but at some tectonic parts which is controlled by fold, $MnCO_3$ shoot ore which has industrial value. Secondary shoot of Jingdian plumb-zonic deposit in Yunnan province is destroyed by radial transextension fault in Jingdian domb. It also indicates that the development of deposit is controlled by tectonic evolution [25].

Bending fold refers that rock layer bends under a force vertical to bedding plane. Generally, it results from uplifting of basement block or diaprism. The rule of stress distribution is that, at crest of a fold [20], compressional stress is vertical to rock layer, extensional stress to axial plane of fold and parallel to bedding plane, the directions of shear stress in both limbs is opposite to buckling fold. The mode of material movement is from the crest to the two limbs and forms thin-top and thick -limb shape. When layer is bended under uplifting force, the rule of the elements migration is that elements with bigger ionic radius are easily migrated to limb to be precipated and enriched in shear and overburden zone along layer. So, sack-like orebody is equillistantly spreaded. At the top of bending fold, due to lateral extension,material is emigrated, so the rock layer becomes thin, even produces top structural denudation. Under the co-action of vertical uplifting force and lateral extensional force, deep materal elevates because of the chimney effect, at this time, the triction and the dislocation between the uplifting mass and the surrounding rocks results in shear force, producing shear crack system while the extensional crack appears in the uplifting mass because of decreasing pressure and expanding volume. If the ore-forming fluid appears, it will be deposited in the crack,

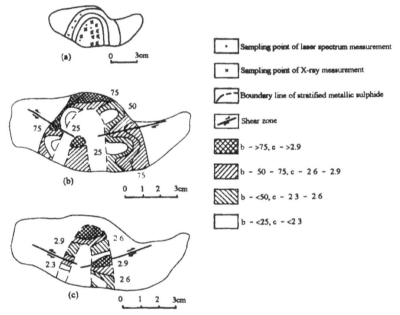

Figure 2. The Contour Line of Concentration of Mineralization Elements in Buckling Fold. a, .fold shape; b, copper content in pyrite (arbitrary assumptive measurement unit) ; c, iron content in terrahedrite.(Simplified after V.I. Starostin, 1994)

forming vein deposits. This suggests that, part of gold deposit of ductile shear zone is formed under regionally extending and bending. To the ductile shear zone taken place at the middle depth, the temperature and pressure condition can lead gold in rock to activate and then move into fluid, however, in ductile shear zone, there does not scale deposit. Laterly, through bending, uplift and lateral extensional force (extending) cause block to elevate and then produce shear displacement and extensional dilation. The former ductile shear zone is destroyed and produce new shear zone and extensional crack system. So ore-forming material is precipitated and enriched to form orebody. In summary, this kind of gold deposits in ductile shear zones is of feature of secondary ore-forming process.

Wu Xueyi studied increase of contents of ore-forming elements resulting from folding by means of modelling [24]. The sampler of experiments is natural ore from Shilu iron ore and he did the experiment on high temperture and high pressure machine. The result indicated that during the formation of fold. with the increase of temperature and pressure, ferroan quartzite and phyllite makes disilocation and iron is enriched. Meanwhile. hematite is deformed plastically and aggregated in the axis. So. it's folding that enriches and thickens the ore layer in Shilu ore.

THE FAULT CONTROLS ON CONCENTRATION AND DISPERSION OF ORE-FORMING ELEMENTS

The Fault Controls on Concentration and Dispersion of Chemical Elements
The fault controls on concentration and dispersion of chemical elements result from the change of physical and chemical enviroment through fractures of rock and displacement between rock blocks due to stress processes. and then make the chemical elements reallot.

Generally, compressional and transpressional faults are in relatively closer reduction enviroments. Along the direction of main compressional stress, compaction of rock causes the compression and mineral deformation or rotation, they are parallel to strike of fault and arrange orientedly, thus cause the permeability along fault belt to be higher than along other directions. Fluid in which there are K^+ and Na^+ etc. can extract from country rock chemical elements and then migrate toward extensional and transextensional space along the direction of decreasing pressure and temperature at the form of complex, chelate or true solution fluid can migrate long distance along compressional fractural belt. Along the main fault plane and the neighbour, contents of Si and Fe often increase due to bleeding of K and Na.The stable order of petrogenic elements from compressional fault to the two plates generally is Si-Fe-Mg-Ca-Al-K-Na, on the whole, it is in accordance with the order of ionic radius from small to big. In compressional and transpressional faults, the element with high stability is relatively enriched and ion is often at the low valence form. It is easy to form high sulfur and low oxygen deposit[12,28]. So it is common that, in compressional and transpressional fault belt, the elements display the horizontal zonation along fault dip and enrichment along strike of fault.

Fig. 3 is a profile of compressional fracture belt developed in green schist of Bailaimiao group in late proterozoic era. It illustrates that the contents of tectonic rock and main petrogenic elements K and Na and main elements of metallogenic Cu, Pb and Zn appear to be zonation and correlation. The zonation of tectonic rock of fault belt is lensing mylonite--mylonitization mortar rock--cataclastic rock--breecia, and the contents of Cu, Pb and Zn gradually become lower from the center of fault to the two sides, while the contents of K and Na increase.

Extensional fault is generally in the oxidization enviroment. Due to extension and dilatancy, the porosity and permeability enhance. So, it is useful for that ore-forming solution stays and aggregates and due to the sudden change of physical and chemical enviroment, the complex of metallogenic elements resolute and precipitate to form ore.

Because of the persistence of the crustal tectonic force process, after the shear fracture of rock, the tectonic force will continuously act on the fractural block and plane of fault, the heterogeneities of force and feature of geological body will cause original shear fractural plane change its single shear feature to be compressional and extensional. The characteristic of concentration and dispersion of chemical elements controlled by transpressional and transextensional fault is simular to that of compressional and extensional fault. The translation of shear fault determines that the chemical elements

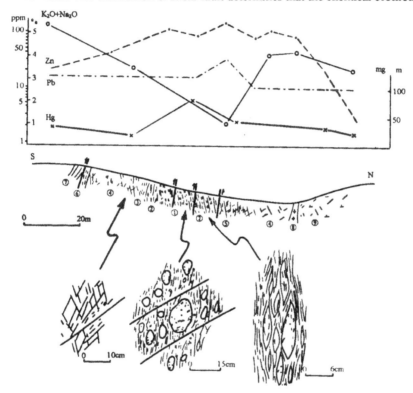

Figure 3. The Profile of the Tectonic-geochemistry in Some Compressional Fracture Zones.①, mylonitized porphyroclastic rock; ②, lentiform prophyroclastic rock;③, cataclastic rock; ④, scattered rhombus block;⑤, cataclactic rock with lens;⑥, late mylonite; ⑦, quartzite; ⑧, granite vein; ⑨, quartz diorite.

mainly disperse along fault belt, at the induced subsidary structure and the place where the occurrence of fault plane is changed, the metalogenic elements easily precipitate and then form orebody.

It is necessary to point out that, because of the complexity of geology, the above rules of tectonic controls on concentrition and dispersion of chemical elements have a little limitation. It is necessary to be sure not to mechanically apply present model to complicated analyses of ore-forming and ore-controlling. It is important, with the view of history of tectonic development, to start with the characteristic of geometry, kinetics and dynamics of tectonics. and combine them with the theory and method of tectonics-geochemistry and use many means to study comprehensively and analyse gynthetically, Therefore, the rule of tectonic controls on the concentration and dispersion can be concluded,and then indentified by its effectiveness in ore-searching and exploration.

Acknowledgements

We have benefited greatly from discussions with V.I. Strarotin, Qingxuan Cheng,Shengqin Cui, Yadong Zheng and An Yan. A grant from Yi Mao and Hongxing Cheng for their helpful reviews and translation is gratefully acknowledged.

REFERENCES

1.   S.B. Bodon and R.K.Valenta. Primary and Tectonic Features of the Currawong Zn-Cu-Pb(-Au)Massive Sulfide Deposit, Benambra, Victoria: Implications for Ore Genesis. Economic Geol. 90, 1694-1721(1995).

2.   D.R. Burrows,E.T.C. Spooner,P.C. Wood, and J.A. Jemielita. Structural Contols on Formation of the Hollonger-McIntyre Au Quartz Vein system in the Hollinger Shear Zone, Timmins, Southern Abitibi Greenstone Belt, Ontario. Economic Geol. 88, 1643-1663(1993).

3.   Guoda Cheng. The Method of Metallogenic Tectonics Research(second   edition). Beijing: Geological publishing house(1986). Guoda Cheng. The Method of Metallogenic Tectonics Research(second   edition). Beijing: Geological publishing house(1986).

4.   S.F. Cox, S.S.Sun, M.A. Etheridge, V.J. Wall, and T.F. Potter. Structural and Geochemical Controls on the Development of Turbidite-Hosted Gold Quartz Vein Deposits, Wattle Gully Mine, Central Victoria, Australia.Economic Geol.90, 1722-1746(1995).

5.   B. Dube and J. Guha.Relationship between Northeast-Trending Regional Faults and Archean Mesothermal Gold-Copper Mineralization: Coode Mine, Aabitibi Greenstone Belt, Quebec, Canada. Economic Geol.87, 1525-1540(1992).

6.   W.E. Elston. Evolution of Volcanic and Tectonic Features in Caldera Settings and Their Imporeance in the Licalization of Ore Deposits. Economic Geol.89, 1662-1686(1994).

7.   R.A. Glen. Thrusts and Thrust-Associated Mineralization in the Lachlan Orogen. Economic Geol.90, 1402-1429(1995).

8.   R.A. Glen, J.L. Walshe, M. Bouffler, T. Ho, and J.A. Jean.Syn-and Post-tectonic Mineralization in the Woodlawn Deposit, New South Wales, Australia.Economic Geol. 90, 1857-1864(1995).

9.   J. Kutina. The Role of Deep Structure of The Lithosphere in Meallogeny-Investigating the Role of Transregional Discontinuities. The 9th Symposium of International Association on the Genesis of Ore Deposits(abstracts)1, 12(1994).

10. J. Lacroix, R. daigneault, F. Chartrand and J. Guha. Structural Evolution of the Grevet Zn-Cu Massive Sulfide Deposit, Lebel-Sur-Quevillon Area, Abitibi Subprovince, Quebec.Economic Geol.88, 1559-1577(1993).

11. J.S. Lee. An Introduction to Geomechanics.Beijing: Science press(1973).

12. Xun Liu. Several Geochemical Problems about the Ore-Control of Tectonic System. Bulletin of Chinese Academic of Geological sciences press, 18(1988).

13. Guxian Lu. Jiaodong Lianlong-Jiaojia type Golden deposit Geology. Beijing:Science press(1993).

14. V.S.E. Mapani and Christopher.J.L. Wilson. Structural Evolution and Gold Mineralization in the Scotchmans Fault Zone. Magdala Gold Mine, Stawell, Western Victoria, Australia. Economic Geol. 89. 566-583(1994).

15. M. Piche, J. Guha and R. Daigneault. Stratigraphic and Structural Aspects of the Volcanic Rocks of the Matagami Mining Camp, Quebec: Implications for the Norita Ore Deposit. Economic Geol. 88. 1542-1558(1993).

16. V.I. Starostin and A.L. Dergachv and K.Hrcorich. The Analysis of Structure of deposit-petrophysics.Moscow:Moscow university press(1994).

17. G. Tourigny, D. Doucet and A.. Bourget. Geology of the Bousquet 2 Mine: An Example of a Deformed, Gold-Bearing, Polymetallic Sulfide Deposit.Economic Geol. 88, 1578-1597(1993).

18. R. Valenta.Syntectonic Discordant Copper Mineralization in the Hilton Mine, Mount Isa. Economic Geol. 89, 1031-1052(1994).

19. J.R. Vearncombe, Dentith, S. Dorling, A. Reed, R. Cooper, J. Hart, P. Muhling, D. Windrim and G. Woad. Regional- and Prospect-Scale Fault Controls on Mississippi Valley-Type Zn-Pb Mineralization at Blendevale, Canning Basin, Western Australia. Economic Geol. 90,181-186(1995).

20. J.L. Walshe, P.S. Heithersay and G.W. Morrison. Toward an Underrstanding of the Metallogeny of the Tasman Fold Belt System. Economic Geol. 90, 1382-1401(1995).

21. J. Windh. Saddle Reef and Related Gold Mineralization. Hill End Gold Field, Australia: Evolution of an Auriferous Vein System during Progressive Deformation. Economic Geol. 90, 1764-1775(1995).

22. Jiaying Wang. An Introduction to Stress mineralogy. Beijing: Geological publishing house(1978).

23. L.A. Woodward. Structural Control of Lode Gold Deposits in the Pone Mining District, Tobacco Root Mountains. Montana. Economic Geol. 88, 1850-1861(1993).

24. Xueyi Wu and Deliang Zhong. The Tecto-Geochemical Simulation Experiment of High Temperature and Pressure and Initial Results. Geotectonica et Metallogenia. Beijing: Geological publishing house(1984).

25. Ganguo Wu and Xidong Wu. The Study of Tectonic Development and the Rule of Mineallogenic Concentrition of Jingdian Plumb-Zonic Deposit in Yunnan Province. Earth science-Jouranl of Chinese university of Geoscience. 14:5(1989).

26. Ganguo Wu. Several Rules of Ore-Controls of Structure. Geology of deposit. 13(supplement)(1994).

27. Kaiqing Yang. The Problems About Ore-control and Rock-control of Tectonics, Collected Essays of Geomechanics, 5(1982).

28. Guoqing Yang. Tecto-Geochemstry. Guilin: Kuangxi normal university press(1990).

Proc. 30th Int'l. Geol. Congr., Vol. 14, pp. 304-311
Zheng *et al.* (Eds)
© VSP 1997

Geomechanics and Site Investigation for Underground Works in Japan

KAZUO HOSHINO
Engineering Advancement Association of Japan (ENAA),
Nishi-Shinbashi 1 - 4 - 6, Minato - ku, Tokyo 105, JAPAN

Abstract

Japan is located at the boundary between three continental and oceanic plates,and there is a great variety and complexity of geological features as a result of tectonic acitivities. In order to ensure both geological stability and engineering feasibility of excavating the large underground tunnels at a great depth in such "fragile" Japanese rocks, comprehensive geological studies was carried out over the whole country. As a result, three sites, two in Cretaceous granites,and one in Miocene volcanics, were selected as suitable for the excavation of the underground caverns for storage of crude oil. After detailed site investigation in five years, construction in these three sites was successfully completed in 1994. Two sites have a total capacity of 1.75million kl of crude oil in ten caverns. The largest cavern completed is 550m in length and 20×30m in cross-section.
The important condition for the large excavation and opening in the orogenic areas that is required for storage of a huge amount of crude oil, liquefied petroleum gas, or compressed air energy is both geomechanical and hydrogeological stability of the rock-mass. Such stabilities are strongly dependent on behavior of the faults and fractures in the rock-mass.

Keywords : underground petroleum storage, LPG storage, strength of rocks,
geomechanics, granite, Kuji, Kikuma, Kushikino,

INTRODUCTION

Geologically, Japan belongs to the circum Pacific orogenic zone, and it is an island arc between Eurasian continent and the Pacific ocean. Some geologists are suspicious about the possibility of excavating large underground tunnels (caverns)in the seemingly "fragile" Japanese rocks in the orogenic areas. In recent years,the national projects on the large underground caverns for storage of huge amount of crude oil or liquefied petroleum gas are in progress. The largest storage caverns for the crude oil is 550m in length and 20×30m in cross-section. The construction of such large and deep caverns raised many geological problems, especially in geomechaincs and hydrogeology. In order to ensure both geological stability and engineering feasibility for the construction, the Ministry of International Trade and Industry (MITI) made the comprehensive geological studies over the whole of Japan, and conducted the preliminary experimentation by excavating a cavern with 112m long and 250m² in cross-section in the Cretaceous granites in Kikuma in the early years of the national projects, [3,4,9,12].

In this report, the author intends to describe the tectonic situation of Japan and geomechanical behavior of the representative Japanese-rocks, and to discuss the geological problems on the site investigation for the excavation of large cavernsinto the hard rock-mass in the orogenic areas.

GEOTECTONICS OF JAPAN

The great variety of Japanese geology is doubtlessly the result of multifold tectonic activities along the boundary between the continental and oceanic plates, [6,9]. This is reflected by the Quaternary stress fields as indicated in the in-situ horizontal stress distribution map around Japan,(Fig.1).

It seems that there are three main directions in the compressional stress systems: east-west, northwest-southeast, and north-south. The east-west stress is distributed mainly in northern and partly in central Japan. On the other hand the other two stress systems are developed mainly in western Japan. The distribution of the north-south stress is limited in some places along the Pacific coast in western Japan. It is likely that the east-west stress is caused by the activity of the Pacific plate (PAC) and the other stress systems are related to the movements of the Philippine plate (PHS). The subduction of the two oceanic plates under the continental Eurasian plate (EUR) is possibly the cause of the violent seismic activities and volcanism around the Japanese islands.

Considering geotectonic effects on underground construction, seismic and volcanic activities are very important. Major damaging earthquakes are believed to occur along the subduction zones of the oceanic plates. The

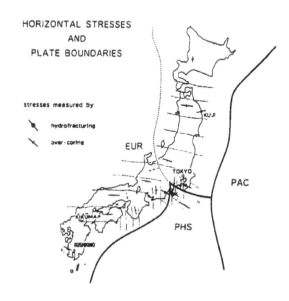

Figure 1. Distribution of in-situ horizontal stresses and plate boundaries. Three sites of underground storage for crude oil: Kuji, Kikuma, and Kushikino are shown.

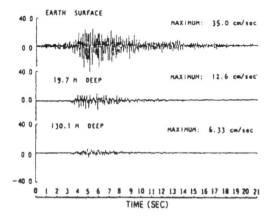

Figure 2. Acceleration of a natural earthquake, recorded in Kikuma.

other large earthquakes are generated by movements of certain active faults. We have recognized more than 1000 active faults on the land [12], which are distributed in most areas, although the largest ones are concentrated in central Japan. Therefore, the areas or zones close to the active faults is unfavorable for the construction [9].However, generally speaking, the underground places are more stable for earthquakes than the surface and the shallow places. Figure 2 shows the acceleration of the natural earthquake, 6.8 in magnitude, that was recorded in the experimental cavern in Kikuma at different depths from the surface to 130m [3,12]. It is clearly shown that the acceleration decreases considerably in deeper places; that is, 35.0cm/sec² at the surface decreases to 12.6 at depth of 19.7m and decreases to 6.33 at 130.1m.

In Japan, about 10% of the land are covered by 250 Quaternary volcanoes [12], of which 80 are still active. The volcanoes are distributed in two belts: east and west volcanic belts. The east volcanic belt was caused by subduction of the Pacific plate, while the west was caused by that of the Philippine plate. The areas in and near active volcanoes are also not suitable for the site of the construction [9].

GEOECHANICAL PROPERTIES OF THE JAPANESE ROCKS

Because of the active tectonics, the rock-mass in Japan is faulted and jointed in many areas. However, if we measure the non-weathered, non-fractured samples in the laboratory, they show the same mechanical properties as the intact rocks [1,2,5]. Fig.3 shows the strength and elastic wave velocity of the Japanese representative plutonic rocks [5]. In the left figure, strength of 13 kinds of the plutonic rocks ranging in SiO_2 component from ultrabasic, gabbro, diorite, to granite are shown at confining pressures of 1, 500, and 1000 bars. The strength at the atmospheric pressure is about 2000kg/cm² for all. The right figure shows P and S wave velocity of the plutonic rocks. The velocity decreases from ultrabasic, diorite to granite. The granite are 5 to 4 km/s in P wave velocity. The strength and elastic wave velocity are the most fundamental parameters for geomechanical properties.

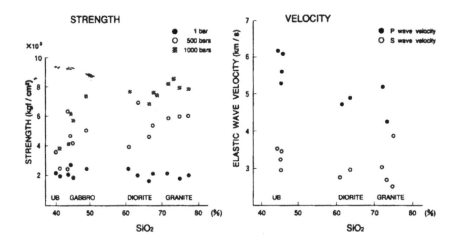

Figure 3. Strength and elastic velocity of the Japanese representative plutonic rocks.

The mechanical properties become weak as the rocks are weathered, fractured or affected by alternation.

Porosity is seemingly a good parameter to indicate the degree of weakening. In Fig.4, relation of uniaxial strength to porosity is shown for the granites. Non-weathered, intact cores are generally less than 1 % in porosity, accordingly they are harder than 50 MPa or 500kg/cm² in strength.

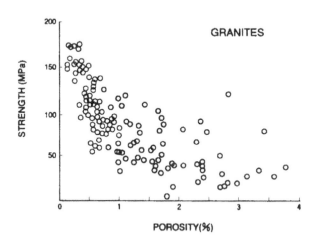

Figure 4. Strength versus porosity of the granites.

SITE INVESTIGATION

In early stages of the project, we conducted comprehensive site investigation over the whole of Japan, which was made by means of analysis of previous data.

The fundamental conditions for the study were that the suitable areas should be (1) composed of hard and uniform rocks, (2) located far away from active faults and volcanoes, and (3) within 10 km of the coast line.

As a result, 170 sites in 57 areas were selected in 1977 to be geologically stable and suitable for underground construction [4,9,10], as shown in Fig.5.

The most preferable areas are in the granitic rocks, because these are physically hard and homogeneous. The next most suitable rocks are Neogene volcanic and pre-Tertiary rocks. The former are pyroclastic volcanic rocks, which are found in many places along the west and east Japan volcanic belts. The latter are thick and massive sandstones in the Mesozoic accretionary complex of southwest Japan. Tertiary sedimentary rocks and basaltic rocks are favorable if considerable thickness of massive and homogeneous layers are present.

For the underground storage of crude oil with a total capacity of millions kl, three sites, Kuji, Kikuma, and Kushikino were selected from the above 57 areas.

The locations of the three sites are shown in Fig.1. Kuji and Kikuma are in Cretaceous granites and Kushikino is in Miocene andesitic pyroclastic rocks. In these places, the feasibility study was carried out for 5 years from 1981-86, purpose of which is to obtain geological and geomechanical data. In the geological survey, we intended to know the distribution of the intact rocks in extent and depth by the following items.
(1) surface route survey.
(2) aerophotographic study of the fracture system.
(3) borings (20 holes).
(4) well logging: density, electrical, caliper, and temperature.
(5) hydrogeological survey : Lugeon test, ground water table, JFT.

The following items were con
ducted to obtain the geomechanical data.

(6) laboratory measurement: density, porosity, elastic wave velocity, elastic constants,
strength at atmospheric and
confining pressure.

(7) in-situ measurement : geo-
stress, shearing stress, deformation and creep.

(8) analytical studies : structural
stability of caverns, hydraulic
behavior around caverns,water
balance and water level.

The seismic prospecting was conducted in the granitic sites, and
proved useful for preliminary
mapping of the weathered zone.
The brittleness determined by
triaxial deformation experimentation is closely related with ability
of fracturing of the rocks. The
triaxial experimentation up to
1000 bars confining pressure or
more is also useful to estimate the
angle of friction and shearing
stress in in-situ condition, if it is
combined with the porosity data.

The construction of the three sites
for crude oil was started in 1986
and completeded in 1991 [4,11].
The arrangement in Kushikino site
are shown in Fig. 6, for example.
The main parts of the Kushikino
consist of ten large caverns used

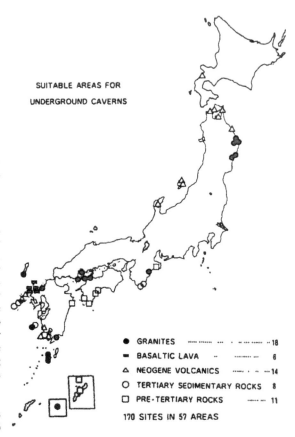

SUITABLE AREAS FOR
UNDERGROUND CAVERNS

● GRANITES	....	18
▬ BASALTIC LAVA	.. ..	6
△ NEOGENE VOLCANICS	....	14
○ TERTIARY SEDIMENTARY ROCKS		8
▢ PRE-TERTIARY ROCKS	....	11

170 SITES IN 57 AREAS

Figure 5. Location and geology of 57 areas selected as
suitable for underground construction

for storage of crude oil, which occupy an area of about 500 × 600m. Each cavern is 555m
long, 18m wide and 22m high. The distance to the earth's surface from the top of the caverns
is roughly 200-100m. The water-table is at an elevation of approximately 120-50m above the
sea level. As the caverns are 20-40m below the sea level, the underground water near the
caverns has a hydrostatic pressure of about 100-150m in water head. The ten caverns were
excavated in the autobrecciated lava and volcanic conglomerate, products of Miocene volcanism, which are geomechanically ductile and very little in permiability.

The Kushikino site has a total capacity of 1.75 million kl of crude oil, [7,9].

FURTHER PROBLEMS

Since the underground caverns of crude oil is successfully completed in the above three areas, the next project of LPG (Liquefied Petroleum Gas) storage was started in granitic areas

in western Japan. In the caverns of LPG, gas pressure is kept larger than 7 kg/cm².

It means that the LPG caverns should be excavated at a depth of 100m or more, that is much deeper than the oil caverns. Therefore, geological condition for site selection is more serious and difficult. When the geological target for underground engineering is directed to the soft sedimentary rocks, apart from granitic or volcanic rocks, different consideration is probably required.

The cross section along Tokyo Bay as shown in Fig. 7 was made as a basic map for the utilization of underground spaces in the metropolitan Tokyo, [2,8,10]. The sediments in Tokyo sedimentary basin ranging from Quaternary to Pliocene in geological age consist of mud and sand. As the mud behaves at normal trend in the compaction at deeper level, porosity decreases to 40% for late Pliocene from 65% for late Pleistocene and the uniaxial strength increases from 20 Kg/cm² for early Pleistocene to 200 Kg/cm² for early Pliocene. This cross section map with the iso-strength lines provides us important information concerning the suitable sites of the underground open spaces for the compressed air energy.

In conclusion, even in active orogenic areas like Japan, most areas involve some feasible places for underground excava-

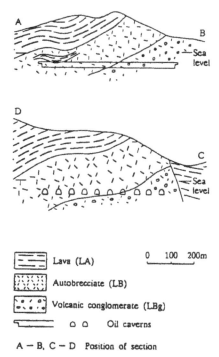

☰ Lava (LA)	0 100 200m
▨ Autobrecciate (LB)	
▦ Volcanic conglomerate (LBg)	
⊐ ◌ ◌ Oil caverns	

A — B, C — D   Position of section

Figure 6. The Kushikino site, geological sections and position of the petroleum storage caverns. A-B is parallel and C-D is vertical to the long axis of the caverns.

tion, including geomechanics and hydrogeology. As for the condition of geomechanical stability, the rock-mass should be (1) hard in strength, (2) compact in permeability, and (3) uniform in mechanical properties. As for the condition of hydrogeology, the underground waterlevel and hydrostatic pressure should be maintained constant. Such geological features are much influenced by geotectonic circumstances and strongly dependent on behavior of the faults and fractures in the rock-mass.

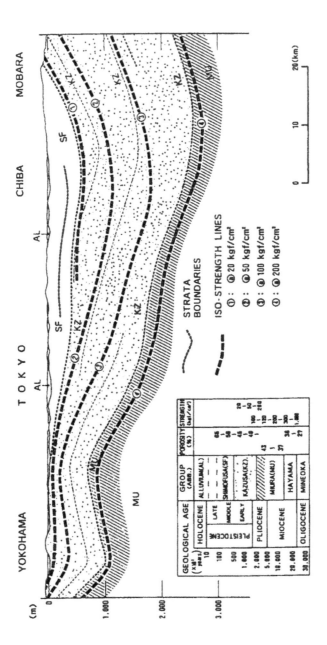

Figure 7. Cross section along Tokyo Bay to show relation between geological structure and the iso-strength lines. the table at lower-left is summary of geology and the development in mechanical properties.

Vertical exaggreation   10 : 1

Acknowledgements

The national projects for the petroleum and LPG storage have been carried out by MITI and the Japan National Oil Corporation (JNOC). The construction of the three sites for underground petroleum storage was conducted by the Japan Underground Oil Storage Company (JUOSC). The author is very grateful for permission to publish this report.

REFERENCES

1. K. Hoshino, H. Koide, K. Inami, S. Iwamura, and S. Mitsui. Mechanical properties of Japanese Tertiary sedimentary rocks under high confining pressures,
  · *Geological Survey Japan, Special Report* No.244 (1972).
2. K. Hoshino. Consolidation and strength of the soft sedimentary rocks, *Proc.Int.Symp. Weak Rocks,ISRM* (Tokyo), 149-153 (1981).
3. K. Hoshino. The Kikuma experimentation on underground petroleum storage, *Chishitsu News* (Geol.Surv.Japan Publ), **350**: 6-18; **353**: 48-62. (1983). (**)
4. K. Hoshino. Present status and some problems of the underground petroleum storage, *Underground Spaces. Min.Mater.Proc.Inst.Japan.* **5-8**. (1983). (**)
5. K. Hoshino. Effect of geological factors on mechanical properties of rocks, *Proc. 6th Japan Symp. Rock Mechanics.* 145-150 (1984).
6. K. Hoshino. Neotectonics of southern Fossa-Magna, a study based on stress measurement and active faults, *Quaternary Research* **23**, 117-128 (1984). (*)
7. K. Hoshino and T. Makita. Construction of large underground caverns for the storage of crude oil in ductile rocks of Miocene volcanism, *Proc. 6th Int.Ass.Engineering Geology* (Amsterdam). 2587-2592 (1990).
8. K. Hoshino. Geological evolution from the soil to the rocks: mechanism of lithification and change of mechanical properties, *Geotechnical Engineering of Hard Soils-Soft Rocks* (Balkema), 131-138 (1993).
9. K. Hoshino. Construction of underground caverns for petroleum storage in orogenic areas: geological stability, *Engineering Geology*, **35**, 199-205 (1993).
10. K. Hoshino. Role of geologic data for the utilization of underground space in great depth, *Chishitsu News* (Geol.Surv.Japan Publ), **492**:55-63 (1995). (**)
11. T. Makita, Y. Miyanaga, K. Iguchi, and T. Hatano. Underground oil facilities in Japan, *Engineering Geology*, **35**, 191-198 (1993).
12. K. Ono and K. Hoshino. Studies on crustal stability in Japan: earthquakes, volcanic eruptions and underground construction. In : *Regional Crustal Stability and Geological Hazards.* C.Qingxuan (editor), 30-44 (1989).

  (*) written in Japansese with English abstract.
  (**) written in Japanese.